Optics, Sound
and Waves

BOOKS BY M. NELKON

Published by Heinemann

ADVANCED LEVEL PHYSICS (SI) (*with P. Parker*)
SCHOLARSHIP PHYSICS (SI)
OPTICS, SOUND, WAVES (SI)
MECHANICS AND PROPERTIES OF MATTER (SI)
ADVANCED LEVEL PRACTICAL PHYSICS (SI) (*with J. Ogborn*)
PRINCIPLES OF ATOMIC PHYSICS AND ELECTRONICS (SI)
REVISION NOTES IN ADVANCED LEVEL PHYSICS (SI)
Book I. Mechanics, Electricity, Atomic Physics
Book II. Optics, Waves, Sound, Heat & Properties of Matter
GRADED EXERCISES AND WORKED EXAMPLES IN PHYSICS (SI)
TEST PAPERS IN PHYSICS (SI)
SOLUTIONS TO ADVANCED LEVEL PHYSICS QUESTIONS (SI)
SOLUTIONS TO ORDINARY LEVEL PHYSICS QUESTIONS (SI)
ELEMENTARY PHYSICS, Book I and II (*with A. F. Abbott*) (SI)
AN INTRODUCTION TO THE MATHEMATICS OF PHYSICS (SI)
(*with J. H. Avery*)
GENERAL SCIENCE PHYSICS
ELECTRONICS AND RADIO (SI)
REVISION BOOK IN ORDINARY LEVEL PHYSICS (SI)

Published by Edward Arnold

ELECTRICITY (Advanced Level, SI)

Published by Chatto & Windus

FUNDAMENTALS OF PHYSICS (O-Level, SI)
EXERCISES IN ORDINARY LEVEL PHYSICS (SI)
C.S.E. PHYSICS (SI)
REVISION BOOK IN C.S.E. PHYSICS (SI)
SI UNITS: AN INTRODUCTION FOR A-LEVEL

Published by Blackie

HEAT (Advanced Level, SI)

Optics, Sound and Waves

(Previously entitled Light and Sound)

Fifth edition with SI units

M. NELKON, M.Sc. (Lond.), F.Inst.P., A.K.C.

*formerly Head of the Science Department,
William Ellis School, London*

HEINEMANN EDUCATIONAL
BOOKS · LONDON

Heinemann Educational Books Ltd

LONDON EDINBURGH MELBOURNE AUCKLAND TORONTO
SINGAPORE HONG KONG KUALA LUMPUR
IBADAN NAIROBI JOHANNESBURG
NEW DELHI

ISBN 0 435 68613 5

© M. Nelkon 1973

First published entitled *Light and Sound* 1950
Reprinted three times
Second Edition 1955
Reprinted five times
Third Edition 1965 (reset)
Reprinted 1966
Fourth Edition (SI) 1969
Reprinted 1970
Fifth edition entitled *Optics, Sound and Waves* 1973

Published by
Heinemann Educational Books Ltd
48 Charles Street, London W1X 8AH

Text set in 10/11 pt. Monotype Times New Roman, printed by photolithography,
and bound in Great Britain at The Pitman Press, Bath

Contents

COVER PHOTOGRAPH

The cover photograph shows a pattern of vibration of a tenor viol, mapped out by the time-averaged technique of laser photography. The instrument was vibrating steadily at a resonance frequency of 285 Hz while hologram was recorded. The interference pattern is a contour map of the amplitude of vibration for this particular resonance, the bright fringes representing the stationary nodal region.

This photograph was obtained by Dr Karl A. Stepson while he was working in England at the National Physical Laboratory. It is published by kind permission of the Director of N.P.L. and is Crown Copyright reserved.

Preface

This textbook deals with the principles of Optics, Sound and their associated Waves to a General Certificate of Education Advanced Level, and assumes an Ordinary level knowledge of the subject. The main part of the work comprises the Optics and Sound section of *Advanced Level Physics*, SI edition, and the chapter and page numbers have been retained except for the final two chapters. These additional new chapters cover further topics in optics and electromagnetic waves. The whole work modernises and replaces the author's textbook *Light and Sound*.

Geometrical Optics has been discussed first. Here the optical functions of mirrors, prisms and lenses have been considered and applied in optical instruments. The sign convention adopted in formulae is the *real is positive, virtual is negative* convention, and examples of its applications, including the case of the virtual object, have been given in the text. (A text using the alternative *New Cartesian* convention is given in a separate edition.)

In Sound, an account has been given of the physical principles of plane-progressive and stationary waves, and their applications to the vibrations in pipes, strings and rods. An introduction to sound recording and reproduction has also been included in the text.

Physical or Wave Optics begins with an account of the wave theory of light, followed by an introduction to interference, diffraction and polarisation. In view of their importance, diffraction, resolving power and interference are discussed further in the next chapter, and a final chapter contains an account of the propagation and properties of electromagnetic waves and their application in the polarisation of light. It is hoped that the last two chapters will be particularly useful for students taking the Nuffield Advanced course.

Throughout the text numerical examples, taken from past examination papers, have been included in illustration of the subject matter and the calculus has been used only where necessary.

The author is indebted to M. V. Detheridge, William Ellis School, London, and R. P. T. Hills, St. John's College, Cambridge, for reading the final section and their valuable advice. He is also indebted to L. J. Beckett, William Ellis School, London, for the opportunity to participate in teaching the new Nuffield Advanced course.

ACKNOWLEDGEMENTS

For assistance in past editions, I am particularly grateful to Professor W. C. Price, F.R.S., King's College, London University; Dr. R. S. Longhurst, senior lecturer, Chelsea College of Science and Technology;

J. H. Avery, senior science master, Stockport Grammar School; L. G. Mead, Wellington School, Somerset; and Professor H. T. Flint, formerly of Bedford College, London University.

I am also indebted to the following Examination Boards for their kind permission to reproduce questions set in past examinations:

University of London (L.)
Joint Matriculation Board (N.)
Oxford and Cambridge Joint Board (O. & C.)
Cambridge Local Examinations Syndicate (C.)
Oxford Local Examinations Delegacy (O.)

OPTICS

chapter fifteen

Introduction

IF you wear spectacles you will appreciate particularly that the science of Light, or *Optics* as it is often called, has benefited people all over the world. The illumination engineer has developed the branch of Light dealing with light energy, and has shown how to obtain suitable lighting conditions in the home and the factory, which is an important factor in maintaining our health. Microscopes, used by medical research workers in their fight against disease; telescopes, used by seamen and astronomers; and a variety of optical instruments which incorporate lenses, mirrors, or glass prisms, such as cameras, driving mirrors, and binoculars, all testify to the scientist's service to the community.

In mentioning the technical achievements in Light, however, it must not be forgotten that the science of Light evolved gradually over the past centuries; and that the technical advances were developed from the experiments and theory on the *fundamental* principles of the subject made by scientists such as NEWTON, HUYGENS, and FRESNEL.

Light Travels in Straight Lines. Eclipses and Shadows

When sunlight is streaming through an open window into a room, observation shows that the edges of the beam of light, where the shadow begins, are straight. This suggests that *light travels in straight lines*, and on this assumption the sharpness of shadows and the formation of eclipses can be explained. Fig. 15.1 (i) illustrates the eclipse of the sun, S,

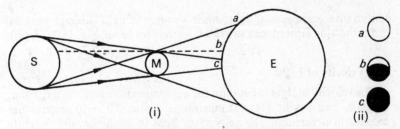

FIG. 15.1. Eclipse of Sun (*not to scale*).

our natural source of light, when the moon, M, passes between the sun and the earth, E. The moon is a non-luminous object which does not allow light to pass through it, and hence the boundaries of the shadows formed by M on the earth are obtained by drawing lines from S which

touch the edge of M. Consequently there is a total eclipse of the sun at
c on the earth, a partial eclipse at b, and no eclipse at a. Fig. 15.1 (ii)
illustrates the appearance of the sun in each case.

Light Rays and Beams

Light is a form of energy. We know this is the case because plants and
vegetables grow when they absorb sunlight. Further, electrons are ejected
from certain metals when light is incident on them, showing that there
was some energy in the light; this phenomenon is the basis of the
photoelectric cell (p. 1077). Substances like wood or brick which allow
no light to pass through them are called "opaque" substances; unless
an opaque object is perfectly black, some of the light falling on it is
reflected (p. 391). A "transparent" substance, like glass, is one which
allows some of the light energy incident on it to pass through, the
remainder of the energy being absorbed and (or) reflected.

A ray of light is the direction along which the light energy travels;
and though rays are represented in diagrams by straight lines, in practice
a ray has a finite width. A **beam** of light is a collection of rays. A searchlight emits a *parallel beam* of light, and the rays from a point on a very
distant object like the sun are substantially parallel, Fig. 15.2 (i). A lamp

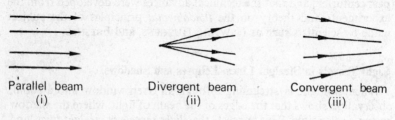

Parallel beam   Divergent beam   Convergent beam
     (i)   (ii)   (iii)

FIG. 15.2. Beams of light.

emits a *divergent beam* of light; while a source of light behind a lens, as
in a projection lantern, can provide a *convergent beam*, Fig. 15.2 (ii), (iii).

The Velocity of Light

The velocity of light is constant for a given medium, such as air, water,
or glass, and has its greatest magnitude, about $3 \cdot 0 \times 10^8$ metres per
second, in a vacuum. The velocity of light in air differs only slightly
from its velocity in a vacuum, so that the velocity in air is also about
$3 \cdot 0 \times 10^8$ metres per second. The velocity of light in glass is about
$2 \cdot 0 \times 10^8$ metres per second; in water it is about $2 \cdot 3 \times 10^8$ metres per
second. On account of the difference in velocity in air and glass, light
changes its direction on entering glass from air (see *Refraction*, p. 679).
Experiments to determine the velocity of light are discussed later, p. 551.

The Human Eye

When an object is seen, light energy passes from the object to the observer's eyes and sets up the sensation of vision. The eye is thus sensitive to light (or luminous) energy. The eye contains a *crystalline lens*, L, made of a gelatinous transparent substance, which normally throws an image of the object viewed on to a sensitive "screen" R at the back of the eye-ball, called the *retina*, Fig. 15.3. Nerves on the retina

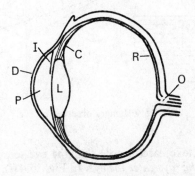

FIG. 15.3. The eyeball.

are joined to the *optic nerve*, O, which carries the sensation produced by the image to the brain. The *iris*, I, is a diaphragm with a circular hole in the middle called the *pupil*, P, which contracts when the light received by the eye is excessive and painful to the eye. The colour of a person's eyes is the colour of the iris; the pupil is always black because no light returns from the interior of the eye-ball. A weak salt solution, called the *aqueous humour*, is present on the left of the lens L, and between L and the retina is a gelatinous substance called the *vitreous humour*. The transparent spherical bulge D in front of L is made of tough material, and is known as the *cornea*.

The *ciliary muscles*, C, enable the eye to see clearly objects at different distances, a property of the eye known as its "power of accommodation". The ciliary muscles are attached to the lens L, and when they contract, the lens' surfaces are pulled out so that they bulge more; in this way a near object can be focused clearly on the retina and thus seen distinctly. When a very distant object is observed the ciliary muscles are relaxed, and the lens' surfaces are flattened.

The use of two eyes gives a three-dimensional aspect of the object or scene observed, as two slightly different views are imposed on the retinæ; this gives a sense of distance not enjoyed by a one-eyed person.

Direction of Image seen by Eye

When a fish is observed in water, rays of light coming from a point such as O on it pass from water into air, Fig. 15.4 (i). At the boundary of the water and air, the rays OA, OC proceed along new directions AB, CD respectively and enter the eye. Similarly, a ray OC from an object O observed in a mirror is reflected along a new direction CD and enters the eye, Fig. 15.4 (ii). These phenomena are studied more fully later, but

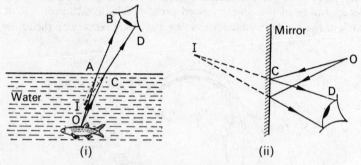

FIG. 15.4. Images observed by eye.

the reader should take careful note that the eye sees an object *in the direction in which the rays enter the eye*. In Fig. 15.4 (i), for example, the object O is seen in the water at I, which lies on BA and DC produced slightly on the right of O; in Fig. 15.4 (ii), is seen behind the mirror at I, which lies on DC produced. In either case, all rays from O which enter the eye appear to come from I, which is called the image of O.

Reversibility of Light

If a ray of light is directed along DC towards a mirror, experiment shows that the ray is reflected along the path CO, Fig. 15.4 (ii). If the ray is incident along OC, it is reflected along CD, as shown. Thus if a light ray is reversed it always travels along its original path, and this is known as *the principle of the reversibility of light*. In Fig. 15.4 (i), a ray BA in air is refracted into the water along the path AO, since it follows the reverse path to OAB. We shall have occasion to use the principle of the reversibility of light later in the book.

chapter sixteen

Reflection at plane surfaces

HIGHLY-POLISHED metal surfaces reflect about 80 to 90 per cent of the light incident on them; *mirrors* in everyday use are therefore usually made by depositing silver on the back of glass. In special cases the front of the glass is coated with the metal; for example, the largest reflector in the world is a curved mirror nearly 5 metres across, the front of which is coated with aluminium (p. 544). Glass by itself will also reflect light, but the percentage reflected is small compared with the case of a silvered surface; it is about 5 per cent for an air-glass surface.

Laws of Reflection

If a ray of light, AO, is incident on a plane mirror XY at O, the angle AON made with the *normal* ON to the mirror is called the "angle of incidence", *i*, Fig. 16.1. The angle BON made by the reflected ray OB with the normal is called the "angle of reflection", *r;* and experiments with a ray-box and a plane mirror, for example, show that:

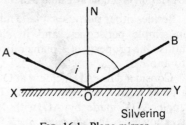

FIG. 16.1. Plane mirror.

(1) *The reflected ray, the incident ray, and the normal to the mirror at the point of incidence all lie in the same plane.*

(2) *The angle of incidence = the angle of reflection.*

These are called the two *laws of reflection*, and they were known to PLATO, who lived about 400 B.C.

Regular and Diffuse Reflection

In the case of a plane mirror or glass surface, it follows from the laws of reflection that a ray incident at a given angle on the surface is reflected in a definite direction. Thus a parallel beam of light incident on a plane mirror in the direction AO is reflected as a parallel beam in the direction OB, and this is known as a case of *regular reflection*, Fig. 16.2 (i). On the other hand, if a parallel beam of light is incident on a sheet of paper in a direction AO, the light is reflected in all different directions from the paper: this is an example of *diffuse reflection*, Fig. 16.2 (ii). Objects in everyday life, such as flowers, books, people, are seen by light diffusely reflected from them. The explanation of the diffusion of light is that the surface of paper, for example, is not perfectly smooth like a mirrored surface; the "roughness" in a paper surface can be seen with a micro-

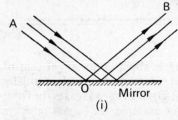

FIG. 16.2 (i). Regular reflection.

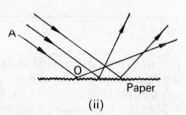

FIG. 16.2 (ii). Diffuse reflection.

scope. At each point on the paper the laws of reflection are obeyed, but the angle of incidence varies, unlike the case of a mirror.

Deviation of Light at Plane Mirror Surface

Besides other purposes, plane mirrors are used in the sextant (p. 395), in simple periscopes, and in signalling at sea. These instruments utilise the property of a plane mirror to deviate light from one direction to another.

Consider a ray AO incident at O on a plane mirror XY, Fig. 16.3 (i). The angle AOX made by AO with XY is known as the *glancing angle*, g, with the mirror; and since the angle of reflection is equal to the angle of incidence, the glancing angle BOY made by the reflected ray OB with the mirror is also equal to g.

FIG. 16.3 (i).
Deviation of light at at plane mirror.

The light has been deviated from a direction AO to a direction OB. Thus if OC is the extension of AO, the *angle of deviation*, d, is angle COB. Since angle COY = vertically opposite angle XOA, it follows that

$$d = 2g \qquad . \qquad . \qquad . \qquad . \qquad (1);$$

so that, in general, *the angle of deviation of a ray by a plane surface is twice the glancing angle.*

Deviation of Reflected Ray by Rotated Mirror

Consider a ray AO incident at O on a plane mirror M_1, α being the glancing angle with M_1, Fig. 16.3 (ii). If OB is the reflected ray, then, as shown above, the angle of deviation COB = $2g$ = 2α.

Suppose the mirror is rotated through an angle θ to a position M_2, the direction of the incident ray AO being *constant*. The ray is now reflected from M_2, in a direction OP, and the glancing angle with M_2 is $(\alpha + \theta)$. Hence the new angle of deviation COP $= 2g = 2(\alpha + \theta)$. The reflected ray

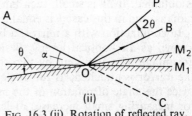

FIG. 16.3 (ii). Rotation of reflected ray.

has thus been rotated through an angle BOP when the mirror rotated through an angle θ; and since

$$\angle BOP = \angle COP - \angle COB,$$
then $$\angle BOP = 2(\alpha + \theta) - 2\alpha = 2\theta.$$

Thus, *if the direction of an incident ray is constant, the angle of rotation of the reflected ray is twice the angle of rotation of the mirror.* If the mirror rotates through 4°, the reflected ray turns through 8°, the direction of the incident ray being kept unaltered.

Optical Lever in Mirror Galvanometer

In a number of instruments a beam of light is used as a "pointer", which thus has a negligible weight and is sensitive to deflections of the moving system inside the instrument. In a mirror galvanometer, used for measuring very small electric currents, a small mirror M_1 is rigidly attached to a system which rotates when a current flows in it, and a beam of light from a fixed lamp L shines on the mirror, Fig. 16.4. If the light is incident normally on the mirror at A, the beam is reflected directly back, and a spot of light is obtained at O on a graduated scale S

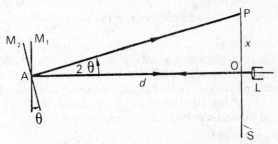

FIG. 16.4. Optical lever principle.

placed just above L. Suppose that the moving system, to which the mirror is attached, undergoes a rotation θ. The mirror is then rotated through this angle to a position M_2, and the spot of light is deflected through a distance x, say to a position P on the scale.

Since the direction OA of the incident light is constant, the rotation of the reflected ray is twice the angle of rotation of the mirror (p. 393). Thus angle OAP $= 2\theta$. Now $\tan 2\theta = x/d$, where d is the distance OA. Thus 2θ can be calculated from a knowledge of x and d, and hence θ

is obtained. If 2θ is small, then $\tan 2\theta$ is approximately equal to 2θ in radians, and in this case θ is equal to $x/2d$ radians.

In conjunction with a mirror, a beam of light used as a "pointer" is known as an "optical lever". Besides a negligible weight, it has the advantage of magnifying by two the rotation of the system to which the mirror is attached, as the angle of rotation of the reflected light is twice the angle of rotation of the mirror. An optical lever can be used for measuring small increases of length due to the expansion or contraction of a solid.

Deviation by Successive Reflections at Two Inclined Mirrors

Before we can deal with the principle of the sextant, the deviation of light by successive reflection at two inclined mirrors must be discussed.

Consider two mirrors, XO, XB, inclined at an angle θ, and suppose AO is a ray incident on the mirror XO at a glancing angle α, Fig. 16.5 (i). The reflected ray OB then also makes a glancing angle α with OX, and from our result on p. 392, the angle of deviation produced by XO in a clockwise direction (angle LOB) = 2α.

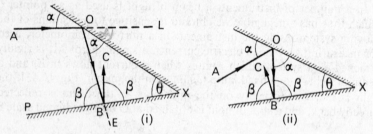

Fig. 16.5. Successive reflection at two plane mirrors.

Suppose OB is incident at a glancing angle β on the second mirror XB. Then, if the reflected ray is BC, the angle of deviation produced by this mirror (angle EBC) = 2β, in an anti-clockwise direction. Thus the net deviation D of the incident ray AO produced by both mirrors = $2\beta - 2\alpha$, in an anti-clockwise direction.

Now from triangle OBX,

$$\text{angle PBO} = \text{angle BOX} + \text{angle BXO},$$

i.e., $$\beta = \alpha + \theta$$

Thus $\theta = \beta - \alpha$, and hence

$$D = 2\beta - 2\alpha = 2\theta.$$

But θ is a *constant* when the two mirrors are inclined at a given angle. *Thus, no matter what angle the incident ray makes with the first mirror, the deviation D after two successive reflections is constant and equal to twice the angle between the mirrors.*

Fig. 16.5 (ii) illustrates the case when the ray BC reflected at the second mirror travels in an opposite direction to the incident ray AO, unlike the case in Fig. 16.5 (i). In Fig. 16.5 (ii), the net deviation, D, after two successive reflections in a clockwise direction is $2\alpha + 2\beta$. But $\alpha + \beta = 180° - \theta$. Hence $D = 2\alpha + 2\beta = 360° - 2\theta$. Thus the deviation, D, in an anti-clockwise direction is 2θ, the same result as obtained above.

Principle of the Sextant

The *sextant* is an instrument used in navigation for measuring the angle of elevation of the sun or stars. It consists essentially of a fixed glass B, silvered on a vertical half, and a silvered mirror O which can be rotated about a horizontal axis. A small fixed telescope T is directed towards B, Fig. 16.6.

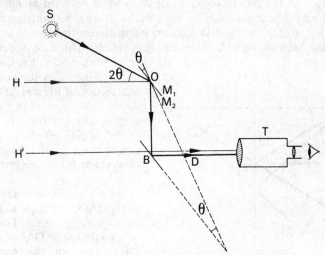

FIG. 16.6. Sextant principle.

Suppose that the angle of elevation of the sun, S, is required. Looking through T, the mirror O is turned until the view H′ of the horizon seen directly through the unsilvered half of B, and also the view of it, H, seen by successive reflection at O and the silvered half of B, are coincident. The mirror O is then parallel to B in the position M_1, and the ray HO is reflected along OB and BD to enter the telescope T. The mirror O is now rotated to a position M_2 until the image of the sun S, seen by successive reflections at O and B, is on the horizon H′, and the angle of rotation, θ, of the mirror is noted, Fig. 16.6.

The ray SO from the sun is now reflected in turn from O and B so that it travels along BD, the direction of the horizon, and the angle of deviation of the ray is thus angle SOH. But the angle between the mirrors M_2 and B is θ. Thus, from our result for successive reflections at two inclined mirrors, angle SOH = 2θ. Now the angle of elevation of the sun, S, is angle SOH. Hence *the angle of elevation is twice the angle of rotation of the mirror O*, and can thus be easily measured from a scale (not shown) which measures the rotation of O.

Since the angle of deviation after two successive reflections is independent of the angle of incidence on the first mirror (p. 394), the image of the sun S through T will continue to be seen on the horizon once O is adjusted, no matter how the ship pitches or rolls. This is an advantage of the sextant.

Images in Plane Mirrors

So far we have discussed the deviation of light by a plane mirror. We have now to consider the *images* in plane mirrors.

Suppose that a **point object** A is placed in front of a mirror M, Fig. 16.7. A ray AO from A, incident on M, is reflected along OB in such a way that angle AON = angle BON, where ON is the normal at O to the mirror. A ray AD incident normally on the mirror at D is reflected back along DA. Thus the rays reflected from M appear to

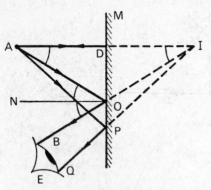

come from a point I *behind* the mirror, where I is the point of intersection of BO and AD produced. As we shall prove shortly, any ray from A, such as AP, is also reflected as if it comes from I, and hence an observer at E sees the image of A at I.

Since angle AON = alternate angle DAO, and angle BON = corresponding angle DIO, it follows that angle DAO = angle DIO. Thus in the triangles ODA, ODI, angle DAO = angle

FIG. 16.7. Image in plane mirror.

DIO, OD is common, and angle ADO = 90° = angle IDO. The two triangles are hence congruent, and therefore AD = ID. For a given position of the object, A and D are fixed points. Consequently, since AD = ID, the point I is a fixed point; and hence *any* ray reflected from the mirror must pass through I, as was stated above.

We have just shown that *the object and image in a plane mirror are at equal perpendicular distances from the mirror*. It should also be noted that AO = OI in Fig. 16.7, and hence the object and image are at equal distances from any point on the mirror.

Image of Finite-sized Object. Perversion

If a right-handed batsman observes his stance in a plane mirror, he appears to be left-handed. Again, the words on a piece of blotting-paper become legible when the paper is viewed in a mirror. This

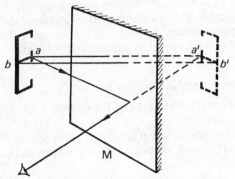

FIG. 16.8. Perverted (laterally inverted) image.

phenomenon can be explained by considering an E-shaped object placed in front of a mirror M, Fig. 16.8. The image of a point a on the object is at a' at an equal distance behind the mirror, and the image of a point b on the left of a is at b', which is on the *right* of a'. The left-hand side of the image thus corresponds to the right-hand side of the object, and vice-versa, and the object is said to be *perverted*, or *laterally inverted* to an observer.

Virtual and Real Images

As was shown on p. 396, an object O in front of a mirror has an image I behind the mirror. The rays reflected from the mirror do not actually pass through I, but only *appear* to do so, and the image cannot be received on a screen because the image is behind the mirror, Fig. 16.9 (i). This type of image is therefore called an unreal or **virtual** image.

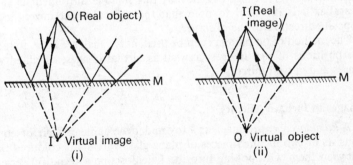

FIG. 16.9. Virtual and real image in plane mirror.

It must not be thought, however, that only virtual images are obtained with a plane mirror. If a *convergent* beam is incident on a plane mirror M, the reflected rays pass through a point I *in front of* M, Fig. 16.9 (ii). If the incident beam converges to the point O, the latter is termed a "virtual" object; I is called a **real** image because it can be received on a screen. Fig. 16.9 (i) and (ii) should now be compared. In the former, a real object (divergent beam) gives rise to a virtual image; in the latter, a virtual object (convergent beam) gives rise to a real image. In each case the image and object are at equal distances from the mirror.

Location of Images by No Parallax Method

A virtual image can be located by the *method of no parallax*, which we shall now describe.

Suppose O is a pin placed in front of a plane mirror M, giving rise to a virtual image I behind M, Fig. 16.10. A pin P behind the mirror is then

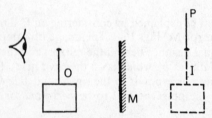

FIG. 16.10. Location of image by no parallax method.

moved towards or away from M, each time moving the head from side to side so as to detect any relative motion between I and P. When the latter appear to move together they are coincident in position, and hence P is at the place of the image I, which is thus located. When P and I do not coincide, they appear to move relative to one another when the observer's head is moved; this relative movement is called "parallax". It is useful to note that the nearer object moves in the opposite direction to the observer.

The method of no parallax can be used, as we shall see later, to locate the positions of real images, as well as virtual images, obtained with lenses and curved mirrors.

Images in Inclined Mirrors

A *kaleidoscope*, produced as a toy under the name of "mirrorscope", consists of two inclined pieces of plane glass with some coloured tinsel between them. On looking into the kaleidoscope a beautiful series of coloured images of the tinsel can often be seen, and the instrument is used by designers to obtain ideas on colouring fashions.

Suppose OA, OB are two mirrors inclined at an angle θ, and P is a point object between them, Fig. 16.11 (i). The image of P in the mirror OB is B_1, and $OP = OB_1$ (see p. 396). B_1 then forms an image B_2 in the mirror OA, with $OB_2 = OB_1$, B_2 forms an image B_3 in OB, and so on. All the images thus lie on a circle of centre O and radius OP. Another set of images, A_1, A_2, A_3 ..., have their origin in the image A_1 formed by P in the mirror OA. When the observer looks into the mirror OB he sees

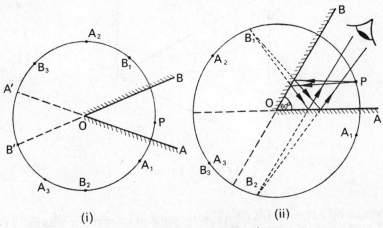

FIG. 16.11. Images in inclined mirrors.

the series of images B_1, A_2, B_3 ...; when he looks into the mirror OA he sees the series of images A_1, B_2, A_3 ... A finite series of images is seen in either mirror, the last image (not shown) being the one formed on the arc A'B', because it is then *behind* the silvering of the next mirror.

When the mirrors are inclined at an angle of 60°, the final images of P, A_3, B_3, of each series coincide, Fig. 16.11 (ii). The total number of images is now 5, as the reader can verify. Fig. 16.11 (ii) illustrates the cone of light received by the pupil of the eye when the image B_2 is observed, reflection occurring successively at the mirrors. The drawing is started by joining B_2 to the boundary of the eye, then using B_1, and finally using P.

EXAMPLE

State the laws of reflection of light and describe how you would verify these laws.

A man 2 m tall, whose eye level is 1·84 m above the ground, looks at his image in a vertical mirror. What is the minimum vertical length of the mirror if the man is to be able to see the whole of himself? Indicate its position accurately in a diagram.

First part. See text. A ray-box can be used to verify the laws.

Second part. Suppose the man is represented by HF, where H is his head and F is his feet; suppose that E represents his eyes, Fig. 16.12. Since the man

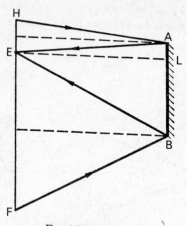

FIG. 16.12. Example.

sees his head H, a ray HA from H to the top A of the mirror is reflected to E. Thus A lies on the perpendicular bisector of HE, and hence $AL = \frac{1}{2} HE = 0.08$ m, where L is the point on the mirror at the same level as E. Since the man sees his feet F, a ray FB from F to the bottom B of the mirror is also reflected to E. Thus the perpendicular bisector of EF passes through B, and hence $BL = \frac{1}{2} FE = \frac{1}{2} \times 1.84$ m $= 0.92$ m.

∴ length of mirror $= AL + LB = 0.08$ m $+ 0.92$ m $= 1$ m.

EXERCISES 16

1. Prove the relation between the angle of rotation of a mirror and the angle of deflection of a reflected ray, when the direction of the incident ray is constant.

2. Two plane mirrors are inclined at an angle of 35°. A ray of light is incident on one mirror at 60°, and undergoes two successive reflections at the mirrors. Show by accurate drawing that the angle of deviation produced is 70°.

Repeat with an angle of incidence of 45°, instead of 60°, and state the law concerning the angle of deviation.

3. Two plane mirrors are inclined to each other at a fixed angle. If a ray travelling in a plane perpendicular to both mirrors is reflected first from one and then from the other, show that the angle through which it is deflected does not depend on the angle at which it strikes the first mirror.

Describe and explain the action of *either* a sextant *or* a rear reflector on a bicycle. (*L.*)

4. State the laws of reflection of light. Two plane mirrors are parallel and face each other. They are *a* cm apart and a small luminous object is placed *b* cm from one of them. Find the distance from the object of an image produced by four reflections. Deduce the corresponding distance for an image produced by 2*n* reflections. (*L.*)

5. Two vertical plane mirrors A and B are inclined to one another at an angle *a*. A ray of light, travelling horizontally, is reflected first from A and then from B. Find the resultant deviation and show it is independent of the original direction of the ray. Describe an optical instrument that depends on the above proposition. (*N.*)

6. State the laws of reflection for a parallel beam of light incident upon a plane mirror.

Indicate clearly by means of diagrams (*a*) how the position and size of the image of an extended object may be determined by geometrical construction, in the case of reflection in a plane; (*b*) how the positions of the images of a small lamp, placed unsymmetrically between parallel reflecting planes, may be graphically determined. (*W.*)

7. Describe the construction of the *sextant* and the *periscope*. Illustrate your answer by clear diagrams and indicate the optical principles involved. (*L.*)

chapter seventeen

Reflection at curved mirrors

CURVED mirrors are reputed to have been used thousands of years ago. Today motor-cars and other vehicles are equipped with driving mirrors which are curved, searchlights have curved mirrors inside them, and the largest telescope in the world utilises a huge curved mirror (p. 544).

Convex and Concave Mirrors. Definitions

In the theory of Light we are mainly concerned with curved mirrors which are parts of *spherical* surfaces. In Fig. 17.1 (*a*), the mirror APB is part of a sphere whose centre C is in front of the reflecting surface; in Fig. 17.1 (*b*), the mirror KPL is part of a sphere whose centre C is behind its reflecting surface. To a person in front of it APB curves inwards and is known as a **concave** mirror, while KPL bulges outwards and is known as a **convex** mirror.

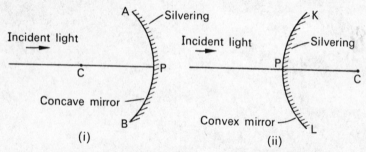

FIG. 17.1. Concave (converging) and convex (diverging) mirrors.

The mid-point, P, of the mirror is called its *pole*; C, the centre of the sphere of which the mirror is part, is known as the *centre of curvature*; and AB is called the *aperture* of the mirror. The line PC is known as the *principal axis*, and plays an important part in the drawing of images in the mirrors; lines parallel to PC are called *secondary axes*.

Narrow and Wide Beams. The Caustic

When a very narrow beam of rays, parallel to the principal axis and close to it, is incident on a concave mirror, experiment shows that all the reflected rays converge to a point F on the principal axis, which is therefore known as the *principal focus* of the mirror, Fig. 17.2 (i). On this account a concave mirror is better described as a "converging" mirror. An image of the sun, whose rays on the earth are parallel, can hence be received on a screen at F, and thus a concave mirror has a *real* focus.

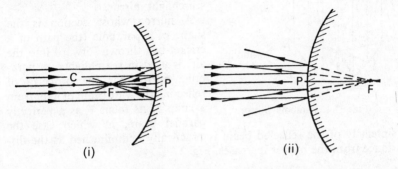

FIG. 17.2. Foci of concave and convex mirrors.

If a narrow beam of parallel rays is incident on a convex mirror, experiment shows that the reflected rays form a divergent beam which appear to come from a point F *behind* the mirror, Fig. 17.2 (ii). A convex mirror has thus a *virtual* focus, and the image of the sun cannot be received on a screen using this type of mirror. To express its action on a parallel beam of light, a convex mirror is often called a "diverging" mirror.

When a *wide* beam of light, parallel to the principal axis, is incident on a concave spherical mirror, experiment shows that reflected rays do not pass through a single point, as was the case with a narrow beam. The reflected rays appear to touch a surface known as a *caustic surface*, S, which has an apex, or cusp, at F, the principal focus, Fig. 17.3. Similarly, if a wide beam of parallel light is incident on a convex mirror, the reflected rays do not appear to diverge from a single point, as was the case with a narrow beam.

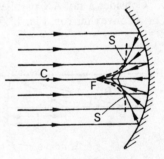

FIG. 17.3. Caustic surface.

Parabolic Mirrors

If a small lamp is placed at the focus, F, of a concave mirror, it follows from the principle of the reversibility of light (p. 390) that rays striking the mirror round a small area about the pole are reflected parallel. See Fig. 17.2 (i). But those rays from the lamp which strike the mirror at points well away from P will be reflected in different directions, because a *wide* parallel beam is not brought to a focus at F, as shown in Fig. 17.3. The beam of light reflected from the mirror thus diminishes

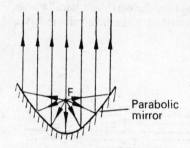

in intensity as its distance from the mirror increases, and a concave spherical mirror is hence useless as a searchlight mirror.

A mirror whose section is the shape of a parabola (the path of a cricket-ball thrown forward into the air) is used in searchlights. A parabolic mirror has the property of reflecting the wide beam of light from a lamp at its focus F as a perfectly parallel beam, in which case the intensity of the reflected beam is practically undiminished as the distance from the mirror increases, Fig. 17.4.

FIG. 17.4. Parabolic mirror.

Focal Length (f) and Radius of Curvature (r)

From now onwards we shall be concerned with curved spherical mirrors of small aperture, so that a parallel incident beam will pass through the focus after reflection. The diagrams which follow are exaggerated for purposes of clarity.

The distance PC from the pole to the centre of curvature is known as the *radius of curvature* (r) of a mirror; the distance PF from the pole to the focus is known as the *focal length* (f) of the mirror. As we shall now prove, there is a simple relation between f and r.

Consider a ray AX parallel to the principal axis of either a concave or a convex mirror, Fig. 17.5 (i), (ii). The normal to the mirror at X is

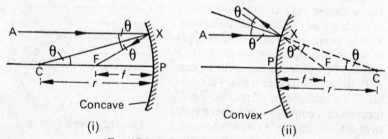

FIG. 17.5. Relation between f and r.

CX, because the radius of the spherical surface is perpendicular to the surface, and hence the reflected ray makes an angle, θ, with CX equal to the incident angle θ. Taking the case of the concave mirror, angle AXC = angle XCP, alternate angles, Fig. 17.5 (i). Thus triangle FXC is isosceles, and FX = FC. As X is a point very close to P we assume to a very good approximation that FX = FP.

$$\therefore \quad FP = FC, \text{ or } FP = \tfrac{1}{2} CP.$$

$$\therefore \quad f = \frac{r}{2} \quad . \quad . \quad . \quad . \quad . \quad . \quad (1)$$

This relation between f and r is the same for the case of the convex mirror, Fig. 17.5 (ii), as the reader can easily verify.

Images in Concave Mirrors

Concave mirrors produce images of different sizes; sometimes they are inverted and real, and on other occasions they are erect (the same way up as the object) and virtual. As we shall see, the nature of the image formed depends on the distance of the object from the mirror.

Consider an object of finite size OH placed at O perpendicular to the principal axis of the mirror, Fig. 17.6 (i). The image, R, of the top point

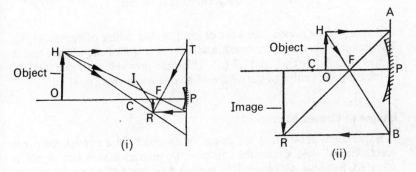

FIG. 17.6. Images in concave mirrors.

H can be located by the intersection of two reflected rays coming initially from H, and the rays usually chosen are two of the following: (1) The ray HT parallel to the principal axis, which is reflected to pass through the focus, F, (2) the ray HC passing through the centre of curvature, C, which is reflected back along its own path because it is a normal to the mirror, (3) the ray HF passing through the focus, F, which is reflected parallel to the principal axis. Since the mirror has a small aperture, and we are considering a narrow beam of light, the mirror must be represented in accurate image drawings by a *straight* line. Thus PT in Fig. 17.6 (i) represents a perfect mirror.

When the object is a very long distance away (at infinity), the image is small and is formed inverted at the focus (p. 402). As the object approaches the centre of curvature, C, the image remains real and inverted, and is formed in front of the object, Fig. 17.6 (i). When the object is between C and F, the image is real, inverted, and larger than the object; it is now further from the mirror than the object, Fig. 17.6 (ii).

As the object approaches the focus, the image recedes further from the mirror, and when the object is at the focus, the image is at infinity. When the object is nearer to the mirror than the focus the image IR becomes *erect* and *virtual*, as shown in Fig. 17.7 (i). In this case the image is *magnified*, and the concave mirror can thus be used as a shaving mirror.

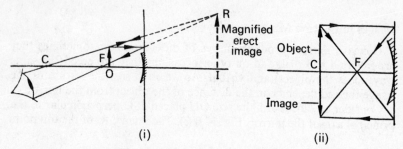

(i)   (ii)

FIG. 17.7. Images in concave mirrors.

A special case occurs when the object is at the centre of curvature, C. The image is then real, inverted, and the same size of the object, and it is also situated at C, Fig. 17.7 (ii). This case provides a simple method of locating the centre of curvature of a concave mirror (p. 413).

Images in Convex Mirrors

Experiment shows that the image of an object in a convex mirror is erect, virtual, and diminished in size, no matter where the object is situated. Suppose an object OH is placed in front of a convex mirror, Fig. 17.8 (i). A ray HM parallel to the principal axis is reflected as if it

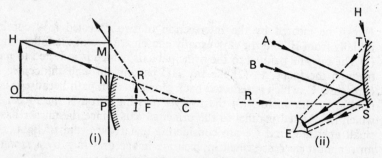

(i)   (ii)

FIG. 17.8. Images in convex mirrors.

appeared to come from the virtual focus, F, and a ray HN incident towards the centre of curvature, C, is reflected back along its path. The two reflected rays intersect *behind* the mirror at R, and IR is a virtual and erect image.

Objects well outside the principal axis of a convex mirror, such as A, B in Fig. 17.8 (ii), can be seen by an observer at E, whose *field of view* is that between HT and RS, where T, S are the edges of the mirror. Thus in addition to providing an erect image the convex mirror has a wide field of view, and is hence used as a driving mirror.

Formulæ for Mirrors. Sign Convention

Many of the advances in the uses of curved mirrors and lenses have resulted from the use of *optical formulæ*, and we have now to consider the relation which holds between the object and image distances in mirrors and their focal length. In order to obtain a formula which holds for both concave and convex mirrors, a *sign rule* or *convention* must be obeyed, and we shall adopt the following:

A real object or image distance is a positive distance.

A virtual object or image distance is a negative distance.

In brief, "real is positive, virtual is negative". The focal length of a concave mirror is thus a positive distance; the focal length of a convex mirror is a negative distance.

Concave Mirror

Consider a point object O on the principal axis of a concave mirror. A ray OX from O is reflected in the direction XI making an equal angle θ with the normal CX; a ray OP from O, incident at P, is reflected back along PO, since CP is the normal at P. The point of intersection, I, of the two rays is the image of O. Fig. 17.9.

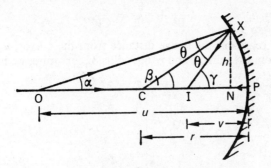

FIG. 17.9. Mirror formula.

Suppose α, β, γ are the angles made by OX, CX, IX respectively with the axis. Since we are considering a mirror of small aperture these angles are small in practice, Fig. 17.9 being exaggerated. As β is the exterior angle of triangle CXO, we have $\beta = \alpha + \theta$.

$$\therefore \theta = \beta - \alpha \quad . \quad . \quad . \quad . \quad \text{(i)}$$

Since γ is the exterior angle of triangle IXC, we have $\gamma = \beta + \theta$.

$$\therefore \theta = \gamma - \beta \quad . \quad . \quad . \quad . \quad \text{(ii)}$$

From (i) and (ii), it follows that

$$\beta - \alpha = \gamma - \beta$$
$$\therefore \alpha + \gamma = 2\beta \quad . \quad . \quad . \quad . \quad \text{(iii)}$$

We can now substitute for α, β, γ in terms of h, the height of X above the axis, and the distances OP, CP, IP. In so doing (a) we assume N is practically coincident with P, as X is very close to P in practice, (b) the appropriate sign, $+$ or $-$, must precede all the numerical values of the distances concerned. Also, as $\alpha = \tan \alpha$ in radians when α is very small, we have

$$\alpha = \frac{XN}{ON} = \frac{XN}{+ OP} = \frac{h}{+ OP},$$

where OP is the distance of the real object O from the mirror in centimetres, say, and XN $= h$. Similarly,

$$\beta = \frac{XN}{CN} = \frac{XN}{+ CP} = \frac{h}{+ CP},$$

as CP, the radius of curvature of the concave mirror, is real.

Also,
$$\gamma = \frac{XN}{IN} = \frac{XN}{+ IP} = \frac{h}{+ IP},$$

where IP is the distance of the real image I from the mirror. Substituting for α, β, γ in (iii),

$$\frac{h}{(+ IP)} + \frac{h}{(+ OP)} = 2\frac{h}{(+ CP)}.$$

Dividing by h,
$$\frac{1}{IP} + \frac{1}{OP} = \frac{2}{CP}.$$

If we let v represent the image distance from the mirror, u the object distance from the mirror, and r the radius of curvature, we have

$$\frac{1}{v} + \frac{1}{u} = \frac{2}{r} \qquad . \qquad . \qquad . \qquad . \qquad (2)$$

Further, since $f = r/2$, then $\dfrac{2}{r} = \dfrac{1}{f}$.

$$\therefore \frac{1}{v} + \frac{1}{u} = \frac{1}{f} \qquad . \qquad . \qquad . \qquad . \qquad (3)$$

The relations (2), (3) are general formulae for curved spherical mirrors; and when they are used the appropriate sign for v, u, f, or r must always precede the corresponding numerical value.

Convex Mirror

We now obtain a relation for object distance (u), image distance (v), and focal length (f) of a convex mirror. In this case the incident rays OX, OP are reflected as if they appear to come from the point I behind the mirror, which is therefore a virtual image, and hence the image distance IP is *negative*, Fig. 17.10. Further, CP is negative, as the centre of curvature of X, a convex mirror, is behind the mirror.

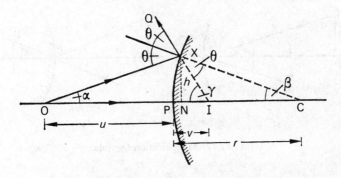

FIG. 17.10. Mirror formula.

Since θ is the exterior angle of triangle COX, $\theta = \alpha + \beta$. As γ is the exterior angle of triangle CIX, $\gamma = \theta + \beta$, or $\theta = \gamma - \beta$.

$$\therefore \gamma - \beta = \alpha + \beta$$
$$\therefore \gamma - \alpha = 2\beta \qquad \cdot \qquad \cdot \qquad \cdot \qquad \cdot \qquad \text{(i)}$$

Now $\gamma = \dfrac{h}{IN} = \dfrac{h}{(-IP)}$, as I is virtual; $\alpha = \dfrac{h}{ON} = \dfrac{h}{(+OP)}$, as O is real,

$\beta = \dfrac{h}{NC} = \dfrac{h}{(-PC)}$, as C is virtual. Substituting in (i),

$$\therefore \quad \frac{h}{(-IP)} - \frac{h}{(+OP)} = \frac{2h}{(-CP)}$$
$$\therefore \quad \frac{1}{IP} + \frac{1}{OP} = \frac{2}{CP}$$
$$\therefore \frac{1}{v} + \frac{1}{u} = \frac{2}{r}$$

and
$$\therefore \frac{1}{v} + \frac{1}{u} = \frac{1}{f}.$$

Thus, using the sign convention, the same formula holds for concave and convex mirrors.

Formula for Magnification

The lateral magnification, m, produced by a mirror is defined by

$$m = \frac{height\ of\ image}{height\ of\ object}.$$

Suppose IR is the image of an object OH in a concave or convex mirror. Fig. 17.11 (i), (ii). Then a ray HP from the top point H of the

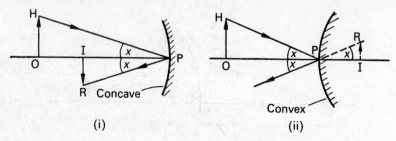

FIG. 17.11. Magnification formula.

object passes through the top point R of the image after reflection from the mirror. Now the normal to the mirror at P is the principal axis, OP. Thus angle OPH = angle IPR, from the law of reflection.

$$\therefore \quad \tan OPH = \tan IPR$$

$$\text{i.e.,} \quad \frac{OH}{OP} = \frac{IR}{IP}$$

$$\therefore \quad \frac{IR}{OH} = \frac{IP}{OP}$$

But IP = image distance = v, and OP = object distance = u

$$\therefore \quad \frac{IR}{OH} = \frac{v}{u}$$

$$\therefore \quad m = \frac{v}{u} \quad . \quad . \quad . \quad . \quad (4)$$

Thus if the image distance is half the object distance, the image is half the length of the object.

Since

$$\frac{1}{v} + \frac{1}{u} = \frac{1}{f}$$

$$\therefore \frac{v}{v} + \frac{v}{u} = \frac{v}{f},$$

multiplying throughout by v.

$$1 + \frac{v}{u} = \frac{v}{f}$$

$$\therefore 1 + m = \frac{v}{f}$$

$$\therefore \quad m = \frac{v}{f} - 1$$

Some Applications of Mirror Formulæ

The following examples will assist the reader to understand how to apply the formulæ $\dfrac{1}{v} + \dfrac{1}{u} = \dfrac{1}{f}$ and $m = \dfrac{v}{u}$ correctly:

1. An object is placed 10 cm in front of a concave mirror of focal length 15 cm. Find the image position and the magnification.

Since the mirror is concave, $f = +15$ cm. The object is real, and hence $u = +10$ cm. Substituting in $\dfrac{1}{v} + \dfrac{1}{u} = \dfrac{1}{f}$,

$$\frac{1}{v} + \frac{1}{(+10)} = \frac{1}{(+15)}$$

$$\therefore \frac{1}{v} = \frac{1}{15} - \frac{1}{10} = -\frac{1}{30}$$

$$\therefore v = -30$$

Since v is negative in sign the image is *virtual*, and it is 30 cm from the mirror. See Fig. 17.7 (i). The magnification, $m = \dfrac{v}{u} = \dfrac{30}{10} = 3$, so that the image is three times as high as the object.

2. The image of an object in a convex mirror is 4 cm from the mirror. If the mirror has a radius of curvature of 24 cm, find the object position and the magnification.

The image in a convex mirror is always virtual (p. 406). Hence $v = -4$ cm. The focal length of the mirror $= \frac{1}{2} r = 12$ cm; and since the mirror is convex, $f = -12$ cm. Substituting in $\dfrac{1}{v} + \dfrac{1}{u} = \dfrac{1}{f}$

$$\frac{1}{(-4)} + \frac{1}{u} = \frac{1}{(-12)}$$

$$\therefore \frac{1}{u} = -\frac{1}{12} + \frac{1}{4} = \frac{1}{6}$$

$$\therefore u = 6$$

Since u is positive in sign the object is real, and it is 6 cm from the mirror. The magnification, $m = \dfrac{v}{u} = \dfrac{4}{6} = \dfrac{2}{3}$, and hence the image is two-thirds as high as the object. See Fig. 17.8 (i).

3. An erect image, three times the size of the object, is obtained with a concave mirror of radius of curvature 36 cm. What is the position of the object? If x cm is the numerical value of the distance of the object from the mirror,

the image distance must be $3x$ cm, since the magnification $m = \dfrac{\text{image distance}}{\text{object distance}}$
$= 3$. Now an *erect* image is obtained with a concave mirror only when the image is *virtual* (p. 406).

$$\therefore \quad \text{image distance}, v = -3x$$

$$\text{Also, object distance}, u = +x$$
$$\text{and focal length}, f = \tfrac{1}{2}r = +18 \text{ cm}.$$

Substituting in $\dfrac{1}{v} + \dfrac{1}{u} = \dfrac{1}{f}$,

$$\therefore \frac{1}{(-3x)} + \frac{1}{(+x)} = \frac{1}{(+18)}$$

$$\therefore -\frac{1}{3x} + \frac{1}{x} = \frac{1}{18}$$

$$\therefore \frac{2}{3x} = \frac{1}{18}$$

$$\therefore \quad x = 12$$

Thus the object is 12 cm from the mirror.

Virtual Object and Convex Mirror

We have already seen that a convex mirror produces a virtual image of an object in front of it, which is a real object. A convex mirror may sometimes produce a real image of a *virtual* object.

As an illustration, consider an incident beam of light bounded by AB, DE, converging to a point O *behind* the mirror, Fig. 17.12. O is

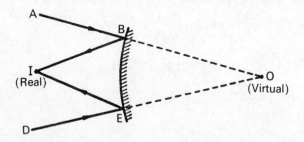

Fig. 17.12. Real image in convex mirror.

regarded as a virtual object, and if its distance from the mirror is 10 cm, then the object distance, $u = -10$. Suppose the convex mirror has a focal length of 15 cm, i.e., $f = -15$.

Since
$$\frac{1}{v} + \frac{1}{u} = \frac{1}{f},$$

$$\frac{1}{v} + \frac{1}{(-10)} = \frac{1}{(-15)}$$

$$\therefore \ \frac{1}{v} = -\frac{1}{15} + \frac{1}{10} = +\frac{1}{30}$$

$$\therefore \ v = +30$$

The point image, I, is thus 30 cm from the mirror, and is *real*. The beam reflected from the mirror is hence a convergent beam, Fig. 17.12; a similar case with a plane mirror is shown in Fig. 16.9 (ii).

Object at Centre of Curvature of Concave Mirror

Suppose an object is placed at the centre of curvature of a concave mirror. Then $u = +r$, where r is the numerical value of the radius of curvature. Substituting in $\frac{1}{v} + \frac{1}{u} = \frac{2}{r}$ to find the image distance v,

$$\frac{1}{v} + \frac{1}{(+r)} = \frac{2}{(+r)}$$

$$\therefore \ \frac{1}{v} = \frac{2}{r} - \frac{1}{r} = \frac{1}{r}$$

$$\therefore \ v = r$$

The image is therefore also formed at the centre of curvature. The magnification in this case is given by $m = \frac{v}{u} = \frac{r}{r} = 1$, and hence the object and image are the same size. This case is illustrated in Fig. 17.7 (ii), to which the reader should now refer.

SOME METHODS OF DETERMINING FOCAL LENGTH AND RADIUS OF CURVATURE OF MIRRORS

Concave Mirror

Method 1. A pin O is placed above a concave mirror M so that an inverted image of the pin can be seen, Fig. 17.13. If the pin is moved up and down with its point on the axis of the mirror, and an observer E moves his eye perpendicularly to the pin at the same time, a position of O is reached when the image I remains perfectly in line with O as E moves; i.e., there is no parallax (no relative displacement) between

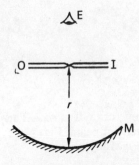

FIG. 17.13. Centre of curvature of concave mirror.

pin and image. The pin is now at exactly the same place as its image I, Fig. 17.13. Since an object and image coincide in position at the centre of curvature of a concave mirror (p. 413), the distance from the point of the pin to the mirror is equal to its radius of curvature, r. The focal length, f, which is $\frac{r}{2}$, is then easily obtained.

If an illuminated object is available instead of a pin, the object is moved to or from the mirror until a clear image is obtained beside the object. The distance of the object from the mirror is then equal to r.

In general the method of no parallax, using a pin, gives a higher degree of accuracy in locating an image.

Method 2. By using the method of no parallax, or employing an illuminated object, several, say six, values of the image distance, v, can be obtained with the concave mirror, corresponding to six different values of the object distance u. Substituting for u, v in the formula $\frac{1}{v} + \frac{1}{u} = \frac{1}{f}$, six values of f can be calculated, and the average value taken.

A better method of procedure, however, is to plot the magnitudes of

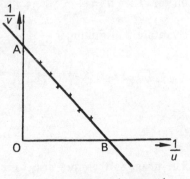

FIG. 17.14. Graph of $\frac{1}{v}$ against $\frac{1}{u}$.

$\frac{1}{v}$ against $\frac{1}{u}$; a straight line BA can be drawn through the points thus obtained, Fig. 17.14. Now $\frac{1}{v} + \frac{1}{u} = \frac{1}{f}$. Hence, when $\frac{1}{v} = 0$, $\frac{1}{u} = \frac{1}{f}$.

But
$$OB = \frac{1}{u} \text{ when } \frac{1}{v} = 0$$

$$\therefore OB = \frac{1}{f}, \quad \text{i.e., } f = \frac{1}{OB}.$$

Thus the focal length can be determined from the reciprocal value of the intercept OB on the axis of $\frac{1}{u}$.

From $\frac{1}{v} + \frac{1}{u} = \frac{1}{f}$, $\frac{1}{v} = \frac{1}{f}$ when $\frac{1}{u} = 0$. Thus $OA = \frac{1}{f}$, Fig. 17.14, and hence $f = \frac{1}{OA}$. It can thus be seen that (i) f can also be calculated from the reciprocal of the intercept OA on the axis of $\frac{1}{v}$, (ii) OAB is an isos-

celes triangle if the same scale is employed for $\frac{1}{v}$ and $\frac{1}{u}$.

Convex Mirror

Method 1. By using a convex lens, L, a real image of an object O can be formed at a point C on the other side of L, Fig. 17.15. The convex mirror, MN, is then placed between L and C with its reflecting face facing the lens, so that a convergent beam of rays is incident on the mirror. When the latter is moved along the axis OC a position will be reached when the beam is incident *normally* on the mirror, in which case the rays are reflected back along the incident path. *A real inverted image is then formed at O.*

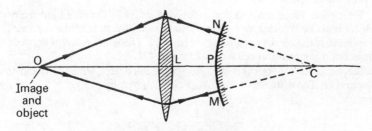

FIG. 17.15. Convex mirror measurement.

Since the rays incident on the mirror, for example at N or M, are normal to the mirror surface, they will, if produced, pass through the centre of curvature, C, of the mirror. Thus the distance PC = r, the radius of curvature. Since PC = LC − LP, this distance can be obtained from measurement of LP and LC, the latter being the image distance from the lens when the mirror is taken away.

Method 2. A more difficult method than the above consists of positioning a pin O in front of the convex mirror, when a virtual image, I, is formed, Fig. 17.16. A small plane mirror M is then moved between O and P until the image I' of the lower part of O *in M* coincides in position with the upper part of the image of O in the convex mirror. The distances OP, MP are then measured.

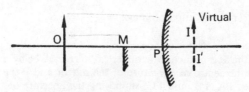

FIG. 17.16. Convex mirror measurement.

Since M is a plane mirror, the image I' of O in it is such that OM = MI'. Thus PI = MI' − MP = OM − MP, and hence PI can be calculated. But PI = v, the image distance of O in the convex mirror, and OP = u, the object distance. Substituting for the virtual distance v, and u, in $\dfrac{1}{v} + \dfrac{1}{u} = \dfrac{1}{f}$, the focal length of the convex mirror can be found.

EXAMPLES

1. Show that a concave spherical mirror can produce a focused image of an object when certain conditions are observed, and prove the usual relation between the object and image distances. A linear object, 10 cm long, lies along the axis of a concave mirror whose radius of curvature is 30 cm, the near end of the object lying 18 cm from the mirror. Find the magnification of the image. (*W.*)

First part. The condition for a focused image is that the light from the object must be incident as a narrow beam round the pole of the mirror. This implies that the object must be small (p. 402). The usual relation between the object and image distances is proved on p. 407.

Second part. Suppose Q is the near end of the object, which is 18 cm from the mirror. The distance of the image of Q is given by

$$\frac{1}{v} + \frac{1}{(+18)} = \frac{1}{(+15)}$$

since $f = \dfrac{r}{2} = \dfrac{30}{2} = 15$ cm

$$\therefore \frac{1}{v} = \frac{1}{15} - \frac{1}{18}$$

from which $v = 90$ cm

The other end P of the object is (10 + 18) cm from the mirror, or 28 cm The image of P is given by

$$\frac{1}{v} + \frac{1}{(+28)} = \frac{1}{(+15)}$$

$$\therefore \frac{1}{v} = \frac{1}{15} - \frac{1}{28}$$

from which $v = 32 \cdot 3$ cm

$$\therefore \text{ length of image} = 90 - 32 \cdot 3 = 57 \cdot 7 \text{ cm}$$

$$\therefore \text{ magnification of image} = \frac{57 \cdot 7}{10} = 5 \cdot 8$$

2. PBCA is the axis of a concave spherical mirror, A being a point object, B its image, C the centre of curvature of the mirror and P the pole. Find a relation between PA, PB, and PC, supposing the aperture of the mirror to be small. A concave mirror forms, on a screen, a real image of twice the linear dimensions of the object. Object and screen are then moved until the image is three times the size of the object. If the shift of the screen is 25 cm determine the shift of the object and the focal length of the mirror. (*N.*)

First part. See text.

Second part. Suppose *v* is the distance of the screen from the mirror when the image is twice the length of the object. Since the magnification, *m*, is given by

$$m = \frac{v}{f} - 1 \qquad \cdots \qquad \cdots \qquad \text{(i)}$$

where *f* is the focal length of the mirror (see p. 410),

$$\therefore 2 = \frac{v}{f} - 1 \qquad \cdots \qquad \cdots \qquad \text{(ii)}$$

When *m* = 3, the image distance is (*v* + 25) cm. Substituting in (i),

$$\therefore 3 = \frac{v + 25}{f} - 1 \qquad \cdots \qquad \cdots \qquad \text{(iii)}$$

Subtracting (ii) from (iii), we have

$$1 = \frac{25}{f}$$
$$\therefore f = 25 \text{ cm.}$$

From (ii), $2 = \dfrac{v}{25} - 1$, or *v* = 75 cm. The object distance, *u*, is thus given by $\dfrac{1}{75} + \dfrac{1}{u} = \dfrac{1}{25}$, from which $u = 37\frac{1}{2}$ cm. From (iii), *v* + 25 = 100 cm.

The object distance, *u*, is then given by $\dfrac{1}{100} + \dfrac{1}{u} = \dfrac{1}{25}$, from which $u = 33\frac{1}{3}$ cm. Thus the shift of the object $= 37\frac{1}{2} - 33\frac{1}{3} = 4\frac{1}{6}$ cm.

EXERCISES 17

1. An object is placed (i) 10 cm, (ii) 4 cm from a concave mirror of radius curvature 12 cm. Calculate the image position in each case, and the respective magnifications.

2. Repeat Q. 1 by accurate drawings to scale. (*Note.*—The mirror must be represented by a straight line.)

3. An object is placed 15 cm from a convex mirror of focal length 10 cm. Calculate the image distance and the magnification produced. Draw an accurate diagram to scale, and verify your drawing from the calculated results.

4. Explain with the aid of diagrams why a curved mirror can be used (i) as a driving mirror, (ii) in a searchlight, (iii) as a shaving mirror. Why is a special form of mirror required in the searchlight?

5. Describe and explain a method of finding the focal length of (*a*) a concave mirror, (*b*) a convex mirror.

6. A pole 4 m long is laid along the principal axis of a convex mirror of focal length 1 m The end of the pole nearer the mirror is 2 m from it. Find the length of the image of the pole.

7. Deduce a formula connecting *u*, *v* and *r*, the distances of object, image and centre of curvature from a spherical mirror.

A mirror forms an erect image 30 cm from the object and twice its height. Where must the mirror be situated? What is its radius of curvature? Assuming the object to be real, determine whether the mirror is convex or concave. (*L.*)

objet between mirror between mirror and f.

8. Establish the formula $\dfrac{1}{v} + \dfrac{1}{u} = \dfrac{1}{f}$ for a concave mirror.

In an experiment with a concave mirror the magnification *m* of the image is measured for a series of values of *v*, and a curve is plotted between *m* and *v*. What curve would you expect to obtain, and how would you use it to deduce the focal length of the mirror? (*C.*)

9. Derive an approximate relation connecting the distances of an object and its image from the surface of a convex spherical mirror.

A small object is placed at right angles to the axis of a concave mirror so as to form (*a*) a real, (*b*) a virtual image, twice as long as the object. If the radius of curvature of the mirror is *R* what is the distance between the two images? (*L.*)

10. Deduce a formula connecting the distances of object and image from a spherical mirror. What are the advantages of a concave mirror over a lens for use in an astronomical telescope?

A driving mirror consists of a cylindrical mirror of radius 10 cm and length (over the curved surface) of 10 cm. If the eye of the driver be assumed at a great distance from the mirror, find the angle of view. (*O. & C.*)

11. Find the relation connecting the focal length of a convex spherical mirror with the distances from the mirror of a small object and the image formed by the mirror.

A convex mirror, radius of curvature 30 cm, forms a real image 20 cm from its surface. Explain how this is possible and find whether the image is erect or inverted. (*L.*)

12. What conditions must be satisfied for an optical system to form an image of an object? Show how these conditions are satisfied for a convex spherical mirror when a small object is placed on its axis and derive a relationship showing how the position of the image depends on the position of the object and the radius of curvature of the mirror.

A millimetre scale is placed at right angles to the axis of a convex mirror of radius of curvature 12 cm. This scale is 18 cm away from the pole of the mirror. Find the position of the image of the scale. What is the size of the divisions of the image? What is the ratio of the angle subtended by the image to that subtended by the object at a point on the axis 25 cm away from the object on the side remote from the mirror? (*O. & C.*)

13. Describe an experiment to determine the radius of curvature of a convex mirror by an optical method. Illustrate your answer with a ray diagram and explain how the result is derived from the observations.

A small convex mirror is placed 60 cm from the pole and on the axis of a large concave mirror, radius of curvature 200 cm. The position of the convex mirror is such that a real image of a distant object is formed in the plane of a hole drilled through the concave mirror at its pole. Calculate (*a*) the radius of curvature of the convex mirror, (*b*) the height of the real image if the distant object subtends an angle of 0·50° at the pole of the concave mirror. Draw a ray diagram to illustrate the action of the convex mirror in producing the image of a non-axial point of the object and suggest a practical application of this arrangement of mirrors. (*N*.)

chapter eighteen

Refraction at plane surfaces

Laws of Refraction

WHEN a ray of light AO is incident at O on the plane surface of a glass medium, observation shows that some of the light is reflected from the surface along OC in accordance with the laws of reflection, while the rest of the light travels along a new direction, OB, in the glass, Fig. 18.1.

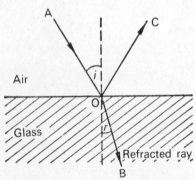

On account of the change in direction the light is said to be "refracted" on entering the glass; and the *angle of refraction*, r, is the angle made by the refracted ray OB with the normal at O.

Historical records reveal that the astronomer PTOLEMY, who lived about A.D. 140, measured numerous values of the angle of incidence, i, and the angle of refraction, r, for glass as the angle of incidence was varied. However, he was unable to discover any relation between i and r.

FIG. 18.1. Refraction at plane surface.

Later scientists were equally unsuccessful, until centuries later SNELL, a Dutch professor, discovered in 1620 that the sines of the angles bear a constant ratio to each other. The *laws of refraction* are:

1. *The incident and refracted rays, and the normal at the point of incidence, all lie in the same plane.*

2. *For two given media,* $\dfrac{\sin i}{\sin r}$ *is a constant, where i is the angle of incidence and r is the angle of refraction* (Snell's law).

Refractive Index.

The constant ratio $\dfrac{\sin i}{\sin r}$ is known as the **refractive index** for the two given media; and as the magnitude of the constant depends on the colour of the light, it is usually specified as that obtained for yellow light. If the medium containing the incident ray is denoted by 1, and that containing the refracted ray by 2, the refractive index can be denoted by $_1n_2$.

Scientists have drawn up tables of refractive indices when the incident ray is travelling *in vacuo* and is then refracted into the medium concerned, for example, glass or water. The values thus obtained are known as the *absolute* refractive indices of the media; and as a vacuum is always the first medium, the subscripts for the absolute refractive index, symbol n, can be dropped. The magnitude of n for glass is about 1·5, n for water is about 1·33, and n for air at normal pressure is about 1·00028. As the magnitude of the refractive index of a medium is only very slightly altered when the incident light is in air instead of a vacuum, experiments to determine the absolute refractive index n are usually performed with the light incident from air on to the medium; thus we can take $_{air}n_{glass}$ as equal to $_{vacuum}n_{glass}$ for most practical purposes.

We have already mentioned that light is refracted because it has different velocities in different media. The Wave Theory of Light, discussed on p. 679, shows that the refractive index $_1n_2$ for two given media 1 and 2 is given by

$$_1n_2 = \frac{velocity\ of\ light\ in\ medium\ 1}{velocity\ of\ light\ in\ medium\ 2} \qquad . \qquad . \qquad . \qquad (1)$$

and this is a *definition* of refractive index which can be used instead of the ratio $\dfrac{\sin i}{\sin r}$. An alternative definition of the absolute refractive index, n, of a medium is thus

$$n = \frac{velocity\ of\ light\ in\ a\ vacuum}{velocity\ of\ light\ in\ medium} \qquad . \qquad . \qquad . \qquad (2)$$

In practice the velocity of light in air can replace the velocity in vacuo in this definition.

Relations Between Refractive Indices

(1) Consider a ray of light, AO, refracted from *glass to air* along the direction OB; observation then shows that the refracted ray OB is bent away from the normal, Fig. 18.2. The refractive index from glass to air, $_gn_a$, is given by $\sin x/\sin y$, by definition, where x is the angle of incidence in the glass and y is the angle of refraction in the air.

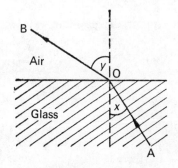

FIG. 18.2. Refraction from glass to air.

From the principle of the reversibility of light (p. 390), it follows that a ray travelling along BO in air is refracted along OA in the glass. The refractive index from air to glass, $_an_g$, is then given by $\sin y/\sin x$, by definition. But $_gn_a = \dfrac{\sin x}{\sin y}$, from the previous paragraph.

$$\therefore \quad _gn_a = \frac{1}{_an_g} \qquad . \qquad . \qquad . \qquad . \qquad . \qquad (3)$$

If $_an_g$ is 1.5, then $_gn_a = \dfrac{1}{1.5} = 0.67$. Similarly, if the refractive index

from air to water is $4/3$, the refractive index from water to air is $3/4$.

(**2**) Consider a ray AO incident in air on a plane glass boundary, then refracted from the glass into a water medium, and finally emerging along a direction CD into air. *If the boundaries of the media are parallel, experiment shows that the emergent ray CD is parallel to the incident ray AO*, although there is a relative displacement. Fig. 18.3. Thus the angles made with the normals by AO, CD are equal, and we shall denote them by i_a.

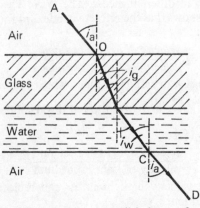

FIG. 18.3. Refraction at parallel plane surfaces.

Suppose i_g, i_w are the angles made with the normals by the respective

rays in the glass and water media. Then, by definition, $_gn_w = \dfrac{\sin i_g}{\sin i_w}$.

But
$$\frac{\sin i_g}{\sin i_w} = \frac{\sin i_g}{\sin i_a} \times \frac{\sin i_a}{\sin i_w},$$

and
$$\frac{\sin i_g}{\sin i_a} = {_gn_a}, \text{ and } \frac{\sin i_a}{\sin i_w} = {_an_w}$$

$$\therefore \quad {_gn_w} = {_gn_a} \times {_an_w} \quad . \qquad . \qquad . \qquad . \qquad . \quad \text{(i)}$$

Further, as $_gn_a = \dfrac{1}{_an_g}$, we can write

$$_gn_w = \frac{_an_w}{_an_g}.$$

Since $_an_w = 1.33$ and $_an_g = 1.5$, it follows that $_gn_w = \dfrac{1.33}{1.5} = 0.89$.

From (i) above, it follows that in general

$$_1n_3 = {_1n_2} \times {_2n_3} \quad . \qquad . \qquad . \qquad . \qquad . \quad \text{(4)}$$

The order of the suffixes enables this formula to be easily memorised.

General Relation Between n and Sin i

From Fig. 18.3, $\sin i_a / \sin i_g = {}_an_g$

$$\therefore \sin i_a = {}_an_g \sin i_g \qquad . \qquad . \qquad . \qquad . \qquad . \qquad \text{(i)}$$

Also, $\sin i_w / \sin i_a = {}_wn_a = 1/{}_an_w$

$$\therefore \sin i_a = {}_an_w \sin i_w \qquad . \qquad . \qquad . \qquad . \qquad . \qquad \text{(ii)}$$

Hence, from (i) and (ii),

$$\sin i_a = {}_an_g \sin i_g = {}_an_w \sin i_w$$

If the equations are re-written in terms of the absolute refractive indices of air (n_a), glass (n_g), and water (n_w), we have

$$n_a \sin i_a = n_g \sin i_g = n_w \sin i_w$$

since $n_a = 1$. This relation shows that when a ray is refracted from one medium to another, *the boundaries being parallel*,

$$n \sin i = \text{a constant} \qquad . \qquad . \qquad . \qquad . \qquad \text{(5)}$$

where n is the absolute refractive index of a medium and i is the angle made by the ray with the normal in that medium.

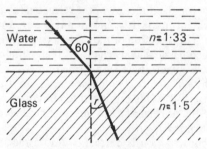

Fig. 18.4. Refraction from water to glass.

This relation applies also to the case of light passing directly from one medium to another. As an illustration of its use, suppose a ray is incident on a water-glass boundary at an angle of 60°, Fig. 18.4. Then, applying "$n \sin i$ is a constant", we have

$$1 \cdot 33 \sin 60° = 1 \cdot 5 \sin r, \qquad . \qquad . \qquad . \qquad \text{(iii)}$$

where r is the angle of refraction in the glass, and $1 \cdot 33$, $1 \cdot 5$ are the respective values of n_w and n_g. Thus $\sin r = 1 \cdot 33 \sin 60°/1 \cdot 5 = 0 \cdot 7679$, from which $r = 50 \cdot 1°$.

Multiple Images in Mirrors

If a candle or other object is held in front of a plane mirror, a series of faint or "ghost" images are observed in addition to one bright image.

424   OPTICS, SOUND AND WAVES

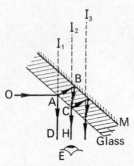

FIG. 18.5. Multiple images.

Suppose O is an object placed in front of a mirror with silvering on the back surface M, as shown in Fig. 18.5. A ray OA from O is then reflected from the front (glass) surface along AD and gives rise to a faint image I_1, while the remainder of the light energy is refracted at A along AB. Reflection then takes place at the silvered surface, and after refraction into the air along CH a bright image is observed at I_2. A small percentage of the light is reflected at C, however, and re-enters the glass again, thus forming a faint image at I_3. Other faint images are formed in the same way. Thus a series of *multiple images* is obtained, the brightest being I_2. The images lie on the normal from O to the mirror, the distances depending on the thickness of the glass and its refractive index and the angle of incidence.

Drawing the Refracted Ray by Geometrical Construction

Since $\sin i/\sin r = n$, the direction of the refracted ray can be calculated when a ray is incident in air at a known angle i on a medium of given refractive index n. The direction of the refracted ray can also be obtained by means of a geometrical construction. Thus suppose AO is a ray incident in air at a given angle i on a medium of refractive index n, Fig. 18.6 (i). With O as centre, two circles, a, b, are drawn whose radii are in the ratio $1:n$, and AO is produced to cut circle a at P. PN is then drawn parallel to the normal at O to intersect circle b at Q. *OQ is then the direction of the refracted ray.*

To prove the construction is correct, we note that angle OPN $= i$, angle OQN $= r$. Thus $\sin i/\sin r =$ ON/OP $\div$ ON/OQ $=$ OQ/OP. But OQ/OP $=$ radius of circle b/radius of circle $a = n$, from our drawing

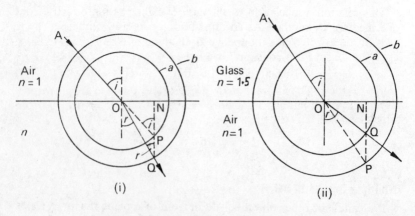

FIG. 18.6. Drawing of refracted rays.

of the circles. Hence $\sin i/\sin r = n$. Thus OQ must be the refracted ray. Although we have taken the case of the incident ray in air, the same construction will enable the refracted ray to be drawn when the incident ray is in any other medium. The radii of the circles are then in the ratio of the absolute refractive indices.

Fig. 18.6 (ii) illustrates the drawing in the case of a ray AO refracted from a dense medium such as glass ($n = 1\cdot5$) into a less dense medium such as air ($n = 1$). Circles a, b are again drawn concentric with O, their radii being in the ratio $1 : n$. The incident ray AO, however, is produced to cut the *larger* circle b this time, at P, and a line PN is then drawn parallel to the normal at O to intersect the circle a at Q. OQ is then the direction of the refracted ray. The proof for the construction follows similar lines to that given for Fig. 18.6 (i), and it is left as an exercise for the reader.

The direction of the incident ray AO in Fig. 18.6 (ii) may be such that the line PN does not intersect the circle a. In this case, which is important and is discussed shortly, no refracted ray can be drawn. See *Total Internal Reflection*, p. 430.

Refractive Index of a Liquid by Using a Concave Mirror

We are now in a position to utilise our formulæ in refraction, and we shall first consider a simple method of determining roughly the refractive index, n, of a small quantity of transparent liquid.

If a small drop of the liquid is placed on a concave mirror S, a position H can be located by the no parallax method where the image of a pin held over the mirror coincides in position with the pin itself, Fig. 18.7. The rays from the pin must now be striking the mirror *normally*, in which case the rays are reflected back along the incident path and form an image at the same place as the object. A ray HN close to the axis HP is refracted at N along ND in the liquid, strikes the mirror normally at D, and is reflected back along the path DNH. Thus if DN is produced it passes through the centre of curvature, C, of the concave mirror.

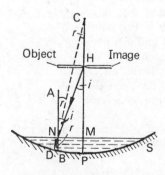

FIG. 18.7. (Depth of liquid exaggerated.)

Let ANB be the normal to the liquid surface at N. Then angle ANH = angle NHM = i, the angle of incidence, and angle BND = angle ANC = angle NCM = r, the angle of refraction in the liquid. From triangles HNM, CNM respectively, sin i = NM/HN and sin r = NM/CN. The refractive index, n, of the liquid is thus given by

$$n = \frac{\sin i}{\sin r} = \frac{NM/HN}{NM/CN} = \frac{CN}{HN}.$$

Now since HN is a ray very close to the principal axis CP, HN = HM and CN = CM, to a very good approximation. Thus $n = \frac{CM}{HM}$. Further, if the depth MP of the liquid is very small compared with HM and CM, CM = CP and HM = HP approximately. Hence, approximately,

$$n = \frac{CP}{HP}.$$

HP can be measured directly. CP is the radius of curvature of the mirror, which can be obtained by the method shown on page 414. The refractive index, n, of the liquid can thus be calculated.

Apparent Depth

Swimmers in particular are aware that the bottom of a pool of water appears nearer the surface than is actually the case; the phenomenon is due to the refraction of light.

Consider an object O at a distance below the surface of a medium such as water or glass, which has a refractive index n, Fig. 18.8. A ray OM from O perpendicular to the surface passes straight through into the air along MS. A ray ON very close to OM is refracted at N into the air away from the normal, in a direction NT; and an observer viewing O directly overhead sees it in the position I, which is the point of intersection of SM and TN produced. Though we have only considered two rays in the air, a *cone* of rays, with SM as the axis, actually enters the observer's eye.

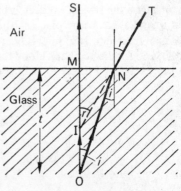

FIG. 18.8. (Inclination of ON to OM exaggerated.)

Suppose the angle of incidence in the glass is i, and the angle of refraction in the air is r. Then, since "n sin i" is a constant (p. 423), we have

$$n \sin i = 1 \times \sin r \qquad . \qquad . \qquad . \qquad . \qquad \text{(i)}$$

where n is the refractive index of glass; the refractive index of air is 1.

Since $i =$ angle NOM, and $r =$ MIN, $\sin i =$ MN/ON and $\sin r =$ MN/IN. From (i), it follows that

$$n \frac{MN}{ON} = \frac{MN}{IN}$$

$$\therefore n = \frac{ON}{IN} \quad . \quad . \quad . \quad . \quad \text{(ii)}$$

Since we are dealing with the case of an observer directly above O, the rays ON, IN are *very* close to the normal OM. Hence to a very good approximation, ON = OM and IN = IM. From (ii),

$$\therefore n = \frac{ON}{IN} = \frac{OM}{IM}.$$

Since the real depth of the object O = OM, and its apparent depth = IM,

$$\therefore \quad n = \frac{\text{real depth}}{\text{apparent depth}} \quad . \quad . \quad . \quad \text{(6)}$$

If the real depth, OM, $= t$, the apparent depth $= \dfrac{t}{n}$, from (6). The *displacement*, OI, of the object, which we shall denote by d, is thus given by $t - \dfrac{t}{n}$, i.e.,

$$d = t \left(1 - \frac{1}{n} \right) \quad . \quad . \quad . \quad . \quad \text{(7)}$$

If an object is 6 cm below water of refractive index, $n = 1\frac{1}{3}$, it appears to be displaced upward to an observer in air by an amount, d $= 6 \left(1 - \dfrac{1}{1\frac{1}{3}} \right) = 1\frac{1}{2}$ cm.

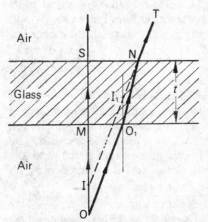

FIG. 18.9. Object below glass block.

Object Below Parallel-sided Glass Block

Consider an object O placed some distance in air below a parallel-sided glass block of thickness t, Fig. 18.9. The ray OMS normal to the surface emerges along MS, while the ray OO_1 close to the normal is refracted along O_1N in the glass and emerges in air along NT in a direction parallel to OO_1 (see p. 422). An observer (not shown) above the glass thus sees the object at I, the point of intersection of TN and SM.

Suppose the normal at O_1 intersects IN at I_1. Then, since O_1I_1 is parallel to OI and IT is parallel to OO_1, OII_1O_1 is a parallelogram. Thus OI $= O_1I_1$. But OI is the displacement of the object O. Hence O_1I_1 is equal to the displacement. Since the apparent position of an object at O_1 is at I_1 (compare Fig. 18.8), we conclude that *the displacement OI of O is independent of the position of O below the glass*, and is given by

$$OI = t\left(1 - \frac{1}{n}\right), \text{ see p. 427.}$$

Measurement of Refractive Index by Apparent Depth Method

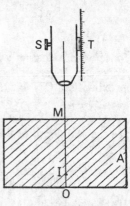

FIG. 18.10. Refractive index by apparent depth.

The formula for the refractive index of a medium in terms of the real and apparent depths is the basis of a very accurate method of measuring refractive index. A *travelling microscope*, S (a microscope which can travel in a vertical direction and which has a fixed graduated scale T beside it) is focused on lycopodium particles at O on a sheet of white paper, and the reading on T is noted, Fig. 18.10. Suppose it is c cm. If the refractive index of glass is required, a glass block A is placed on the paper, and the microscope is raised until the particles are refocused at I. Suppose the reading on T is b cm. Some lycopodium particles are then sprinkled at M on the top of the glass block, and the microscope is raised until they are focused, when the reading on T is noted. Suppose it is a cm.

Then   real depth of O = OM = $(a - c)$ cm.

and   apparent depth = IM = $(a - b)$ cm.

$$\therefore n = \frac{\text{real depth}}{\text{apparent depth}} = \frac{a - c}{a - b}$$

The high accuracy of this determination of n lies mainly in the fact that the objective of the microscope collects only those rays near to its axis, so that the object O, and its apparent position I, are seen by rays very close to the normal OM. The experiment thus fulfils the theoretical conditions assumed in the proof of the formula $n = \dfrac{\text{real depth}}{\text{apparent depth}}$ p. 427.

The refractive index of water can also be obtained by an apparent depth method. The block A is replaced by a dish, and the microscope is focused first on the bottom of the dish and then on lycopodium powder sprinkled on the surface of water poured into the dish. The apparent position of the bottom of the dish is also noted, and the refractive index of the water n_w is calculated from the relation

$$n_w = \frac{\text{real depth of water}}{\text{apparent depth of water}}$$

General formula for real and apparent depth. So far we have considered the rays refracted from a medium like glass into air. As a more general case, suppose an object O is in a medium of refractive index n_1 and the rays from it are refracted at M, N into a medium of refractive index n_2, Fig. 18.11. The image of O to an observer in the latter medium is then at I.

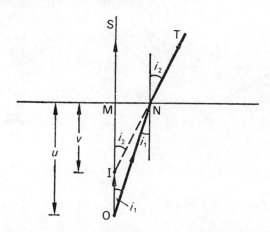

FIG. 18.11. Relation between real and apparent depth.

Suppose i_1 is the angle of incidence at N, and i_2 is the angle of refraction as shown. Then, since "$n \sin i$" is a constant,

$$n_1 \sin i_1 = n_2 \sin i_2.$$

Now angle MON $= i_1$, and angle MIN $= i_2$.

$$\therefore n_1 \frac{\text{MN}}{\text{ON}} = n_2 \frac{\text{MN}}{\text{IN}}$$

$$\therefore \frac{n_1}{\text{ON}} = \frac{n_2}{\text{IN}} \qquad \cdot \qquad \cdot \qquad \cdot \qquad \cdot \qquad \text{(i)}$$

If we consider rays very close to the normal, then IN = IM = v, say, and ON = OM = u, say. Substituting in (i),

$$\therefore \frac{n_1}{u} = \frac{n_2}{v} \qquad \cdot \qquad \cdot \qquad \cdot \qquad \cdot \qquad \text{(ii)}$$

This formula can easily be remembered, as the refractive index of a medium is divided by the corresponding distance of the object (or image) *in* that medium.

Total Internal Reflection. Critical Angle

If a ray AO in glass is incident at a small angle a on a glass–air plane boundary, observation shows that part of the incident light is reflected along OE in the glass, while the remainder of the light is refracted away from the normal at an angle β into the air. The reflected ray OE is weak, but the refracted ray OL is bright, Fig. 18.12 (i). This means that most of the incident light energy is transmitted, and a little is reflected.

When the angle of incidence, a, in the glass is increased, the angle of emergence, β, is increased at the same time; and at some angle of incidence c in the glass the refracted ray OL travels along the glass–air boundary, making the angle of refraction of 90°, Fig. 18.12 (ii). The reflected ray OE is still weak in intensity, but as the angle of incidence in the glass is increased slightly the reflected ray suddenly becomes bright, and no refracted ray is then observed, Fig. 18.12 (iii). Since *all* the

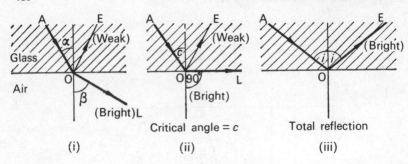

FIG. 18.12. Total internal reflection.

incident light energy is now reflected, **total reflection** is said to take place
in the glass at O.

When the angle of refraction in air is 90°, a critical stage is reached
at the point of incidence O, and the angle of incidence *in the glass* is
accordingly known as the **critical angle** for glass and air, Fig. 18.12 (ii).
Since "$n \sin i$" is a constant (p. 423), we have

$$n \sin c = 1 \times \sin 90°,$$

where n is the refractive index of the glass. As $\sin 90° = 1$, then

$$n \sin c = 1,$$

or, $$\sin c = \frac{1}{n} \quad . \quad . \quad . \quad . \quad (8)$$

Crown glass has a refractive index of about 1·51 for yellow light, and
thus the critical angle for glass to air is given by $\sin c = 1/1·51 = 0·667$.
Consequently $c = 41·5°$. Thus if the incident angle in the glass is
greater than c, for example 45°,
total reflection occurs, Fig. 18.12
(iii). The critical angle between two
media for blue light is less than for
red light, since the refractive index
for blue light is greater than that for
red light (see p. 458).

The phenomenon of total reflection may occur when light in glass
($n_g = 1·51$, say) is incident on a
boundary with water ($n_w = 1·33$).
Applying "$n \sin i$ is a constant" to
the critical case, Fig. 18.13, we have

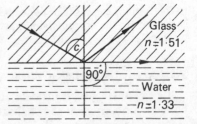

FIG. 18.13. Critical angle for
water and glass.

$$n_g \sin c = n_w \sin 90°,$$

where c is the critical angle. As $\sin 90° = 1$

$$n_g \sin c = n_w$$

$$\therefore \sin c = \frac{n_w}{n_g} = \frac{1·33}{1·51} = 0·889$$

$$\therefore c = 63°.$$

Thus if the angle of incidence in the glass exceeds 63°, total internal reflection occurs.

It should be carefully noted that the phenomenon of total internal reflection can occur only when light travels from one medium to another which has a *smaller* refractive index, i.e., which is optically less dense. The phenomenon cannot occur when light travels from one medium to another optically denser, for example from air to glass, or from water to glass, as a refracted ray is then always obtained.

SOME APPLICATIONS OF TOTAL INTERNAL REFLECTION

1. *Reflecting prisms* are pieces of glass of a special shape which are used in prism binoculars and in certain accurate ranging instruments such as submarine periscopes. These prisms, discussed on p. 449, act as reflectors of light by total internal reflection.

2. *The mirage* is a phenomenon due to total reflection. In the desert the air is progressively hotter towards the sand, and hence the density of the air decreases in the direction *bcd*, Fig. 18.14 (i). A downward ray OA from a tree or the sky is thus refracted more and more away from the normal; but at some layer of air *c*, a critical angle is reached, and the ray begins to travel in an upward direction along *cg*. A distant observer P thus sees the object O at I, and hence an image of a palm tree, for example, is seen below the actual position of the tree. As an image of part of the sky is also formed by total reflection round the image of the tree, the whole appearance is similar to that of a pool of water in which the tree is reflected.

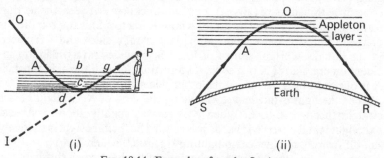

FIG. 18.14. Examples of total reflection.

3. *Total reflection of radio waves.* A radio wave is an example of an *electromagnetic wave* because it comprises electric and magnetic forces. Light waves also are electromagnetic waves (p. 983). Light waves and radio waves are therefore the same in nature, and a close analogy can be made between the refraction of light and the refraction of a radio wave when the latter enters a medium containing electric particles.

In particular, the phenomenon of total reflection occurs when radio waves travel from one place, S, on the earth, for example England, to another place, R, on the other side of the earth, for example America, Fig. 18.14 (ii). A layer of considerable density of electrons exists many miles above the earth (at night this is the *Appleton layer*), and when a radio wave SA from a transmitter is sent skyward it is refracted away from the normal on entering the electron layer. At some height, corresponding to O, a critical angle is reached, and the wave then begins to be refracted downward. After emerging from the electron layer it returns to R on the earth, where its presence can be detected by a radio receiver.

Measurement of Refractive Index of a Liquid by an Air-cell Method

The phenomenon of total internal reflection is utilised in many methods of measuring refractive index. Fig. 18.15 (i) illustrates how the

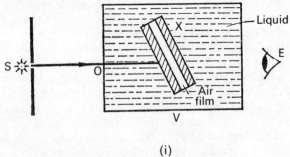

(i)

FIG. 18.15 (i). Air-cell method.

refractive index of a liquid can be determined. Two thin plane-parallel glass plates, such as microscope slides, are cemented together so as to contain a thin film of air of constant thickness between them, thus forming an air-cell, X. The liquid whose refractive index is required is placed in a glass vessel V having thin plane-parallel sides, and X is placed in the liquid. A bright source of light, S, provides rays which are incident on one side of X in a constant direction SO, and the light through X is observed by a person on the other side at E.

When the light is incident normally on the sides of X, the light passes straight through X to E. When X is rotated slightly about a vertical axis, light is still observed; but as X is rotated farther, the light is suddenly cut off from E, and hence no light now passes through X, Fig. 18.15 (i).

Fig. 18.15 (ii) illustrates the behaviour of the light when this happens. The ray SO is refracted along OB in the glass, but at B *total internal reflection begins*. Suppose i_1 is the angle of incidence in the liquid, i_2 is the angle of incidence in the glass, and n, n_g are the corresponding refractive indices. Since the boundaries of the media are parallel we can apply the relation "$n \sin i$ is a constant", and hence

$$n \sin i_1 = n_g \sin i_2 = 1 \times \sin 90°, \quad . \quad . \quad . \quad (i)$$

the last product corresponding to the case of refraction in the air-film.

$$\therefore \quad n \sin i_1 = 1 \times \sin 90° = 1 \times 1 = 1$$

$$\therefore n = \frac{1}{\sin i_1} \quad . \quad . \quad . \quad . \quad . \quad . \quad (9)$$

It should now be carefully noted that i_1 is the angle of incidence in the *liquid* medium, and is thus determined by measuring the rotation of X from its position when normal to SO to the position when the light

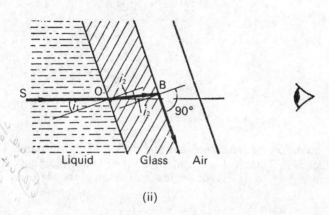

(ii)

FIG. 18.15 (ii). Air-cell theory.

is cut off. In practice, it is better to rotate X in opposite directions and determine the angle θ between the *two* positions for the extinction of the light. The angle i_1 is then half the angle θ, and hence $n = 1/\sin\dfrac{\theta}{2}$.

From equations (i) and (9), it will be noted that i_1 is the critical angle between the liquid and air, and i_2 is the critical angle between the glass and air. We cannot measure i_2, however, as we can i_1, and hence the method provides the refractive index of the *liquid*.

The source of light, S, in the experiment should be a monochromatic source, i.e., it should provide light of one colour, for example yellow light. The extinction of the light is then sharp. If white light is used, the colours in its spectrum are cut off at slightly different angles of incidence, since refractive index depends upon the colour of the light (p. 458). The extinction of the light is then gradual and ill-defined.

Pulfrich Refractometer

A *refractometer* is an instrument which measures refractive index by making use of total internal reflection. PULFRICH designed a refractometer enabling the refractive index of a liquid to be easily obtained, which consists of a block of glass G with a polished and vertical face. On top of G is cemented a circular glass tube V, Fig. 18.16. The liquid L is placed in V, and a convergent beam of monochromatic light is directed so that the liquid–glass interface is illuminated. On observing the light refracted through G by a telescope T, a light and dark field of view are seen.

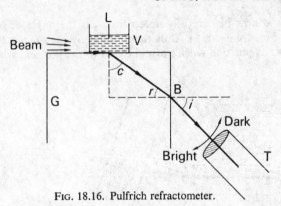

FIG. 18.16. Pulfrich refractometer.

The boundary between the light and dark fields corresponds to the ray which is incident just horizontally on the liquid-glass boundary, as shown in Fig. 18.16. If c is the angle of refraction in the glass, it follows that

$$n \sin 90° = n_g \sin c \quad . \qquad . \qquad . \qquad . \qquad . \qquad . \qquad \text{(i)}$$

where n, n_g are the refractive indices of the liquid and glass respectively. For refraction at B,

$$n_g \sin r = \sin i \quad . \qquad . \qquad . \qquad . \qquad . \qquad . \qquad \text{(ii)}$$

Also, $\qquad\qquad c + r = 90° \quad . \qquad . \qquad . \qquad . \qquad . \qquad . \qquad \text{(iii)}$

From (i), $\sin c = n/n_g$. Now from (iii), $\sin r = \sin (90° - c) = \cos c$. Substituting in (ii), we have $n_g \cos c = \sin i$, or $\cos c = \sin i/n_g$.

But $\qquad\qquad \sin^2 c + \cos^2 c = 1$

$$\therefore \quad \frac{n^2}{n_g^2} + \frac{\sin^2 i}{n_g^2} = 1$$

$$\therefore \quad n^2 + \sin^2 i = n_g^2$$

$$\therefore \quad n = \sqrt{n_g^2 - \sin^2 i}$$

Thus if i is measured and $n_g = 1 \cdot 51$, n can be calculated. In practice, tables are supplied giving the refractive index in terms of i, and another block is used in place of G for liquids of higher n than $1 \cdot 51$.

Abbe refractometer. Abbe designed a refractometer for measuring the refractive index of liquids whose principle is illustrated in Fig. 18.17. Two similar prisms X, Y are placed on a table A, the prism X being hinged at H so that it could be swung away from Y. A drop of the liquid is placed on the surface a, which is matt, and the prisms are placed together so that the liquid is squeezed into a thin film between them. Light from a suitable source is directed towards the prisms by means of a mirror M, where it strikes the surface a and is scattered by the matt surface into the liquid film. The emergent rays are collected in a telescope T directed towards the prisms, and the field of view is divided into a dark and bright portion. The table A is then turned until the dividing line between the dark and bright fields is on the crosswires of the telescope, which is fixed. The reading on the scale S, which is attached to, and moves with, the table, gives the refractive index of the liquid directly, as explained below.

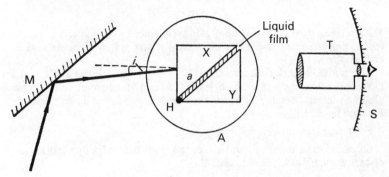

Fig. 18.17. Abbe refractometer principle.

Theory. The dividing line, BQ, between the bright and dark fields corresponds to the case of the ray DA, incident in the liquid L at grazing incidence on the prism Y, Fig. 18.18. The refracted ray AB in the prism then makes the critical angle c with the normal at A, where $n \sin 90° = n_g \sin c$, n and n_g being the respective refractive indices of the liquid and the glass.

$$\therefore \quad n = n_g \sin c \quad \cdots \quad \cdots \quad \text{(i)}$$

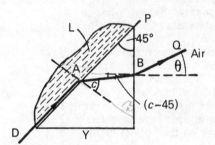

Fig. 18.18. Theory of refractometer.

For simplicity, suppose that Y is a right-angled isosceles prism, so that the angle P of the prism is 45°. The angle of incidence at B in the glass is then $(c - 45°)$, by considering the geometry of triangle PAB, and hence for refraction at B we have

$$n_g \sin (c - 45°) = \sin \theta \quad \cdots \quad \cdots \quad \text{(ii)}$$

where θ is the angle with the normal at B made by the emerging ray BQ. By eliminating c from (i) and (ii), we obtain finally

$$n = \sin 45° (n_g^2 - \sin^2\theta)^{\frac{1}{2}} + \cos 45° \sin \theta$$
$$= \frac{1}{\sqrt{2}} [(n_g^2 - \sin^2\theta)^{\frac{1}{2}} + \sin \theta]$$

since $\sin 45° = 1/\sqrt{2} = \cos 45°$. Thus knowing n_g and θ, the refractive index of the liquid, n, can be evaluated. The scale S, Fig. 18.17, which gives θ, can thus be calibrated in terms of n.

EXAMPLES

1. Describe a method, based on grazing incidence or total internal reflection, for finding the refractive index of water for the yellow light emitted by a sodium flame.

The refractive index of carbon bisulphide for red light is 1·634 and the difference between the critical angles for red and blue light at a carbon bisulphide-air interface is 0° 56′. What is the refractive index of carbon bisulphide for blue light? (*N*.)

First part. See air-cell method, p. 432.

Second part. Suppose n_r and c_r are the refractive index and critical angle of carbon bisulphide for red light.

Then
$$\sin c_r = \frac{1}{n_r} = \frac{1}{1\cdot634} = 0\cdot6119$$
$$\therefore\ c_r = 37°\ 44′$$

The critical angle, c_b, for blue light is less than that for red light.
$$\therefore\ c_b = 37°\ 44′ - 0°\ 56′ = 36°\ 48′$$

The refractive index for blue light, n_b, $= \dfrac{1}{\sin c_b}$
$$\therefore\ n_b = \frac{1}{\sin 36°\ 48′} = 1\cdot669$$

2. Find an expression for the distance through which an object appears to be displaced towards the eye when a plate of glass of thickness t and refractive index n is interposed.

A tank contains a slab of glass 8 cm thick and of refractive index 1·6. Above this is a depth of 4·5 cm of a liquid of refractive index 1·5 and upon this floats 6 cm of water ($n = 4/3$). To an observer looking from above, what is the apparent position of a mark on the bottom of the tank? (*O. & C.*)

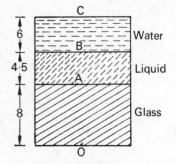

FIG. 18.19.

First part. See text.

Second part. Suppose O is the mark at the bottom of the tank, Fig. 18.19. Then since the boundaries of the media are parallel, the total displacement of O is the sum of the displacements due to each of the media.

For glass, displacement, $d = t\left(1 - \dfrac{1}{n}\right) = 8\left(1 - \dfrac{1}{1\cdot6}\right) = 3$ cm

For liquid, $d = t\left(1 - \dfrac{1}{n}\right) = 4\cdot5\left(1 - \dfrac{1}{1\cdot5}\right) = 1\cdot5$ cm

For water, $d = 6\left(1 - \dfrac{1}{4/3}\right) = 1\cdot5$ cm

$$\therefore\ \text{total displacement} = 3 + 1\cdot5 + 1\cdot5 = 6\text{ cm}$$
$$\therefore\ \text{apparent position of O is 6 cm from bottom.}$$

3. A small object is placed on the principal axis of a concave spherical mirror of radius 20 cm at a distance of 30 cm. By how much will the position

and size of the image alter when a parallel-sided slab of glass, of thickness 6 cm and refractive index 1·5, is introduced between the centre of curvature and the object? The parallel sides are perpendicular to the principal axis. Prove any formula used. (*N.*)

Suppose O is the position of the object before the glass is placed in position, Fig. 18.20. The image position is given by $\dfrac{1}{v} + \dfrac{1}{u} = \dfrac{2}{r}$,

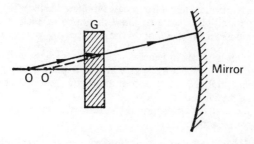

FIG. 18.20.

$$\therefore \qquad \frac{1}{v} + \frac{1}{30} = \frac{2}{20}$$

Solving, $\qquad\qquad v = 15 \text{ cm.}$

The magnification, $m = \dfrac{v}{u} = \dfrac{15}{30} = 0\cdot5.$

When the glass slab G of thickness, t, 6 cm is inserted, the rays from O appear to come from a point O′ whose displacement from O is $t\left(1 - \dfrac{1}{n}\right)$, where n is the glass refractive index. See p. 428. The displacement is thus $6\left(1 - \dfrac{1}{1\cdot5}\right) = 2 \text{ cm.}$ The distance of O′ from the mirror is therefore $(30 - 2)$, or 28 cm. Applying the equation $\dfrac{1}{v} + \dfrac{1}{u} = \dfrac{2}{r}$, we find $v = +\ 15\frac{5}{9}$ cm. The image position changes by $(15\frac{5}{9} - 15)$ or $\frac{5}{9}$ cm. The magnification becomes $15\frac{5}{9} \div 28$, or 0·52.

EXERCISES 18

1. A ray of light is incident at 60° in air on an air–glass plane surface. Find the angle of refraction in the glass by calculation and by drawing (n for glass = 1·5).

2. A ray of light is incident in water at an angle of 30° on a water–air plane surface. Find the angle of refraction in the air by calculation and by drawing (n for water = 4/3).

3. A ray of light is incident in water at an angle of (i) 30°, (ii) 70° on a water–glass plane surface. Calculate the angle of refraction in the glass in each case ($_an_g = 1\cdot5$, $_an_w = 1\cdot33$).

4. (*a*) Describe the apparent depth method of finding the refractive index of glass, and prove the formula used. (*b*) What is the apparent position of an object below a rectangular block of glass 6 cm thick if a layer of water 4 cm thick is on top of the glass (refractive index of glass and water $= 1\frac{1}{2}$ and $1\frac{1}{3}$ respectively)?

5. Describe and explain a method of measuring approximately the refractive index of a small quantity of liquid.

6. Calculate the critical angle for (i) an air–glass surface, (ii) an air–water surface, (iii) a water–glass surface; draw diagrams in each case illustrating the total reflection of a ray incident on the surface ($_a n_g = 1\cdot5$, $_a n_w = 1\cdot33$).

7. Explain what happens in general when a ray of light strikes the surface separating transparent media such as water and glass. Explain the circumstances in which total reflection occurs and show how the critical angle is related to the refractive index.

Describe a method for determining the refractive index of a medium by means of critical reflection. (*L.*)

8. Define *refractive index* of one medium with respect to another and show how it is related to the values of the velocity of light in the two media.

Describe a method of finding the refractive index of water for sodium light, deducing any formula required in the reduction of the observations. (*N.*)

9. Explain carefully why the apparent depth of the water in a tank changes with the position of the observer.

A microscope is focused on a scratch on the bottom of a beaker. Turpentine is poured into the beaker to a depth of 4 cm, and it is found necessary to raise the microscope through a vertical distance of $1\cdot28$ cm to bring the scratch again into focus. Find the refractive index of the turpentine. (*C.*)

10. What is meant by *total reflection* and *critical angle*? Describe two methods of measuring the refractive index of a material by determining the critical angle, one of which is suitable for a solid substance and the other for a liquid. (*L.*)

11. (*a*) State the conditions under which total reflection occurs. Show that the phenomenon will occur in the case of light entering normally one face of an isosceles right-angle prism of glass, but not in the case when light enters similarly a similar hollow prism full of water. (*b*) A concave mirror of small aperture and focal length 8 cm lies on a bench and a pin is moved vertically above it. At what point will image and object coincide if the mirror is filled with water of refractive index 4/3? (*N.*)

12. State the laws of refraction, and define *refractive index*.

Describe an accurate method of determining the refractive index of a transparent liquid for sodium light. Give the theory of the method, and derive any formula you require. Discuss the effect of substituting white light for sodium light in your experiment. (*W.*)

13. Describe an experiment for finding the refractive index of a liquid by measuring its apparent depth.

A vessel of depth $2d$ cm is half filled with a liquid of refractive index n_1, and the upper half is occupied by a liquid of refractive index n_2. Show that the apparent depth of the vessel, viewed perpendicularly, is $d\left(\dfrac{1}{n_1}+\dfrac{1}{n_2}\right)$. (*L.*)

14. The base of a cube of glass of refractive index n_1 is in contact with the surface of a liquid of refractive index n_2. Light incident on one vertical face of the cube is reflected internally from the base and emerges again from the opposite vertical face in a direction making an angle θ with its normal. Assuming that $n_1 > n_2$, show that the light has just been totally reflected internally if $n_2 = \sqrt{(n_1^2 - \sin^2 \theta)}$.

Describe how the above principle may be used to measure the refractive index of a small quantity of liquid. (*N.*)

15. Explain the meaning of critical angle and total internal reflection. Describe fully (*a*) one natural phenomenon due to total internal reflection, (*b*) one practical application of it. Light from a luminous point on the lower face of a rectangular glass slab, 2·0 cm thick, strikes the upper face and the totally reflected rays outline a circle of 3·2 cm radius on the lower face. What is the refractive index of the glass? (*N.*)

16. Describe one experiment in each case to determine the refractive index for sodium light of (*a*) a sample of glass which could be supplied to any shape and size which you specify, (*b*) a liquid of which only a *very* small quantity is available. Show how the result is calculated in each case. (You are *not* expected to derive standard formulae.)

How would you modify the experiment in (*a*) to find how the refractive index varies with the wavelength of the light used? What general result would you expect? (*O. & C*).

17. Explain the meaning of *critical angle*, and describe how you would measure the critical angle for a water–air boundary.

ABCD is the plan of a glass cube. A horizontal beam of light enters the face AB at grazing incidence. Show that the angle θ which any rays emerging from BC would make with the normal to BC is given by $\sin \theta = \cot a$, where a is the critical angle. What is the greatest value that the refractive index of glass may have if any of the light is to emerge from BC? (*N.*)

18. State the *laws of refraction of light*. Explain how you would measure the refractive index of a transparent liquid available only in *small* quantity, i.e., less than 0·5 cm³.

A ray of light is refracted through a sphere, whose material has a refractive index n, in such a way that it passes through the extremities of two radii which make an angle a with each other. Prove that if γ is the deviation of the ray caused by its passage through the sphere

$$\cos \tfrac{1}{2} (a - \gamma) = n \cos \tfrac{1}{2}a. \tag{L.}$$

19. Explain what is meant by the terms *critical angle* and *total reflection*. Describe an accurate method of determining the critical angle for a liquid, indicating how you would calculate the refractive index from your measurements.

A man stands at the edge of the deep end of a swimming bath, the floor of which is covered with square tiles. If the water is clear and undisturbed, explain carefully how the floor of the bath appears to him. (*O. & C.*)

20. Summarize the various effects that may occur when a parallel beam of light strikes a plane interface between two transparent media.

Explain why, when looking at the windows of a railway carriage from inside, one sees by day the country outside and by night the reflection of the inside of the carriage.

An observer looks normally through a thick window of thickness d and refractive index n at an object at a distance e behind the farther surface. Where does the object appear to be, and how can this apparent position be found experimentally? (*O. & C.*)

chapter nineteen

Refraction through prisms

In Light, a *prism* is a transparent object usually made of glass which has two plane surfaces, XDEY, XDFZ, inclined to each other, Fig. 19.1

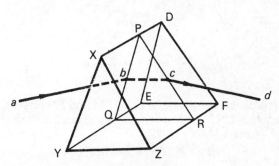

FIG. 19.1. Prism.

Prisms are used in many optical instruments, for example prism binoculars, and they are also utilised for separating the colours of the light emitted by glowing objects, which affords an accurate knowledge of their chemical composition. A prism of glass enables the refractive index of this material to be measured very accurately.

The angle between the inclined plane surfaces XDFZ, XDEY is known as the *angle of the prism*, or the *refracting angle*, the line of intersection XD of the planes is known as the *refracting edge*, and any plane in the prism perpendicular to XD, such as PQR, is known as a *principal section* of the prism. A ray of light *ab*, incident on the prism at *b* in a direction perpendicular to XD, is refracted towards the normal along *bc* when it enters the prism, and is refracted away from the normal along *cd* when it emerges into the air. From the law of refraction (p. 420), the rays *ab*, *bc*, *cd* all lie in the same plane, which is PQR in this case. If the incident ray is directed towards the refracting angle, as in Fig. 19.1, the light is always deviated by the prism towards its base.

Refraction Through a Prism

Consider a ray HM incident in air on a prism of refracting angle A, and suppose the ray lies in the principal section PQR, Fig. 19.2. Then, if i_1, r_1 and i_2, r_2 are the angles of incidence and refraction at M, N as shown, and n is the prism refractive index,

$$\sin i_1 = n \sin r_1 \qquad . \qquad . \qquad . \qquad . \qquad \text{(i)}$$
$$\sin i_2 = n \sin r_2 \qquad . \qquad . \qquad . \qquad . \qquad \text{(ii)}$$

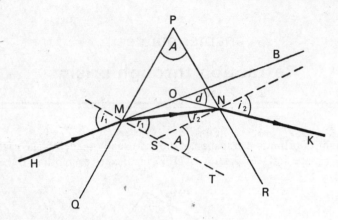

FIG. 19.2. Refraction through prism.

Further, as MS and NS are normals to PM and PN respectively, angle MPN + angle MSN = 180°, considering the quadrilateral PMSN. But angle NST + angle MSN = 180°.

$$\therefore \quad \text{angle NST} = \text{angle MPN} = A$$

$$\therefore \quad A = r_1 + r_2 \quad . \quad . \quad . \quad . \quad \text{(iii)}$$

as angle NST is the exterior angle of triangle MSN.

In the following sections, we shall see that the *angle of deviation, d*, of the light, caused by the prism, is utilised considerably. The angle of deviation at M = angle OMN = $i_1 - r_1$; the angle of deviation at N = angle MNO = $i_2 - r_2$. Since the deviations at M, N are in the same direction, the total deviation, d (angle BOK), is given by

$$d = (i_1 - r_1) + (i_2 - r_2) \quad . \quad . \quad . \quad . \quad \text{(iv)}$$

Equations (i) – (iv) are the general relations which hold for refraction through a prism. In deriving them, it should be noted that the geometrical form of the prism base plays no part.

Minimum Deviation

The angle of deviation, d, of the incident ray HM is the angle BOK in Fig. 19.2. The variation of d with the angle of incidence, i, can be obtained experimentally by placing the prism on paper on a drawing board and using a ray AO from a ray-box (or two pins) as the incident ray, Fig. 19.3 (i). When the direction AO is kept constant and the drawing board is turned so that the ray is always incident at O on the prism, the angle of incidence i is varied; the corresponding emergent rays CE, HK, LM, NP can be traced on the paper. Experiment shows that as the angle of incidence i is increased from zero, the deviation d begins to decrease continuously to some value D, and then increases to a maximum as i is increased further to 90°. A *minimum deviation*, corresponding to the emergent ray NP, is thus obtained. A graph of d plotted against i has the appearance of the curve X, which has a minimum value at R, Fig. 19.3 (ii).

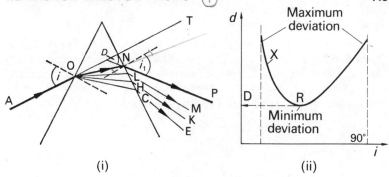

FIG. 19.3. Minimum deviation.

Experiment and theory show that *the minimum deviation, D, of the light occurs when the ray passes symmetrically through the prism.* Suppose this corresponds to the case of the ray AONP in Fig. 19.3 (i). Then the corresponding incident angle, i, is equal to the angle of emergence, i_1, into the air at N for this special case. See also Fig. 19.5 and Fig. 19.9.

A proof of symmetrical passage of ray at minimum deviation. Experiment shows that minimum deviation is obtained at *one* particular angle of incidence. On this assumption it is possible to prove by a *reductio ad absurdum* method that the angle of incidence is equal to the angle of emergence in this case. Thus

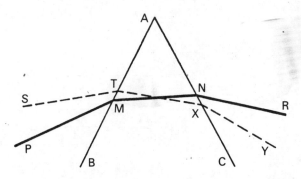

FIG. 19.4. Minimum deviation proof.

suppose that minimum deviation is obtained with a ray PMNR when these angles are *not* equal, so that angle PMB is not equal to angle RNC, Fig. 19.4. It then follows that a ray YX, incident on AC at an angle CXY equal to angle PMB, will emerge along TS, where angle BTS = angle CNR; and from the principle of the reversibility of light, a ray incident along ST on the prism emerges along XY. We therefore have *two* cases of minimum deviation, corresponding to two different angles of incidence. But, from experiment, this is impossible. Consequently our initial assumption must be wrong, and hence the angle of emergence *does* equal the angle of incidence. Thus the ray passes symmetrically through the prism in the minimum deviation case.

Relation Between A, D, and n

A very convenient formula for refractive index, n, can be obtained in the minimum deviation case. The ray PQRS then passes symmetrically through the prism, and the angles made with the normal in the air and in the glass at Q, R respectively are equal, Fig. 19.5. Suppose the angles are denoted by i, r, as shown. Then, as explained on p. 442,

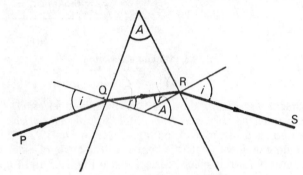

Fig. 19.5. Formula for n of prism.

$$i - r + i - r = D \qquad \qquad \text{(i)}$$

and
$$r + r = A \qquad \qquad \text{(ii)}$$

From (ii),
$$r = \frac{A}{2}$$

Substituting for r in (i),
$$2i = A + D$$
$$\therefore i = \frac{A + D}{2}$$

$$\therefore n = \frac{\sin i}{\sin r} = \frac{\sin \dfrac{A + D}{2}}{\sin \dfrac{A}{2}} \qquad \qquad \text{(1)}$$

The Spectrometer

The spectrometer is an optical instrument which is mainly used to study the light from different sources. As we shall see later, it can be used to measure accurately the refractive index of glass in the form of a prism. The instrument consists essentially of a *collimator*, C, a *telescope*, T, and a *table*, R, on which a prism B can be placed. The lenses in C, T are achromatic lenses (p. 515). The collimator is fixed, but the table and the telescope can be rotated round a circular scale graduated in half-degrees (not shown) which has a common vertical axis with the table, Fig. 19.6. A vernier is also provided for this scale. The *source of light*, S, used in the experiment is placed in front of a narrow slit at one end of the collimator, so that the prism is illuminated by light from S.

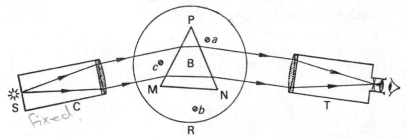

FIG. 19.6. Spectrometer.

Before the spectrometer can be used, however, three adjustments must be made: (1) The collimator C must be adjusted so that parallel light emerges from it; (2) the telescope T must be adjusted so that parallel rays entering it are brought to a focus at cross-wires near its eye-piece; (3) the refracting edge of the prism must be parallel to the axis of rotation of the telescope, i.e., the table must be "levelled".

Adjustments of Spectrometer

The telescope adjustment is made by first moving its eye-piece until the cross-wires are distinctly seen, and then sighting the telescope on to a *distant* object through an open window. The length of the telescope is now altered by a screw arrangement until the object is clearly seen at the same place as the cross-wires, so that parallel rays now entering the telescope are brought to a focus at the cross-wires.

The collimator adjustment. With the prism removed from the table, the telescope is now turned to face the collimator, C, and the slit in C is illuminated by a sodium flame which provides yellow light. The edges of the slit are usually blurred, showing that the light emerging from the lens of C is not a parallel beam. The position of the slit is now adjusted by moving the tube in C, to which the slit is attached, until the edges of the latter are sharp.

"Levelling" the table. If the rectangular slit is not in the centre of the field of view when the prism is placed on the table, the refracting edge of the prism is not parallel to the axis of rotation of the telescope. The table must then be adjusted, or "levelled", by means of the screws *a*, *b*, *c* beneath it. One method of procedure consists of placing the prism on the table with one face MN approximately perpendicular to the line joining two screws *a*, *b*, as shown in Fig. 19.6. The table is turned until MN is illuminated by the light from C, and the telescope T is then moved to receive the light reflected from MN. The screw *b* is then adjusted until the slit appears in the centre of the field of view. With C and T fixed, the table is now rotated until the slit is seen by reflection at the face NP of the prism, and the screw *c* is then adjusted until the slit is again in the middle of the field of view. The screw *c* moves MN in its own plane, and hence the movement of *c* will not upset the adjustment of MN in the perpendicular plane.

Measurement of the Angle, A, of a Prism

The angle of a prism can be measured very accurately by a spectro-meter. The refracting edge, P, of the prism is turned so as to face the collimator lens, which then illuminates the two surfaces containing

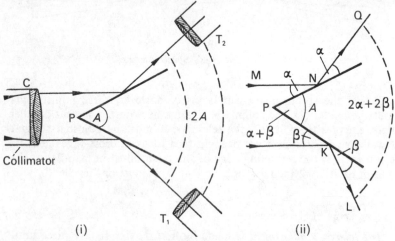

FIG. 19.7. Measurement of angle of prism.

the refracting angle A with parallel light, Fig. 19.7 (i). An image of the collimator slit is hence observed with the telescope in positions T_1, T_2, corresponding to reflection of light at the respective faces of the prism. It is shown below that the angle of rotation of the telescope from T_1 to T_2 is equal to $2A$, and hence the angle of the prism, A, can be obtained.

Proof. Suppose the incident ray MN makes a glancing angle α with one face of the prism, and a parallel ray at K makes a glancing angle β with the other face, Fig. 19.7 (ii). The reflected ray NQ then makes a glancing angle α with the prism surface, and hence the deviation of MN is 2α (see p. 392). Similarly, the deviation by reflection at K is 2β. Thus the reflected rays QN, LK are inclined at an angle equal to $2\alpha + 2\beta$, corresponding to the angle of rotation of the telescope from T_1 to T_2. But the angle, A, of the prism $= \alpha + \beta$, as can be seen by drawing a line through P parallel to MN, and using alternate angles. Hence the rotation of the telescope $= 2\alpha + 2\beta = 2A$.

Measurement of the Minimum Deviation, D

In order to measure the minimum deviation, D, caused by refraction through the prism, the latter is placed with its refracting angle A pointing *away* from the collimator, as shown in Fig. 19.8 (i). The telescope is then turned until an image of the slit is obtained on the cross-wires, corresponding to the position T_1. The table is now slowly rotated so that the angle of incidence on the left side of the prism decreases, and the image of the slit is kept on the cross-wires by moving the telescope at the same time. The image of the slit, and the telescope, then slowly approach the fixed line XY. But at one position, corresponding to T_2, the image of the slit begins to move *away* from XY. If the table is now turned in the opposite direction the image of the slit again moves back when the telescope reaches the position T_2. The angle between the emergent ray CH and the line XY is hence the smallest angle of deviation caused by the prism, and is thus equal to D.

The minimum deviation is obtained by finding the angle between the positions of the telescope (i) at T_2, (ii) at T; the prism is removed in the latter case so as to view the slit directly. Alternatively, the experiment to find the minimum deviation is repeated with the refracting angle pointing

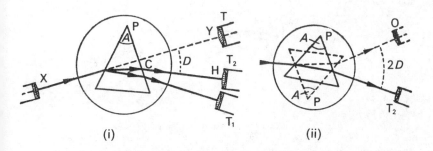

FIG. 19.8. Measurement of minimum deviation.

the opposite way, the prism being represented by dotted lines in this case, Fig. 19.8 (ii). If the position of the telescope for minimum deviation is now O, it can be seen that the angle between the position O and the other minimum deviation position T_2 is $2D$. The value of D is thus easily calculated.

The Refractive Index of the Prism Material

The refractive index, n, of the material of the prism can be easily calculated once A and D have been determined, since, from p. 444.

$$n = \sin \frac{A + D}{2} \Big/ \sin \frac{A}{2}.$$

In an experiment of this nature, the angle, A, of a glass prism was found to be 59° 52′, and the minimum deviation, D, was 40° 30′. Thus

$$n = \sin \frac{59° \; 52′ + 40° \; 30′}{2} \Big/ \sin \frac{59° \; 52′}{2}$$

$$= \sin 50° \; 11′ / \sin 29° \; 56′$$

$$= 1·539$$

The spectrometer prism method of measuring refractive index is capable of providing an accuracy of one part in a thousand. The refractive index of a liquid can also be found by this method, using a hollow glass prism made from thin parallel-sided glass strips.

Grazing Incidence for a Prism

We shall now leave any further considerations of minimum deviation, and shall consider briefly other special cases of refraction through a prism.

When the surface PQ of a prism is illuminated by a source of yellow light placed near Q, the field of view seen through the other surface PR is divided into a bright and dark portion, Fig. 19.9. If NM is the

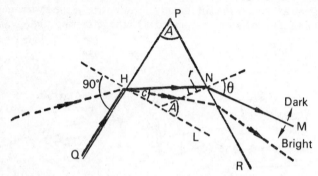

FIG. 19.9. Grazing incidence.

emergent ray corresponding to the incident ray QH which just grazes the prism surface, the dark portion lies above NM and the bright portion exists below NM. The boundary of the light and dark portions is thus NM.

Since the angle of incidence of QH is 90°, angle NHL is equal to c, the critical angle for the glass of the prism. From Fig. 19.9, it follows that $A = c + r$, and hence

$$r = A - c \qquad . \qquad . \qquad . \qquad . \qquad \text{(i)}$$

Further, for refraction at N,

$$\sin \theta = n \sin r.$$

$$\therefore \sin \theta = n \sin (A - c) = n (\sin A \cos c - \cos A \sin c) . \qquad \text{(ii)}$$

But $\sin c = \dfrac{1}{n}$, i.e., $\cos c = \sqrt{1 - \sin^2 c} = \sqrt{1 - \dfrac{1}{n^2}} = \dfrac{1}{n} \sqrt{n^2 - 1}.$

Substituting in (ii) and simplifying, we obtain finally

$$n = \sqrt{1 + \left(\frac{\cos A + \sin \theta}{\sin A}\right)^2} .$$

Thus if A and θ are measured, the refractive index of the prism material can be calculated.

It should be noted that *maximum* deviation by a prism is obtained (i) at grazing incidence, $i = 90°$, (ii) at an angle of incidence θ (Fig. 19.9), corresponding to grazing emergence.

Grazing Incidence and Grazing Emergence

If a ray BM is at grazing incidence on the face of a prism, and the angle A of the prism is increased, a calculation shows that the refracted ray MN in the glass will make a bigger and bigger angle of incidence on the other face PR, Fig. 19.10. This is left as an exercise for the reader. At

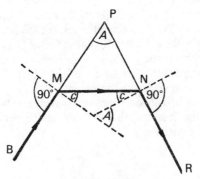

FIG. 19.10. Maximum angle of prism.

a certain value of A, MN will make the critical angle, c, with the normal at N, and the emergent ray NR will then graze the surface PR, as shown in Fig. 19.10. As A is increased further, the rays in the glass strike PR at angles of incidence greater than c, and hence no emergent rays are obtained. Thus Fig. 19.10 illustrates the largest angle of a prism for which emergent rays are obtained, and this is known as the *limiting angle* of the prism. It can be seen from the simple geometry of Fig. 19.10 that $A = c + c$ in this special case, and hence *the limiting angle of a prism is twice the critical angle*. For crown glass of $n = 1\cdot51$ the critical angle c is 41° 30′, and hence transmission of light through a prism of crown glass is impossible if the angle of the prism exceeds 83°.

Total Reflecting Prisms

When a plane mirror silvered on the back is used as a reflector, multiple images are obtained (p. 424). This disadvantage is overcome by using right-angled isosceles prisms as reflectors of light in optical instruments such as submarine periscopes (see p. 540).

Consider a ray OQ incident normally on the face AC of such a prism. Fig. 19.11 (i). The ray is undeviated, and is therefore incident at P in the

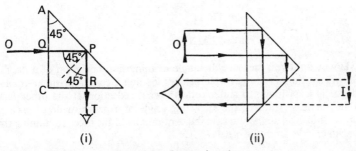

(i)   (ii)

FIG. 19.11. Images in prisms.

glass at an angle of 45° to the normal at P. If the prism is made of crown glass its critical angle is 41° 30′. Hence the incident angle, 45°, in the glass is greater than the critical angle, and consequently the light is *totally* reflected in the glass at P. A bright beam of light thus emerges from the prism along RT, and since the angle of reflection at P is equal to the incident angle, RT is perpendicular to OQ. The prism thus deviates the light through 90°. If the prism is positioned as shown in Fig. 19.11 (ii), an inverted bright virtual image I of the object O is seen by total reflection at the two surfaces of the prism.

There is no loss of brightness when total internal reflection occurs at a surface, whereas the loss may be as much as 10 per cent or more in reflection at a silver surface.

EXAMPLES

1. Describe a good method of measuring the refractive index of a substance such as glass and give the theory of the method. A glass prism of angle 72° and index of refraction 1·66 is immersed in a liquid of refractive index 1·33. What is the angle of minimum deviation for a parallel beam of light passing through the prism? (*L*.)

First part. Spectrometer can be used, p. 444.

Second part.

$$n = \frac{\sin\left(\dfrac{A+D}{2}\right)}{\sin\dfrac{A}{2}}$$

where n is the *relative refractive index* of glass with respect to the liquid.

But

$$n = \frac{1·66}{1·33}$$

$$\therefore \quad \frac{1·66}{1·33} = \frac{\sin\left(\dfrac{72°+D}{2}\right)}{\sin\dfrac{72°}{2}} = \frac{\sin\left(\dfrac{72°+D}{2}\right)}{\sin 36°}$$

$$\therefore \quad \sin\left(\frac{72°+D}{2}\right) = \frac{1·66}{1·33}\sin 36° = 0·7335$$

$$\therefore \quad \frac{72°+D}{2} = 47° \; 11'$$

$$\therefore \quad D = 22° \; 22'$$

2. How would you measure the angle of minimum deviation of a prism? (*a*) Show that the ray of light which enters the first face of a prism at grazing incidence is least likely to suffer total internal reflection at the other face. (*b*) Find the least value of the refracting angle of a prism made of glass of refractive index 7/4 so that no rays incident on one of the faces containing this angle can emerge from the other. (*N*.)

First part. See text.

Second part. (a) Suppose PM is a ray which enters the first face of the prism at grazing incidence, i.e., at an angle of incidence of 90°, Fig. 19.12. The refracted ray MQ then makes an angle of refraction c, where c is the critical angle. Suppose QN is the normal at Q on the other face of the prism. Then since angle BNQ = A, where A is the angle of the prism, the angle of refraction MQN at Q = A − c, from the exterior angle property of triangle MQN. Similarly, if RM is a ray at an angle of incidence i at the first face less than 90°, the angle of refraction at S at the second face = A − r, where r is the angle of refraction BMS.

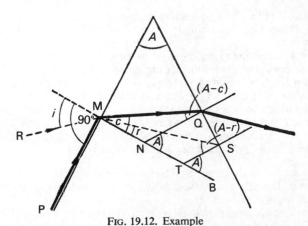

FIG. 19.12. Example

Now c is the *maximum* angle of refraction in the prism.

Hence MQ makes the minimum angle of incidence on the second face, and thus is least likely to suffer total internal reflection.

(b) The least value of the refracting angle of the prism corresponds to a ray at grazing incidence and grazing emergence, as shown in Fig. 19.10.

Thus   minimum angle = 2c
where c is the critical angle (p. 449).

But $\sin c = \dfrac{1}{n} = \dfrac{4}{7} = 0.5714$

$$\therefore \quad c = 34° \, 51'$$

$$\therefore \quad \text{minimum angle} = 2c = 69° \, 42'$$

EXERCISES 19

1. A ray of light is refracted through a prism of angle 70°. If the angle of refraction in the glass at the first face is 28°, what is the angle of incidence in the glass at the second face?

2. (i) The angle of a glass prism is 60°, and the minimum deviation of light through the prism is 39°. Calculate the refractive index of the glass. (ii) The refractive index of a glass prism is 1·66, and the angle of the prism is 60°. Find the minimum deviation.

3. By means of a labelled diagram show the paths of rays from a mono-chromatic source to the eye through a correctly adjusted prism spectrometer.

Obtain an expression relating the deviation of the beam by the prism to the refracting angle and the angles of incidence and emergence.

A certain prism is found to produce a minimum deviation of 51° 0′, while it produces a deviation of 62° 48′ for two values of the angle of incidence, namely 40° 6′ and 82° 42′ respectively. Determine the refracting angle of the prism, the angle of incidence at minimum deviation and the refractive index of the material of the prism. (*L*).

4. A ray of light passing symmetrically through a glass prism of refracting angle *A* is deviated through an angle *D*. Derive an expression for the refractive index of the glass.

A prism of refracting angle about 60° is mounted on a spectrometer table and all the preliminary adjustments are made to the instrument. Describe and explain how you would then proceed to measure the angles *A* and *D*.

PQR represents a right-angled isosceles prism of glass of refractive index 1·50. A ray of light enters the prism through the hypotenuse *QR* at an angle of incidence *i*, and is reflected at the critical angle from *PQ* to *PR*. Calculate and draw a diagram showing the path of the ray through the prism. (Only rays in the plane of *PQR* need be considered.) (*N*.)

5. Give a labelled diagram showing the essential optical parts of a prism spectrometer. Describe the method of adjusting a spectrometer and using it to measure the angle of a prism.

A is the vertex of a triangular glass prism, the angle at *A* being 30°. A ray of light *OP* is incident at *P* on one of the faces enclosing the angle *A*, in a direction such that the angle *OPA* = 40°. Show that, if the refractive index of the glass is 1·50, the ray cannot emerge from the second face. (*L*.)

6. Define refractive index and derive an expression relating the relative refractive index n_{AB} for light travelling out of medium *A* into medium *B* with the velocities of light v_A and v_B respectively in those media.

Draw a diagram showing how a parallel beam of monochromatic light is deviated by its passage through a triangular glass prism. Given that the angle of deviation is a minimum when the angles of incidence and emergence are equal show that the refractive index *n* of the glass is related to the refracting angle a of the prism and the minimum deviation δ by the equation

$$n = \sin \tfrac{1}{2}(a + \delta)/\sin \tfrac{1}{2}a.$$

Describe how you would apply this result to measure the dispersive power of the glass of a given triangular prism. You may assume the availability of sources of light of standard wavelengths. (*O. & C.*)

7. Explain how you would adjust the telescope of a spectrometer before making measurements.

Draw and label a diagram of the optical parts of a prism spectrometer after the adjustments have been completed. Indicate the position of the crosswires and show the paths through the instrument of two rays from a monochromatic source when the setting for minimum deviation has been obtained.

The refracting angle of a prism is 62·0° and the refractive index of the glass for yellow light is 1·65°. What is the smallest possible angle of incidence of a ray of this yellow light which is transmitted without total internal reflection? Explain what happens if white light is used instead, and the angle of incidence is varied in the neighbourhood of this minimum. (*N*.)

8. Explain the meaning of the term *critical angle*. Describe and give the theory of a critical angle method for determining the refractive index of water.

A right-angled prism ABC has angle BAC = angle $ACB = 45°$, and is made of glass of refractive index 1·60. A ray of light is incident upon the hypotenuse face AC so that after refraction it strikes face AB and emerges at minimum deviation. What is the angle of incidence upon AC?

What is the smallest angle of incidence upon AC for which the ray can still emerge at AB? If the angle of incidence upon AC is made zero, what will be the whole deviation of the ray? (*L.*)

9. Draw a labelled diagram of a spectrometer set up for studying the deviation of light through a triangular prism. Describe how you would adjust the instrument and use it to find the refractive index of the prism material.

Indicate briefly how you would show that the radiation from an arc lamp is not confined to the visible spectrum. (*L.*)

10. How would you investigate the way in which the deviation of a ray of light by a triangular glass prism varies with the angle of incidence on the first face of the prism? What result would you expect to obtain?

The deviation of a ray of light incident on the first face of a 60° glass prism at an angle of 45° is 40°. Find the angle which the emergent ray makes with the normal to the second face of the prism and determine, preferably by graphical construction, the refractive index of the glass of the prism. (*L.*)

11. A prism has angles of 45°, 45°, and 90° and all three faces polished. Trace the path of a ray entering one of the smaller faces in a direction parallel to the larger face and perpendicular to the prism edges. Assume 1·5 for the refractive index.

If you had two such prisms how would you determine by a simple pin or ray-box method the refractive index of a liquid available only in small quantity? (*L.*)

12. Under what circumstances does total internal reflection occur? Show that a ray of light incident in a principal section of an equilateral glass prism of refractive index 1·5, can only be transmitted after two refractions at adjacent faces if the angle of incidence on the prism exceeds a certain value. Find this limiting angle of incidence. (*W.*)

13. Draw a graph showing, in a general way, how the deviation of a ray of light when passed through a triangular prism depends on the angle of incidence.

You are required to measure the refractive index of glass in the form of a prism by means of a spectrometer provided with a vertical slit. Explain how you would level the spectrometer table and derive the formula from which you would calculate the refractive index. (You are not required to explain any other adjustment of the apparatus nor to explain how you would find the refracting angle of the prism.) (*L.*)

chapter twenty

Dispersion. Spectra

Spectrum of White Light

IN 1666, NEWTON made a great scientific discovery. He found that sunlight, or white light, was made up of different colours, consisting of red, orange, yellow, green, blue, indigo, violet. Newton made a small hole in a shutter in a darkened room, and received a white circular patch of sunlight on a screen S in the path of the light, Fig. 20.1 (i). But on interposing a glass prism between the hole and the screen he observed

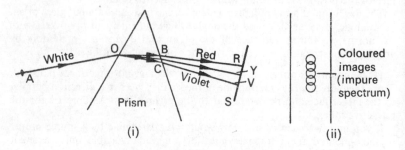

FIG. 20.1. Impure spectrum.

a series of overlapping coloured patches in place of the white patch, the total length of the coloured images being several times their width, Fig. 20.1 (ii). By separating one colour from the rest, Newton demonstrated that the colours themselves could not be changed by refraction through a prism, and he concluded that the colours were not *introduced* by the prism, but were components of the white light. The *spectrum* (colours) of white light consists of red, orange, yellow, green, blue, indigo, and violet, and the separaion of the colours by the prism is known as *dispersion*.

The red rays are the least deviated by the prism, and the violet rays are the most deviated, as shown in the exaggerated sketch of Fig. 20.1 (i). Since the angle of incidence at O in the air is the same for the red and violet rays, and the angle of refraction made by the red ray OB in the glass is greater than that made by the violet ray OC, it follows from $\sin i/\sin r$ that the refractive index of the prism material for red light is less than for violet light. Similarly, the refractive index for yellow light lies between the refractive index values for red and violet light (see also p. 458).

Production of Pure Spectrum

Newton's spectrum of sunlight is an *impure spectrum* because the different coloured images overlap, Fig. 20.1 (ii). A *pure spectrum* is one in which the different coloured images contain light of one colour only, i.e., they are monochromatic images. In order to obtain a pure spectrum (i) the white light must be admitted through a very narrow opening, so as to assist in the reduction of the overlapping of the images, (ii) the beams of coloured rays emerging from the prism must be parallel, so that each beam can be brought to a separate focus.

The spectrometer can be used to provide a pure spectrum. The collimator slit is made very narrow, and the collimator C and the telescope T are both adjusted for parallel light, Fig. 20.2. A bright source

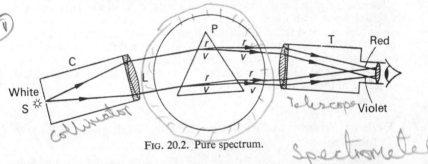

FIG. 20.2. Pure spectrum.

of white light, S, is placed near the slit, and the prism P is usually set in the minimum deviation position for yellow light, although this is not essential. The rays refracted through P are now separated into a number of different coloured parallel beams of light, each travelling in slightly different directions, and the telescope brings each coloured beam to a separate focus. A pure spectrum can now be seen through T, consisting of a series of monochromatic images of the slit.

If only one lens, L, is available, the prism P *must* be placed in the minimum deviation position for yellow light in order to obtain a fairly pure spectrum, Fig. 20.3. The prism is then also approximately in the

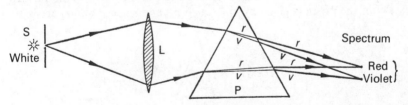

FIG. 20.3. Fairly pure spectrum.

minimum deviation position for the various colours in the incident convergent beam, and hence the rays of one colour are approximately deviated by the same amount by the prism, thus forming an image of the slit S at roughly the same place.

Infra-red and ultra-violet rays. In 1800 HERSCHEL discovered the existence

456 OPTICS, SOUND AND WAVES

of *infra-red rays*, invisible rays beyond the red end of the spectrum. Fundamentally, they are of the same nature as rays in the visible spectrum but having longer wavelengths than the red, and produce a sensation of heat (see p. 344). Their existence may be demonstrated in the laboratory by means of an arc light in place of S in Fig. 20.3, a rocksalt lens at L and a rocksalt prism at P. A phototransistor, such as Mullard OCP71, connected to an amplifier and galvanometer is very sensitive to infra-red light. When this detector is moved into the dark part beyond the red end of the spectrum, a deflection is obtained in the galvanometer. Since they are not scattered by fine particles as much as the rays in the visible spectrum, infra-red rays can penetrate fog and mist. Clear pictures have been taken in mist by using infra-red filters and photographic plates.

About 1801 RITTER discovered the existence of invisible rays beyond the violet end of the visible spectra. *Ultra-violet rays*, as they are known, affect photographic plates and cause certain minerals to fluoresce. They can also eject electrons from metal plates as seen in the *Photoelectric effect*. Ultraviolet rays can be detected in the laboratory by using an arc light in the place of S in Fig. 20.3, a quartz lens at L, and a quartz prism at P. A sensitive detector is a photoelectric cell connected to a galvanometer and battery. When the cell is moved beyond the violet into the dark part of the spectrum a deflection is observed in the galvanometer.

Deviation Produced by Small-angle Prism for Small Angles of Incidence

Before discussing in detail the colour effect produced when white light is incident on a prism, we must derive an expression for the deviation produced by a *small-angle prism*.

Consider a ray PM of monochromatic light incident almost normally on the face TM of a prism of small angle A, so that the angle of incidence, i_1, is small, Fig. 20.4. Then $\sin i_1 / \sin r_1 = n$, where r_1 is the angle of refraction in the prism, and n is the refractive index for the colour of the light. As r_1 is less than i_1, r_1 also is a small angle. Now the sine of a small angle is practically equal to the angle measured in radians. Thus $i_1/r_1 = n$, or

$$i_1 = nr_1 \qquad \qquad \text{(i)}$$

From the geometry of Fig. 20.4, the angle of incidence r_2 on the face TN of the prism is given by $r_2 = A - r_1$; and since A and r_1 are both

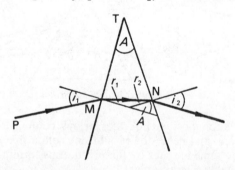

FIG. 20.4. Deviation through small-angle prism.

small, it follows that r_2 is a small angle. The angle of emergence i_2 is thus also small, and since $\sin i_2/\sin r_2 = n$ we may state that $i_2/r_2 = n$, or

$$i_2 = nr_2 \ . \qquad . \qquad . \qquad . \qquad . \qquad . \qquad \text{(ii)}$$

The deviation, d, of the ray on passing through the prism is given by $d = (i_1 - r_1) + (i_2 - r_2)$. Substituting for i_1 and i_2 from (i) and (ii),

$$\therefore \quad d = nr_1 - r_1 + nr_2 - r_2 = n(r_1 + r_2) - (r_1 + r_2)$$
$$\therefore \quad d = (n - 1)(r_1 + r_2)$$

But $\quad r_1 + r_2 = A$

$$\therefore \quad \mathbf{d = (n - 1)\,A} \quad . \qquad . \qquad . \qquad . \qquad . \qquad . \qquad \text{(1)}$$

This is the magnitude of the deviation produced by a *small*-angle prism for *small* angles of incidence. If A is expressed in radians, then d is in radians; if A is expressed in degrees, then d is in degrees. If $A = 6°$ and $n = 1 \cdot 6$ for yellow light, the deviation d of that colour for small angles of incidence is given by $d = (1 \cdot 6 - 1)\,6° = 3 \cdot 6°$. It will be noted that the deviation is independent of the magnitude of the small angle of incidence on the prism.

Dispersion by Small-angle Prism

We have already seen from Newton's experiment that the colours in a beam of white light are separated by a glass prism into red, orange, yellow, green, blue, indigo, violet, so that the emergent light is no longer white but coloured. The separation of the colours by the prism is known generally as the phenomenon of *dispersion*, and the *angular dispersion* between the red and blue emergent rays, for example, is defined as the *angle* between these two rays. Thus, in Fig. 20.5, θ is the angular dispersion between the red and blue rays. Of course, the angular dispersion is also equal to the *difference in deviation* of the two colours produced by the prism; and since we have already derived the expression $d = (n - 1)\,A$ for the deviation of monochromatic light by a small-angle prism we can obtain the angular dispersion between any two colours.

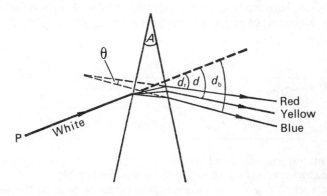

FIG. 20.5. Dispersion.

Suppose d_b, d_r are the respective deviations of the blue and red light when a ray of white light is incident at a small angle on a prism of small angle A, Fig. 20.5. Then, if n_b, n_r are the refractive indices of the prism material for blue and red light respectively,

$$d_b = (n_b - 1)\,A,$$
and
$$d_r = (n_r - 1)\,A.$$

$\therefore$ angular dispersion, $d_b - d_r = (n_b - 1)\,A - (n_r - 1)\,A$

$$\therefore \quad d_b - d_r = (n_b - n_r)\,A \qquad . \qquad . \qquad (2)$$

For a particular crown glass, $n_b = 1\cdot521$, $n_r = 1\cdot510$. Thus if $A = 8°$, the angular dispersion between the blue and red colours =

$$d_b - d_r = (n_b - n_r)\,A = (1\cdot521 - 1\cdot510)\,8° = 0\cdot09°$$

The *mean deviation* of the white light by the prism is commonly chosen as the deviation of the *yellow* light, since this is the colour approximately in the middle of the spectrum; the mean refractive index of a material is also specified as that for yellow light. Now the deviation, d, of monochromatic light is given by $d = (n - 1)\,A$, from equation (1), and unless otherwise stated, the magnitudes of d and n will be understood to be those for yellow light when these symbols contain no suffixes. If $n_b = 1\cdot521$ and $n_r = 1\cdot510$, then *approximately* the refractive index, n, for yellow light is the average of n_b and n_r, or $\frac{1}{2}(1\cdot521 + 1\cdot510)$; thus $n = 1\cdot515$. Hence if the prism has an angle of $8°$, the mean deviation, d, $= (n - 1)\,A = (1\cdot515 - 1)\,8° = 4\cdot1°$.

Dispersive Power

The *dispersive power*, ω, of the material of a small-angle prism for blue and red rays may be defined as the ratio

$$\omega = \frac{\textit{angular dispersion between blue and red rays}}{\textit{mean deviation}} \qquad (3)$$

The dispersive power depends on the material of the prism. As an illustration, suppose that a prism of angle $8°$ is made of glass of a type X, say, and another prism of angle $8°$ is made of glass of a type Y.

	n_b	n_r	n
Crown glass, X	1·521	1·510	1·515
Flint glass, Y	1·665	1·645	1·655

Further, suppose the refractive indices of the two materials for blue red, and yellow light are those shown in the above table.

For a small angle of incidence on the prism of glass X, the angular dispersion

$$= d_b - d_r = (n_b - 1) A - (n_r - 1) A$$
$$= (n_b - n_r) A = (1\cdot521 - 1\cdot510)\ 8° = 0\cdot09° \quad . \quad \text{(i)}$$

The mean deviation, $d = (n - 1) A = (1\cdot515 - 1)\ 8° = 4\cdot1°$. (ii)

∴ dispersive power, $\omega, = \dfrac{0\cdot09}{4\cdot1} = 0\cdot021$ . . . . (iii)

Similarly, for the prism of glass Y,

angular dispersion $= (n_b - n_r) A = (1\cdot665 - 1\cdot645)\ 8 = 0\cdot16°$

and   mean deviation $= (n - 1) A = 1\cdot655 - 1)\ 8 = 5\cdot24°$

∴ dispersive power $= \dfrac{0\cdot16}{5\cdot24} = 0\cdot03$ . . . . (iv)

From (iii) and (iv), it follows that the dispersive power of glass Y is about 1·5 times as great as that of glass X.

General Formula for Dispersive Power

We can now derive a general formula for dispersive power, ω, which is independent of angles. From equation 3, it follows that

$$\omega = \frac{d_b - d_r}{d}$$

as d_b, d_r, d denote the deviations of blue, red, and yellow light respectively.

But $d_b - d_r = (n_b - 1) A - (n_r - 1) A = (n_b - n_r) A$

and $d = (n - 1) A.$

$$\therefore \quad \omega = \frac{d_b - d_r}{d} = \frac{(n_b - n_r) A}{(n - 1) A}$$

$$\therefore \quad \omega = \frac{n_b - n_r}{n - 1} \quad . \quad . \quad . \quad . \quad . \quad (4)$$

From this formula, it can be seen that (i) ω depends only on the material of the prism and is independent of its angle, (ii) ω is a number and has therefore no units. In contrast to "dispersive power", it should be noted that "dispersion" is an angle, and that its magnitude depends on the angle A of the prism and the two colours concerned, for $d_b - d_r = \text{dispersion} = (n_b - n_r) A$.

Achromatic Prisms

We have seen that a prism separates the colours in white light. If a prism is required to deviate white light without dispersing it into colours, *two* prisms of different material must be used to eliminate the dispersion, as shown in Fig. 20.6. The prism P is made of crown glass, and causes dispersion between the red and blue in the incident white light. The prism Q is inverted with respect to P, and with a suitable choice of its angle A' (discussed fully later), the red and blue rays incident on it can be made to emerge in *parallel* directions. If the rays are viewed the eye-lens brings them to a focus at the same place on the retina, and hence the colour effect due to red and blue rays is eliminated. The dispersion of the other colours in white light still remains, but most of the colour effect is eliminated as the red and blue rays are the "outside" (extreme) rays in the spectrum of white light.

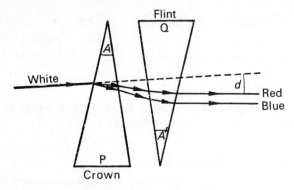

FIG. 20.6. Achromatic prisms.

Prisms which eliminate dispersion between two colours, blue and red say, are said to be *achromatic* prisms for those colours. Suppose n_b, n_r are the refractive indices of crown glass for blue and red light, and A is the angle of the crown glass prism P. Then, from p. 458,

$$\text{dispersion} = (n_b - n_r)\,A \quad . \quad . \quad . \quad . \quad \text{(i)}$$

If n'_b, n'_r are the refractive indices of flint glass for blue and red light and A' is the angle of the flint glass prism Q, then similarly,

$$\text{dispersion} = (n'_b - n'_r)\,A' \quad . \quad . \quad . \quad . \quad \text{(ii)}$$

Now prism P produces its dispersion in a "downward" direction since a prism bends rays towards its base, Fig. 20.6, and prism Q produces its dispersion in an "upward" direction. For achromatic prisms, therefore, the dispersions produced by P and Q must be equal.

$$\therefore (n_b - n_r)\,A = (n'_b - n'_r)\,A' \quad . \quad . \quad . \quad . \quad \text{(5)}$$

Suppose P has an angle of 6°. Then, using the refractive indices for n_b, n_r, n'_b, n'_r in the table on p. 458, it follows from (5) that the angle A' is given by

$$(1\cdot521 - 1\cdot510)\,6° = (1\cdot665 - 1\cdot645)\,A'$$

Thus
$$A' = \frac{0\cdot011}{0\cdot02} \times 6° = 3\cdot3°$$

Deviation Produced by Achromatic Prisms

Although the colour effects between the red and blue rays are eliminated by the use of achromatic prisms, it should be carefully noted that the incident light beam, as a whole, has been deviated. This angle of deviation, d, is shown in Fig. 20.6, and is the angle between the incident and emergent beams. The deviation of the mean or yellow light by prism P is given by $(n - 1)\,A$, and is in a "downward" direction.

Since the deviation of the yellow light by the prism Q is in an opposite direction, and is given by $(n' - 1) A'$, the net deviation, d, is given by

$$d = (n - 1) A - (n' - 1) A'.$$

Using the angles $6°$ and $3·3°$ obtained above, with $n = 1·515$ and $n' = 1·655$,

$$d = (1·515 - 1) 6° - (1·655 - 1) 3·3° = 0·93°.$$

Direct-vision Spectroscope

The direct-vision spectroscope is a simple instrument used for examining the different colours in the spectrum obtained from a glowing gas in a flame or in a discharge tube. It contains several crown and flint prisms cemented together, and contained in a straight tube having lenses which constitute an eye-piece. The tube is pointed at the source of light examined, when various colours are seen on account of the dispersion produced by the prisms, Fig. 20.7.

In practice, the direct-vision spectroscope contains several crown and flint glass prisms, but for convenience suppose we consider two such prisms, as in Fig. 20.7. For "direct vision", the *net deviation of the mean* (yellow) *ray* produced by the prisms must be zero. Thus the mean deviation caused by the crown glass prism in one direction must be equal to that caused by the flint glass prism in the opposite direction. Hence, with the notation already used, we must have

$$(n - 1) A = (n' - 1) A'.$$

Suppose $A = 6°$, $n = 1·515$, $n' = 1·655$. Then A' is given by

$$(1·515 - 1) 6° = (1·655 - 1) A'$$

$$\therefore \qquad A' = \frac{0·515}{0·655} \times 6° = 4·7°$$

The *net dispersion* of the blue and red rays is given by

$$(n_b - n_r) A - (n'_b - n'_r) A'$$
$$= (1·521 - 1·510) 6° - (1·665 - 1·645) 4·7°$$
$$= 0·066 - 0·094 = - 0·028°.$$

The minus indicates that the net dispersion is produced in a "blue-upward" direction, as the dispersion of the flint glass prism is greater than that of the crown glass prism.

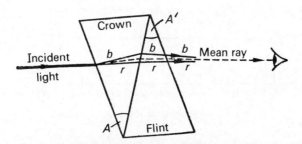

FIG. 20.7. Dispersion; but no deviation of mean ray.

SPECTRA

The Importance of the Study of Spectra

The study of the wavelengths of the radiation from a hot body comes under the general heading of *Spectra*. The number of spectra of elements and compounds which have been recorded runs easily into millions, and it is worth while stating at the outset the main reasons for the interest in the phenomenon.

It is now considered that an atom consists of a nucleus of positive electricity surrounded by electrons moving in various orbits, and that a particular electron in an orbit has a definite amount of energy. In certain circumstances the electron may jump from this orbit to another, where it has a smaller amount of energy. When this occurs radiation is emitted, and the energy in the radiation is equal to the difference in energy of the atom between its initial and final states. The displacement of an electron from one orbit to another occurs when a substance is raised to a high temperature, in which case the atoms present collide with each other very violently. Light of a definite wavelength will then be emitted, and will be characteristic of the electron energy changes in the atom. There is usually more than one wavelength in the light from a hot body (iron has more than 4,000 different wavelengths in its spectrum), and each wavelength corresponds to a change in energy between two orbits. A study of spectra should therefore reveal much important information concerning the structure and properties of atoms.

Every element has a unique spectrum. Consequently a study of the spectrum of a substance enables its composition to be readily determined. *Spectroscopy* is the name given to the exact analysis of mixtures or compounds by a study of their spectra, and the science has developed to such an extent that the presence in a substance of less than a millionth of a milligram of sodium can be detected.

Types of emission spectra. There are three different types of spectra, which are easily recognised. They are known as (*a*) *line spectra*, (*b*) *band spectra*, (*c*) *continuous spectra*.

(*a*) *Line spectra.* When the light emitted by the atoms of a glowing substance (such as vaporised sodium or helium gas) is examined by a prism and spectrometer, lines of various wavelengths are obtained. These lines, it should be noted, are images of the narrow slit of the spectrometer on which the light is incident. The spectra of hydrogen, Fig. 20.8, and helium are line spectra, and it is generally true that line spectra are obtained from *atoms*.

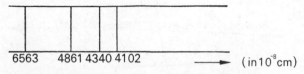

6563   4861 4340 4102   ⟶   (in 10^{-8} cm)

FIG. 20.8. Visible line spectra of hydrogen.

(*b*) *Band spectra.* Band spectra are obtained from *molecules*, and consist of a series of bands each sharp at one end but "fading" at the other end, Fig.

20.9. The term "fluting" is often used to describe the way in which the bands are spaced. Careful examination reveals that the bands are made up of numerous fine lines very close to each other. Two examples of band spectra are those usually obtained from nitrogen and oxygen.

FIG. 20.9. Diagrammatic representation of band spectra.

(c) *Continuous spectra*. The spectrum of the sun is an example of a continuous spectrum, and, in general, the latter are obtained from *solids and liquids*. In these states of matter the atoms and molecules are close together, and electron orbital changes in a particular atom are influenced by neighbouring atoms to such an extent that radiations of all different wavelengths are emitted. In a gas the atoms are comparatively far apart, and each atom is uninfluenced by any other. The gas therefore emits radiations of wavelengths which result from orbital changes in the atom due solely to the high temperature of the gas, and a line spectrum is obtained. When the temperature of a gas is decreased and pressure applied so that the liquid state is approached, the line spectrum of the gas is observed to broaden out considerably.

Production of spectra. In order to produce its spectrum the substance under examination must be heated to a high temperature. There are four main methods of *excitation*, as the process is called, and spectra are classified under the method of their production.

(a) *Flame spectra*. The temperature of a Bunsen flame is high enough to vaporise certain solids. Thus if a piece of platinum wire is dipped into a sodium salt and then placed in the flame, a vivid yellow colour is obtained which is characteristic of the element sodium. This method of excitation can only be used for a limited number of metals, the main class being the alkali and alkaline earth metals such as sodium, potassium, lithium, calcium, and barium. The line spectra produced in each case consist of lines of different colours, but some lines have a greater intensity than others. Thus sodium is characterised by two prominent yellow lines barely distinguishable in a small spectroscope, and lithium by a prominent green line.

(b) *Spark spectra*. If metal electrodes are connected to the secondary of an induction coil and placed a few millimetres apart, a spark can be obtained which bridges the gap. It was discovered that a much more intense and violent spark could be obtained by placing a capacitor in parallel with the gap. This spark is known as a *condensed* spark. The solid under investigation forms one of the electrodes, and is vaporised at the high temperature obtained.

(c) *Arc spectra*. This is the method most used in industry. If two metal rods connected to a d.c. voltage supply are placed in contact with each other and then drawn a few millimetres apart, a continuous spark, known as an arc, is obtained across the gap. The arc is a source of very high temperature, and therefore vaporises substances very readily. In practice the two rods are placed in a vertical position, and a small amount of the substance investigated is placed on the lower rod.

(d) *Discharge-tube spectra*. If a gas is contained at low pressure inside a tube having two aluminium electrodes and a high a.c. or d.c. voltage is applied to the gas, a "discharge" occurs between the electrodes and the gas becomes luminous. This is the most convenient method of examining the spectra of gases. The luminous neon gas in a discharge tube has a reddish colour, while mercury vapour is greenish-blue.

Absorption Spectra. Kirchhoff's Law

The spectra just discussed are classified as *emission spectra*. There is another class of spectra known as *absorption spectra*, which we shall now briefly consider.

If light from a source having a continuous spectrum is examined after it has passed through a sodium flame, the spectrum is found to be crossed by a dark line; this dark line is in the position corresponding to the bright line emission spectrum obtained with the sodium flame alone. The continuous spectrum with the dark line is naturally characteristic of the absorbing substance, in this case sodium, and it is known as an *absorption spectrum*. An absorption spectrum is obtained when red glass is placed in front of sunlight, as it allows only a narrow band of red rays to be transmitted.

KIRCHHOFF'S investigations on absorption spectra in 1855 led him to formulate a simple law concerning the emission and absorption of light by a substance. This states: *A substance which emits light of a certain wavelength at a given temperature can also absorb light of the same wavelength at that temperature*. In other words, a good emitter of a certain wavelength is also a good absorber of that wavelength. From Kirchhoff's law it follows that if the radiation from a hot source emitting a continuous spectrum is passed through a vapour, the absorption spectrum obtained is deficient in those wavelengths which the vapour would emit if it were raised to the same high temperature. Thus if a sodium flame is observed through a spectrometer in a darkened room, a bright yellow line is seen; if a strong white arc light, richer in yellow light than the sodium flame, is placed behind the flame, a dark line is observed in the place of the yellow line. The sodium absorbs yellow light from the white light, and re-radiates it in all directions. Consequently there is less yellow light in front of the sodium flame than if it were removed, and a dark line is thus observed.

Fraunhofer Lines

In 1814 FRAUNHOFER noticed that the sun's spectrum was crossed by many hundreds of dark lines. These *Fraunhofer lines*, as they are called, were mapped out by him on a chart of wavelengths, and the more prominent were labelled by the letters of the alphabet. Thus the dark line in the blue part of the spectrum was known as the *F* line, the dark line in the yellow part as the *D* line, and the dark line in the red part as the *C* line.

The Fraunhofer lines indicate the presence in the sun's atmosphere of certain elements in a vaporised form. The vapours are cooler than the central hot portion of the sun, and they absorb their own characteristic wavelengths from the sun's continuous spectrum. Now every element has a characteristic spectrum of wavelengths. Accordingly, it became possible to identify the elements round the sun from a study of the wavelengths of the Fraunhofer (dark) lines in the sun's spectrum, and it was then found that hydrogen and helium were present. This was how helium was first discovered. The *D* line is the yellow sodium line.

The incandescent gases round the sun can be seen as flames many miles high during a total eclipse of the sun, when the central portion of the sun is cut off from the observer. If the spectrum of the sun is observed just before an eclipse takes place, a continuous spectrum with Fraunhofer lines is obtained, as already stated. At the instant when the eclipse becomes total, however, bright emission lines are seen in exactly the same position as those previously occupied by the Fraunhofer lines, and they correspond to the emission spectra of the vapours alone. This is an illustration of Kirchhoff's law, p. 464.

Measurement of Wavelengths by Spectrometer

As we shall discuss later (p. 690) the light waves produced by different colours are characterised by different *wavelengths*. Besides measuring refractive index, the spectrometer can be adapted for measuring unknown wavelengths, corresponding to the lines in the spectrum of a glowing gas in a discharge tube, for example.

A prism is first placed on the spectrometer table in the minimum deviation position for yellow (sodium) light, thus providing a reference position for the prism in relation to incident light from the collimator. The source of yellow light is now replaced by a helium discharge tube, which contains helium at a very low pressure, glowing as a result of the high voltage placed across the tube. Several bright lines of various colours can now be observed through the telescope (they are differently coloured images of the slit), and the *deviation*, θ, of each of the lines is obtained by rotating the telescope until the image is on the cross-wires, and then noting the corresponding reading on the circular graduated scale. Since the wavelengths, λ, of the various lines in the helium spectrum are known very accurately from tables, a graph can now be plotted between θ and λ. The helium discharge tube can then be replaced by a hydrogen or mercury discharge tube, and the deviations due to other lines of known wavelength obtained. In this way a *calibration curve* for the spectrometer can be obtained, Fig. 20.10.

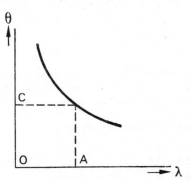

FIG. 20.10. Measurement of wavelength by spectrometer.

The wavelength due to a line Q in the spectrum of an unknown glowing gas can now be easily derived. With the prism still in the minimum deviation position for yellow light, the deviation, θ, of Q is measured. If this angle corresponds to C in Fig. 20.10, the wavelength λ is OA.

EXAMPLES

1. Show that when a ray of light passes nearly normally through a prism of small angle a and refractive index n, the deviation δ is given by $\delta = (n - 1)a$. A parallel beam of light falls normally upon the first face of a prism of small angle. The portion of the beam which is refracted at the second surface is deviated through an angle of 1° 35′, and the portion which is reflected at the second surface and emerges again at the first surface makes an angle of 8° 9′ with the incident beam. Calculate the angle of the prism and the refractive index of the glass. (*C.*)

First part. See text.

Second part. Let θ = angle of prism, n = the refractive index, and RH the ray incident normally on the face AN, striking the second face at K, Fig. 20.11.

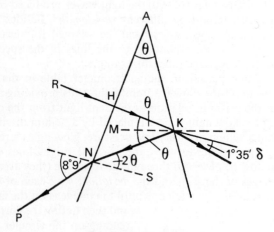

FIG. 20.11. Example.

Then the angle of incidence at K = θ, and angle HKN = 2θ. By drawing the normal NS at N, which is parallel to HK, it can be seen that angle KNS = 2θ. The angle of emergence from the prism = 8° 9′ since the incident beam was normal to AN.

The angle of deviation, δ, of the beam by the prism is given by

$$\delta = (n - 1)\theta$$
$$\therefore \quad 1° 35′ = (n - 1)\theta \quad . \quad . \quad . \quad . \quad \text{(i)}$$

For refraction at N, $n = \dfrac{\sin 8° 9′}{\sin 2\theta}$

Since the angles concerned are small,

$$n = \frac{8° 9′}{2\theta} \quad . \quad . \quad . \quad . \quad \text{(ii)}$$

where θ is in degrees.

From (ii), $\qquad\qquad\qquad \theta = \dfrac{8° 9'}{2n}$; substituting in (i),

$$\therefore \quad 1° 35' = (n - 1)\frac{8° 9'}{2n}$$

$$\therefore \quad \frac{n - 1}{2n} = \frac{95}{489}$$

$$\therefore \quad 489n - 489 = 190n$$

$$\therefore \quad n = 1\cdot63$$

$$\therefore \quad \theta = \frac{8° 29'}{2n} = \frac{8° 29'}{3\cdot26} = 2° 30'$$

2. Define dispersive power. The following table gives the refractive indices of crown and flint glass for three lines of the spectrum.

	C	D	F
Crown	1·514	1·517	1·523
Flint	1·644	1·650	1·664

Calculate the refracting angle of a flint glass prism which, when combined with a crown glass prism of refracting angle 5°, produces a combination that does not deviate the light corresponding to the D line. What separation of the rays corresponding to the C and F lines will such a compound prism produce? (*L.*)

For definition, see text.

The D line corresponds to the mean, or yellow, ray, the F and C lines to the blue and red rays respectively. Let n', n = the refractive indices for crown and flint glass respectively, A', A = the corresponding angles of the prisms.

For no deviation $\quad (n'_D - 1) A' - (n_D - 1) A = 0,$

$$\therefore \ (1\cdot517 - 1) 5° - (1\cdot650 - 1) A = 0$$

$$\therefore \qquad A = \frac{0\cdot517}{0\cdot650} \times 5 = 3\cdot99°$$

The separation of the F and C lines

$$= (n_F - n_C) A - (n'_F - n'_C) A'$$
$$= (1\cdot664 - 1\cdot644) 3\cdot99° - (1\cdot523 - 1\cdot514) 5°$$
$$= 0\cdot0798° - 0\cdot045° = 0\cdot0348°$$

3. Prove that for a prism of small angle A the deviation of a ray of light is $(n - 1) A$, provided that the angle of incidence also is small. A crown glass prism of refracting angle 6° is to be achromatised for red and blue light with a flint glass prism. Using the data below and the formula above find (*a*) the angle of the flint glass prism, (*b*) the mean deviation.

	Crown glass	*Flint glass*
n red	1·513	1·645
n blue	1·523	1·665

(N.)

First part. See text.

Second part. Let A = the angle of the flint prism, n', n = the refractive indices of the crown and flint glass respectively. For achromatism,
$(n'_b - n'_r) \, 6° = (n_b - n_r) \, A$

$$\therefore \quad (1\cdot523 - 1\cdot513) \, 6° = 1\cdot665 - 1\cdot645) \, A$$

$$\therefore \quad A = \frac{0\cdot010}{0\cdot020} \times 6° = 3°$$

The mean refractive index, n', for crown glass $= \dfrac{1\cdot523 + 1\cdot513}{2} = 1\cdot518$

and mean refractive index n, for flint glass $= \dfrac{1\cdot665 + 1\cdot645}{2} = 1\cdot655$

$\therefore$ deviation of mean ray $= (n' - 1) \, 6° - (n - 1) \, 3°$
$$= (1\cdot518 - 1) \, 6° - (1\cdot655 - 1) \, 3° = 1\cdot043°$$

EXERCISES 20

1. Write down the formula for the deviation of a ray of light through a prism of small angle A which has a refractive index n for the colour concerned. Using the following table, calculate the deviation of (i) red light, (ii) blue light, (iii) yellow light through a flint glass prism of refracting angle 4°, and through a crown glass prism of refracting angle 6°.

	Crown glass	Flint glass
n red	1·512	1·646
n blue	1·524	1·666

2. Using the above data, calculate the *dispersive powers* of crown glass and flint glass.

3. Explain how it is possible with two prisms to produce dispersion without mean deviation. A prism of crown glass with refracting angle of 5° and mean refractive index 1·51 is combined with one flint glass of refractive index 1·65 to produce no mean deviation. Find the angle of the flint glass prism. The difference in the refractive indices of the red and blue rays in crown glass is 0·0085 and in flint glass 0·0162. Find the inclination between the red and blue rays which emerge from the composite prism. (*L.*)

4. Draw a ray diagram showing the passage of light of two different wavelengths through a prism spectrometer. Why is it that such a spectrometer is almost invariably used with (*a*) a very narrow entrance slit, (*b*) parallel light passing through the prism, (*c*) the prism set at, or near, minimum deviation?

A spectrometer is used with a *small angle* prism made from glass which has a refractive index of 1·649 for the blue mercury line and 1·631 for the green mercury line. The collimator lens and the objective of the spectrometer both have a focal length of 30 cm. If the angle of the prism is 0·1 radian what is the spacing of the centres of the blue and green mercury lines in the focal plane of the objective, and what maximum slit width may be used without the lines overlapping? The effect of diffraction need not be considered. (*O. & C.*)

5. A glass prism of refracting angle $6.0°$ and of material of refractive index 1.50 is held with its refracting angle downwards alongside another prism of angle $4.0°$ which has its refracting angle pointing upwards. A narrow parallel beam of yellow light is incident nearly normally on the first prism, passes through both prisms, and is observed to emerge parallel to its original direction. Calculate the refractive index of the material of the second prism. If white light were used and the glasses of the two prisms were very different in their power to disperse light, describe very briefly what would be seen on a white screen placed at right angles to the emergent light. (*C.*)

6. A ray of monochromatic light is incident at an angle i on one face of a prism of refracting angle A of glass of refractive index n and is transmitted. The deviation of the ray is D.

Considering only rays incident on the side of the normal away from the refracting angle, sketch graphs on the same set of axes showing how D varies with i when (*a*) A is about $60°$, (*b*) A is very small.

From first principles derive an expression for D when i and A are both very small angles. (*N.*)

7. Distinguish between emission spectra and absorption spectra. Describe the spectrum of the light emitted by (i) the sun, (ii) a car headlamp fitted with yellow glass, (iii) a sodium vapour street lamp.

What are the approximate wavelength limits of the visible spectrum? How would you demonstrate the existence of radiations whose wavelengths lie just outside these limits? (*O. & C.*)

8. State what is meant by *dispersion* and describe, with diagrams, the principle of (i) an achromatic and (ii) a direct-vision prism.

Derive an expression for the refractive index of the glass of a *narrow* angle prism in terms of the angle of minimum deviation and the angle of the prism. If the refractive index of the glass of refracting angle $8°$ is 1.532 and 1.514 for blue and red light respectively, determine the angular dispersion produced by the prism. (*L.*)

9. Describe the processes which lead to the formation of numerous dark lines (Fraunhofer lines) in the solar spectrum. Explain why the positions of these lines in the spectrum differ very slightly when the light is received from opposite ends of an equatorial diameter of the sun. (*N.*)

10. Describe with the aid of diagrams what is meant by *dispersion* and *deviation* by a glass prism. Derive a formula for the deviation D produced by a glass prism of small refracting angle A for small angles of incidence. Sketch the graph showing how the deviation varies with angle of incidence for a beam of light striking such a prism, and on the same axes indicate what would happen with a prism of much larger refracting angle but of material of the same index of refraction. (*C.*)

11. Describe the optical system of a simple prism spectrometer. Illustrate your answer with a diagram showing the paths through the spectrometer of the pencils of rays which form the red and blue ends of the spectrum of a source of white light. (Assume in your diagram that the lenses are achromatic.)

The prism of a spectrometer has a refracting angle of $60°$ and is made of glass whose refractive indices for red and violet are respectively 1.514 and 1.530. A white source is used and the instrument is set to give minimum deviation for red light. Determine (*a*) the angle of incidence of the light on the prism, (*b*) the angle of emergence of the violet light, (*c*) the angular width of the spectrum. (*N.*)

12. Calculate the angle of a crown glass prism which makes an achromatic combination for red and blue light with a flint glass prism of refracting angle 4°. What is the mean deviation of the light by this combination? Use the data given in question 1.

13. Describe and give a diagram of the optical system of a spectrometer. What procedure would you adopt when using the instrument to measure the refractive index of the glass of a prism for sodium light? What additional observations would be necessary in order to determine the dispersive power of the glass?

The refractive index of the glass of a prism for red light is 1·514 and for blue light 1·523. Calculate the difference in the velocities of the red and blue light in the prism if the velocity of light *in vacuo* is 3×10^5 kilometres per second. (*N.*)

14. Explain, with diagrams, how a 'pure' spectrum is produced by means of a spectrometer. What source of light may be used and what readings must be taken in order to find the dispersive power of the material of which the prism is made? (*L.*)

15. (*a*) Explain, giving a carefully drawn, labelled diagram, the function of the various parts of a spectrometer. How is it adjusted for normal laboratory use? (*b*) Distinguish between a continuous spectrum, an absorption spectrum, a band spectrum, and a line spectrum. State briefly how you would obtain each type with a spectrometer. (*W.*)

16. Describe a prism spectrometer and the adjustment of it necessary for the precise observation of the spectrum of light by a gaseous source.

Compare and contrast briefly the spectrum of sunlight and of light emitted by hydrogen at low pressure contained in a tube through which an electric discharge is passing. (*L.*)

chapter twenty-one

Refraction through lenses

A *lens* is a piece of glass bounded by one or two spherical surfaces. When a lens is thicker in the middle than at the edges it is called a *convex* or *converging lens*, Fig. 21.1 (i); when it is thinner in the middle than at the edges it is known as a *concave* or *diverging lens*, Fig. 21.1 (ii). Fig. 21.9, on p. 478, illustrates other types of converging and diverging lenses.

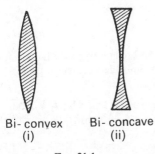

Bi- convex (i)   Bi- concave (ii)

FIG. 21.1.
Converging and diverging lenses.

Lenses were no doubt made soon after the art of glass-making was discovered; and as the sun's rays could be concentrated by these curved pieces of glass they were called "burning glasses". ARISTOPHANES, in 424 B.C., mentions a burning glass. To-day, lenses are used in spectacles, cameras, microscopes, and telescopes, as well as in many other optical instruments, and they afford yet another example of the many ways in which Science is used to benefit our everyday lives.

Since a lens has a curved spherical surface, a thorough study of a lens should be preceded by a discussion of the refraction of light through a curved surface. We shall therefore proceed to consider what happens in this case, and defer a discussion of lenses until later, p. 478.

REFRACTION AT CURVED SPHERICAL SURFACE

Relation Between Object and Image Distances

Consider a curved spherical surface NP, bounding media of refractive indices n_1, n_2 respectively, Fig. 21.2. The medium of refractive index n_1 might be air, for example, and the other of refractive index n_2 might be glass. The centre, C, of the sphere of which NP is part is the centre of curvature of the surface, and hence CP is the radius of curvature, r. The line joining C to the mid-point P of the surface is known as its *principal axis*. P is known as the *pole*.

Suppose a point object O is situated on the axis PC in the medium of refractive index n_1. The image of O by refraction at the curved surface can be obtained by taking two rays from O. A ray OP passes straight through along PC into the medium of refractive index n_2, since OP is normal to the surface, while a ray ON *very close* to the axis is refracted at N along NI towards the normal CN, if we assume n_2 is greater than n_1. Thus at the point of intersection, I, of OP and NI is the image O, and we have here the case of a real image.

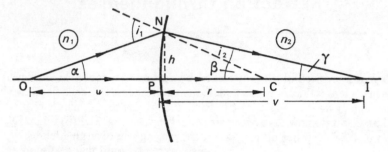

FIG. 21.2. Refraction at curved surface.

Suppose i_1, i_2 are the angles made by ON, NI respectively with the normal, CN, at N, Fig. 21.2. Then, applying "$n \sin i$" is a constant (p. 423),

$$n_1 \sin i_1 = n_2 \sin i_2 \qquad . \qquad . \qquad . \qquad . \qquad \text{(i)}$$

But if we deal with rays from O very close to the axis OP, i_1 is small; and hence $\sin i_1 = i_1$ in radians. Similarly, $\sin i_2 = i_2$ in radians. From (i), it follows that

$$n_1 i_1 = n_2 i_2 \qquad . \qquad . \qquad . \qquad . \qquad \text{(ii)}$$

If a, β, γ are the angles with the axis made by ON, CN, IN respectively, we have

$$i_1 = a + \beta, \text{ from the geometry of triangle ONC,}$$

and $i_2 = \beta - \gamma$, from the geometry of triangle CNI.

Substituting for i_1, i_2 in (ii), we have

$$n_1 (a + \beta) = n_2 (\beta - \gamma)$$
$$\therefore \ n_1 a + n_2 \gamma = (n_2 - n_1) \beta \ . \qquad . \qquad . \qquad \text{(iii)}$$

If h is the height of N above the axis, and N is so close to P that NP is perpendicular to OP,

$$a = \frac{h}{\text{OP}}, \ \gamma = \frac{h}{\text{PI}}, \ \beta = \frac{h}{\text{PC}},$$

From (iii), using our sign convention on p. 407, we have

$$h \left(\frac{n_1}{+\text{OP}} + \frac{n_2}{+\text{PI}} \right) = h \, \frac{(n_2 - n_1)}{\text{PC}},$$

since O is a real object and I is a real image.

$$\therefore \quad \frac{n_1}{\text{OP}} + \frac{n_2}{\text{PI}} = \frac{n_2 - n_1}{\text{PC}}.$$

If the object distance, OP, from P = u, the image distance, IP, from P = v, and PC = r, then

$$\frac{n_1}{u} + \frac{n_2}{v} = \frac{n_2 - n_1}{r} \qquad . \qquad . \qquad \text{(1)}$$

Sign Convention for Radius of Curvature

Equation (1) is the general relation between the object and image distances, u, v, from the middle or pole of the refracting surface, its radius of curvature r, and the refractive indices of the media, n_2, n_1. The quantity $(n_2 - n_1)/r$ is known as the **power** of the surface. If a ray is made to converge by a surface, as in Fig. 21.1, the power will be assumed *positive* in sign; if a ray is made to diverge by a surface, the power will be assumed negative. Since refractive index is a ratio of velocities (p. 421), n_1 and n_2 have no sign. $(n_2 - n_1)$ on the right side of equation (1) will be taken always as a *positive* quantity, and thus denotes the smaller refractive index subtracted from the greater refractive index. The sign convention for the radius of curvature, r, of a spherical surface is now as follows: if the surface is *convex to the less dense* medium, its radius is *positive*; if it is concave to the less dense medium, its radius is negative. We have thus to view the surface from a point in the less dense medium. In Fig. 21.3 (i), the surface A is convex to the less dense

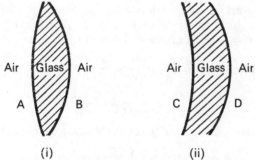

(i)   (ii)

FIG. 21.3. Sign convention for radius of curvature.

medium air, and hence its radius is positive. The surface C is concave to the less dense medium air, and its radius is thus negative, Fig. 21.3 (ii). The radii of the surfaces B and D are both positive.

Special Cases

The general formula $\dfrac{n_1}{u} + \dfrac{n_2}{v} = \dfrac{n_2 \sim n_1}{r}$, can easily be remembered on account of its symmetry. The object distance u corresponds to the refractive index n_1 of the medium in which the object is situated; while the image distance v corresponds to the medium of refractive index n_2 in which the image is situated.

Suppose an object O in air is x cm from a curved spherical surface, and the image I is real and in glass of refractive index n, at a distance of y cm from the surface, Fig. 21.4 (i). Then $u = +x$, $v = +y$, $n_1 = 1$, $n_2 = n$. If the surface is convex to the less dense medium, as shown in Fig. 21.4 (i), the radius of curvature, a cm, is given by $r = +a$.

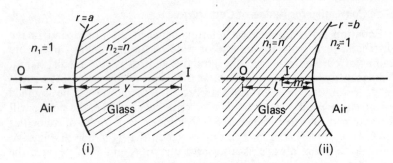

FIG. 21.4. Special cases of refraction formula.

Substituting in
$$\frac{n_1}{u} + \frac{n_2}{v} = \frac{n_2 \sim n_1}{r}$$

$$\therefore \quad \frac{1}{x} + \frac{n}{y} = \frac{n-1}{a}$$

Fig. 21.4 (ii) illustrates the case of an object O in glass of refractive index n, the surface being concave to the less dense surface air. The radius, b cm, is then given by $r = -b$. If the image I is virtual, its distance $v = -m$. If l is the distance of O, then $u = +l$.

Substituting in
$$\frac{n_1}{u} + \frac{n_2}{v} = \frac{n_2 \sim n_1}{r}$$

$$\therefore \quad \frac{n}{l} + \frac{1}{-m} = \frac{n-1}{-b}$$

If a surface is *plane*, its radius of curvature, r, is infinitely large.

Hence $\dfrac{n_2 \sim n_1}{r}$ is zero, whatever different values n_1 and n_2 may have.

Deviation of Light by Sphere

Suppose a ray AO in air is incident on a sphere of glass or a drop of water, Fig. 21.5 (i). The light is refracted at O, then reflected inside B,

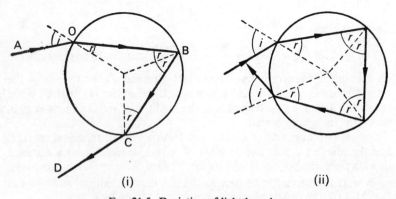

FIG. 21.5. Deviation of light by sphere.

and finally emerges into the air along CD. If i, r are the angles of incidence and refraction at O, the deviation of the light at O and C is $(i - r)$ each time; it is $(180° - 2r)$ at B. The total deviation, δ, in a clockwise direction is thus given by

$$\delta = 2(i - r) + 180° - 2r = 180° + 2i - 4r \quad . \quad . \quad \text{(i)}$$

It can be seen that the deviation at each reflection inside the sphere is $(180° - 2r)$ and that the deviation at each refraction is $(i - r)$. Thus if a ray undergoes two reflections inside the sphere, and two refractions, as shown in Fig. 21.5 (ii), the total deviation in a clockwise direction $= 2(i - r) + 2(180° - 2r) = 360° + 2i - 6r$. After m internal reflections,

$$\text{the total deviation} = 2(i - r) + m(180° - 2r).$$

The Rainbow

The explanation of the colours of the rainbow was first given by Newton about 1667. He had already shown that sunlight consisted of a mixture of colours ranging from red to violet, and that glass could disperse or separate the colours (p. 454). In the same way, he argued, water droplets in the air dispersed the various colours in different directions, so that the colours of the spectrum were seen.

The curved appearance of the rainbow was first correctly explained about 1611. It was attributed to refraction of light at a water drop, followed by reflection inside the drop, the ray finally emerging into the air as shown in Fig. 21.6. The *primary bow* is the rainbow usually seen, and is obtained by two refractions and one reflection at the drops, as in Fig. 21.6. Sometimes a *secondary bow* is seen higher in the sky, and it is formed by rays undergoing two refractions and two reflections at the drop, as in Fig. 21.6.

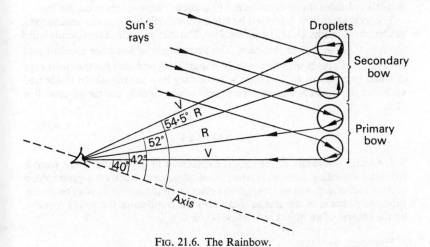

FIG. 21.6. The Rainbow.

The total deviation δ of the light when one reflection occurs in the drop, Fig. 21.6, is given by $\delta = 180° + 2i - 4r$, as proved before. Now the light emerging from the drop will be intense at those angles of incidence corresponding to the *minimum* deviation position, since a considerable number of rays have about the same deviation at the minimum value and thus emerge almost parallel. Now for a minimum value, $\dfrac{d\delta}{di} = 0$.

Differentiating the expression for δ, we have $2 - 4\dfrac{dr}{di} = 0$.

$$\therefore \quad \frac{dr}{di} = \frac{1}{2}$$

But $\quad\quad\quad\quad\quad\quad\quad\quad \sin i = n \sin r,$

where n is the refractive index of water.

$$\therefore \quad \cos i = n \cos r \frac{dr}{di} = \frac{n}{2} \cos r$$

$$\therefore \quad 4 \cos^2 i = n^2 \cos^2 r = n^2 - n^2 \sin^2 r = n^2 - \sin^2 i$$
$$= n^2 - (1 - \cos^2 i)$$

$$\therefore \quad 3 \cos^2 i = n^2 - 1$$

$$\therefore \quad \cos i = \sqrt{\frac{n^2 - 1}{3}} \quad . \quad . \quad . \quad . \quad \text{(ii)}$$

The refractive index of water for red light is $1\cdot331$. Substituting this value in (ii) i can be found, and thus r is obtained. The deviation δ can then be calculated, and the acute angle between the incident and emergent red rays, which is the supplement of δ, is about $42\cdot1°$. By substituting the refractive index of water for violet light in (ii), the acute angle between the incident and emergent violet rays is found to be about $40\cdot2°$. Thus if a shower of drops is illuminated by the sun's rays, an observer standing with his back to the sun sees a brilliant red light at an angle of $42\cdot1°$ with the line joining the sun to him, and a brilliant violet light at an angle of $40\cdot2°$ with this line, Fig. 21.6. Since the phenomenon is the same in all planes passing through the line, the brightly coloured drops form an arc of a circle whose centre is on the line.

The secondary bow is formed by two internal reflections in the water drops, as illustrated in Fig. 21.5 (ii) and Fig. 21.6. The minimum deviation occurs when $\cos i = \sqrt{(n^2 - 1)/8}$ in this case. The acute angle between the incident and emergent red rays is then found to be about $51\cdot8°$, and that for the violet rays is found to be about $54\cdot5°$. Thus the secondary bow has red on the inside and violet on the outside, whereas the primary bow colours are the reverse, Fig. 21.6.

EXAMPLES

1. Obtain a formula connecting the distances of object and image from a spherical refracting surface. A small piece of paper is stuck on a glass sphere of 5 cm radius and viewed through the glass from a position directly opposite. Find the position of the image. Find also the position of the image formed, by the sphere, of an object at infinity. (*O. & C.*)

First part. See text.

Second part. Suppose O is the piece of paper, Fig. 21.7 (i). The refracting surface of the glass is at P, and $u = +10$. Now

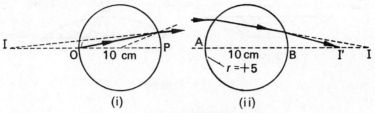

Fig. 21.7. Example

$$\frac{n_2}{v} + \frac{n_1}{u} = \frac{n_2 \sim n_1}{r}$$

where $n_1 = 1\cdot5$, $n_2 = 1$, $r = +5$ p. 472 and v is the image distance from P.

Substituting,
$$\frac{1}{v} + \frac{1\cdot5}{10} = \frac{1\cdot5 - 1}{5}$$

$$\therefore \qquad \frac{1}{v} = 0\cdot1 - 0\cdot15 = -0\cdot05$$

$$\therefore \qquad v = -20 \text{ cm}.$$

Thus the image is virtual, i.e., it is 20 cm from P on the same side as O.

Third part. Suppose I is the position of the image by refraction at the first surface, A, Fig. 21.7 (ii). Now $\frac{n_2}{v} + \frac{n_1}{u} = \frac{n_2 \sim n_1}{r}$, where $u = \infty$, $n_1 = 1$ $n_2 = 1\cdot5$, $r = +5$.

$$\therefore \qquad \frac{1\cdot5}{v} = \frac{1\cdot5 - 1}{5}$$

$$\therefore \qquad v = 15 \text{ cm} = \text{AI, or BI} = 5 \text{ cm}.$$

I is a virtual object for refraction at the curved surface B. Since $u = -\text{BI}$ $= -5$ cm, $n_1 = 1\cdot5$, $n_2 = 1$, $r = +5$, it follows from

$$\frac{n_2}{v} + \frac{n_1}{u} = \frac{n_2 \sim n_1}{r}$$

that
$$\frac{1}{v} + \frac{1\cdot5}{(-5)} = \frac{1\cdot5 - 1}{5}$$

from which
$$v = 2\cdot5 \text{ cm} = \text{BI}'.$$

2. An object is placed in front of a spherical refracting surface. Derive an expression connecting the distances from the refracting surface of the object and the image produced. The apparent thickness of a thick plano-convex lens is measured with (a) the plane face uppermost (b) the convex face uppermost, the values being 2 cm and $2\frac{2}{9}$ cm respectively. If its real thickness is 3 cm, calculate the refractive index of the glass and the radius of curvature of the convex face. (L.)

First part. See text.

Second part. With the plane face uppermost, the image I of the lowest point O is obtained by considering refraction at the plane surface D, Fig. 21.8 (i). Now

FIG. 21.8. Example

$$n = \frac{\text{real depth}}{\text{apparent depth}}$$

$$\therefore \quad n = \frac{3}{2} = 1 \cdot 5$$

With the curved surface uppermost, the image I_1 of the lowest point O_1 is obtained by considering refraction at the curved surface M, Fig. 21.8 (ii). In this case $MI_1 = v$ = apparent thickness = $- 2\frac{2}{9}$ cm, the image I_1 being virtual. Now $u = MO_1 = 3$ cm, $n_2 = 1$, $n_1 = n = 1 \cdot 5$. Substituting in

$$\frac{n_2}{v} + \frac{n_1}{u} = \frac{n_2 \sim n_1}{r}$$

we have

$$\frac{1}{-2\frac{2}{9}} + \frac{1 \cdot 5}{3} = \frac{1 \cdot 5 - 1}{r}.$$

Simplifying, $r = 10$ cm.

REFRACTION THROUGH THIN LENSES

Converging and Diverging Lenses

At the beginning of the chapter we defined a lens as an object, usually of glass, bounded by one or two spherical surfaces. Besides the converging (convex) lens shown in Fig. 21.1 (i) on p. 471, Fig. 21.9 (i) illustrates two other types of converging lenses, which are thicker in the middle than at the edges. Fig. 84 (ii) illustrates two types of diverging (concave) lenses, a diverging lens being also shown in Fig. 21.1 (ii) on p. 471.

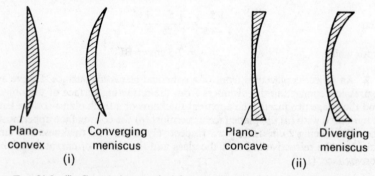

Plano-convex	Converging meniscus	Plano-concave	Diverging meniscus
(i)		(ii)	

FIG. 21.9. (i). Convex (converging) lenses. (ii). Concave (diverging) lenses.

The *principal axis* of a lens is the line joining the centres of curvature of the two surfaces, and passes through the middle of the lens. Experiments with a ray-box show that a thin convex lens brings an incident parallel beam of rays to a *principal focus*, F, on the other side of the lens when the beam is narrow and incident close to the principal axis, Fig. 21.10 (i). On account of the convergent beam contained with it, the convex lens is better described as a "converging" lens. If a similar parallel beam is incident on the other (right) side of the lens, it converges to a focus F′, which is at the same distance from the lens as F when the lens is thin. To distinguish F from F′ the latter is called the "first principal focus"; F is known as the "second principal focus".

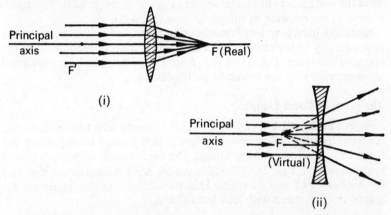

Fig. 21.10. Focus of converging (convex) and diverging (concave) lenses.

When a narrow parallel beam, close to the principal axis, is incident on a thin concave lens, experiment shows that a beam is obtained which appears to diverge from a point F on the same side as the incident beam, Fig. 21.10 (ii). F is known as the principal "focus" of the concave lens. Since a divergent beam is obtained, the concave lens is better described as a "diverging" lens.

Explanation of Effects of Lenses

A thin lens may be regarded as made up of a very large number of *small-angle prisms* placed together, as shown in the exaggerated sketches of Fig. 21.11. If the spherical surfaces of the various truncated prisms are

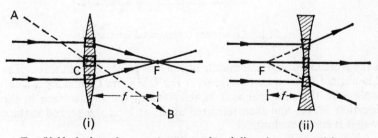

Fig. 21.11. Action of converging (convex) and diverging (concave) lenses.

imagined to be produced, the angles of the prisms can be seen to increase from zero at the middle to a small value at the edge of the lens. Now the deviation, d, of a ray of light by a small-angle prism is given by $d = (n - 1) A$, where A is the angle of the prism, see p. 457. Consequently the truncated prism corresponding to a position farther away from the middle of the lens deviates an incident ray more than those prisms nearer the middle. Thus, for the case of the converging lens, the refracted rays converge to the same point or focus F, Fig. 21.11 (i). It will be noted that a ray AC incident on the middle, C, of the lens emerges parallel to AC, since the middle acts like a rectangular piece of glass (p. 422). This fact is utilised in the drawing of images in lenses (p. 485).

Since the diverging lens is made up of truncated prisms pointing the opposite way to the converging lens, the deviation of the light is in the opposite direction, Fig. 21.11 (ii). A divergent beam is hence obtained when parallel rays are refracted by the lens.

The Signs of Focal Length, f

From Fig. 21.11 (i), it can be seen that a convex lens has a real focus; the focal length, f, of a *converging* lens is thus *positive* in sign. Since the focus of a diverging lens is virtual, the focal length of such a lens is negative in sign, Fig. 21.11 (ii). The reader must memorise the sign of f for a converging and diverging lens respectively, as this is always required in connection with lens formulæ.

Relations Between Image and Object Distances for Thin Lens

We can now derive a relation between the object and image distances when a lens is used. We shall limit ourselves to the case of a *thin* lens, i.e., one whose thickness is small compared with its other dimensions, and consider narrow beams of light incident on its central portion.

Suppose a lens of refractive index n_2 is placed in a medium of refractive index n_1, and a point object O is situated on the principal axis, Fig. 21.12.

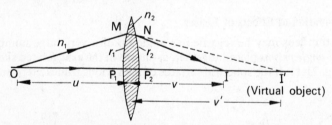

FIG. 21.12. Lens proof (exaggerated for clarity).

A ray from O through the middle of the lens passes straight through as it is normal to both lens surfaces. A ray OM from O, making a *small* angle with the principal axis, is refracted at the first surface in the direction MNI', and then refracted again at N at the second surface so that it emerges along NI.

Refraction at first surface, MP₁. Suppose u is the distance of the

REFRACTION THROUGH LENSES

object from the lens, i.e., $u = OP_1$, and v' is the distance of the image I' by refraction at the first surface, MP_1, of the lens, i.e., $v' = I'P_1$. Then, since I' is situated in the medium of refractive index n_2 (I' is on the ray MN produced), we have, if $n_2 > n_1$,

$$\frac{n_2}{v'} + \frac{n_1}{u} = \frac{n_2 - n_1}{r_1} \qquad . \qquad . \qquad . \qquad . \qquad . \qquad \text{(i)}$$

where r_1 is the radius of the spherical surface MP_1, see p. 472.

Refraction at second surface, NP_2. Since MN and P_1P_2 are the incident rays on the second surface NP_2, it follows that I' is a *virtual object* for refraction at this surface (see p. 412). Hence the object distance $I'P_2$ is negative; and as we are dealing with a thin lens, $I'P_2 = -v'$. The corresponding image distance, IP_2 or v, is positive since I is a real image. Substituting in the formula for refraction at a single spherical surface,

$$\frac{n_2}{-v'} + \frac{n_1}{v} = \frac{n_2 - n_1}{r_2} \qquad . \qquad . \qquad . \qquad . \qquad \text{(ii)}$$

where r_2 is the radius of curvature of the surface NP_2 of the lens.

Lens equation. Adding (i) and (ii) to eliminate v', we have

$$\frac{n_1}{v} + \frac{n_1}{u} = (n_2 - n_1)\left(\frac{1}{r_1} + \frac{1}{r_2}\right);$$

and dividing throughout by n_1,

$$\frac{1}{v} + \frac{1}{u} = \left(\frac{n_2}{n_1} - 1\right)\left(\frac{1}{r_1} + \frac{1}{r_2}\right). \qquad . \qquad . \qquad \text{(iii)}$$

Now parallel rays incident on the lens are brought to a focus. In this case, $u = \infty$ and $v = f$. From (iii),

$$\frac{1}{f} + \frac{1}{\infty} = \left(\frac{n_2}{n_1} - 1\right)\left(\frac{1}{r_1} + \frac{1}{r_2}\right)$$

$$\therefore \quad \frac{1}{f} = \left(\frac{n_2}{n_1} - 1\right)\left(\frac{1}{r_1} + \frac{1}{r_2}\right) \qquad . \qquad . \qquad \text{(2)}$$

Substituting $\dfrac{1}{f}$ for the right-hand side of (iii), we obtain the important

equation $$\qquad \frac{1}{v} + \frac{1}{u} = \frac{1}{f} \qquad . \qquad . \qquad . \qquad . \qquad . \qquad \text{(3)}$$

This is the "lens equation", and it applies equally to converging and diverging lenses if the sign convention is used (see also p. 407).

Focal Length of Lens. Small-angle Prism Method

The focal length f of a lens can also be found by using the deviation formula due to a small-angle prism. Consider a ray PQ parallel to the principal axis at a height h above it. Fig. 21.13 (i). This ray is refracted to the principal focus, and thus undergoes a small deviation through an angle d given by

$$d = \frac{h}{f} \qquad . \qquad . \qquad . \qquad . \qquad . \qquad \text{(i)}$$

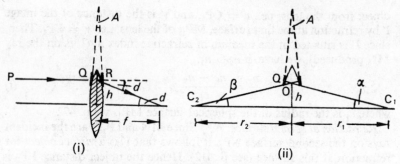

FIG. 21.13. Focal length by small angle prism.

This is the deviation through a prism of small angle A formed by the tangents at Q, R to the lens surfaces, as shown. Now for a small angle of incidence, which is the case for a thin lens and a ray close to the principal axis, $d = (n - 1) A$. See p. 457.

From (i),
$$\frac{h}{f} = (n - 1) A$$

$$\therefore \frac{1}{f} = (n - 1)\frac{A}{h} \qquad . \qquad . \qquad \text{(ii)}$$

The normals at Q, R pass respectively through the centres of curvatures C_1, C_2 of the lens surfaces. From the geometry, angle $ROC_1 = A = \alpha + \beta$, where α, β, are the angles with the principal axis at C_1, C_2 respectively, as shown. But $\alpha = h/r_1$, $\beta = h/r_2$.

$$\therefore \quad A = \alpha + \beta = \frac{h}{r_1} + \frac{h}{r_2} \qquad \text{(iii)}$$

$$\therefore \frac{A}{h} = \frac{1}{r_1} + \frac{1}{r_2}.$$

Substituting in (ii),
$$\frac{1}{f} = (n - 1)\left(\frac{1}{r_1} + \frac{1}{r_2}\right).$$

Focal Length Values

Since $\dfrac{1}{f} = \left(\dfrac{n_2}{n_1} - 1\right)\left(\dfrac{1}{r_1} + \dfrac{1}{r_2}\right)$, it follows that the focal length of a lens depends on the refractive index, n_2, of its material, the refractive index, n_1, of the medium in which it is placed, and the radii of curvature, r_1, r_2, of the lens surfaces. The quantity $\dfrac{n_2}{n_1}$ may be termed the "relative refractive index" of the lens material; if the lens is made of glass of $n_2 = 1\cdot5$, and it is placed in water of $n_1 = 1\cdot33$, then the relative refractive index $= \dfrac{1\cdot5}{1\cdot33} = 1\cdot13$.

In practice, however, lenses are usually situated in air; in which case $n_1 = 1$. If the glass has a refractive index, n_2, equal to n, the relative refractive index, $\dfrac{n_2}{n_1} = \dfrac{n}{1} = n$. Substituting in (22), then

$$\frac{1}{f} = (n - 1) \left(\frac{1}{r_1} + \frac{1}{r_2} \right) \qquad . \qquad . \qquad . \qquad (4)$$

Fig. 21.14 illustrates four different types of glass lenses in air, whose refractive indices, n, are each 1·5. Fig. 21.14 (i) is a biconvex lens, whose radii of curvature, r_1, r_2, are each 10 cm. Since a spherical surface convex to a less dense medium has a positive sign (see p. 473), $r_1 = + 10$ and $r_2 = + 10$. Substituting in (4).

$$\therefore \frac{1}{f} = (1·5 - 1) \left(\frac{1}{(+ 10)} + \frac{1}{(+ 10)} \right) = 0·5 \times \frac{2}{10} = 0·1$$

$$\therefore f = + 10 \text{ cm.}$$

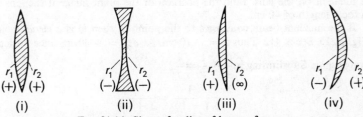

FIG. 21.14. Signs of radius of lens surface.

Fig. 21.14 (ii) is a biconcave lens in air. Since its surfaces are both concave to the less dense medium, $r_1 = - 10$ and $r_2 = - 10$, assuming the radii are both 10 cm. Substituting in (4),

$$\therefore \frac{1}{f} = (1·5 - 1) \left(\frac{1}{(- 10)} + \frac{1}{(- 10)} \right) = 0·5 \times - \frac{2}{10} = - 0·1$$

$$\therefore f = - 10 \text{ cm.}$$

In the case of a plano-convex lens, suppose the radius is 8 cm. Then $r_1 = + 8$ and $r_2 = \infty$, Fig. 21.14 (iii). Hence $\dfrac{1}{f} = (1·5 - 1) \left(\dfrac{1}{(+ 8)} + \dfrac{1}{\infty} \right)$

$= 0·5 \times \dfrac{1}{8} = \dfrac{1}{16}$. Thus $f = + 16$ cm.

In Fig. 21.14 (iv), suppose the radii r_1, r_2 are numerically 16 cm, 12 cm respectively. Then $r_1 = - 16$ and $r_2 = + 12$. Hence

$$\frac{1}{f} = (1·5 - 1) \left(\frac{1}{(- 16)} + \frac{1}{(+ 12)} \right) = 0·5 \left(- \frac{1}{16} + \frac{1}{12} \right) = + \frac{1}{96}.$$

Thus $f = + 96$ cm, confirming that the lens is a converging one.

Some Applications of the Lens Equation

The following examples should assist the reader in understanding how to apply correctly the lens equation $\dfrac{1}{v} + \dfrac{1}{u} = \dfrac{1}{f}$:

1. An object is placed 12 cm from a converging lens of focal length 18 cm. Find the position of the image.

Since the lens is converging, $f = +18$ cm. The object is real, and therefore $= +12$ cm. Substituting in $\dfrac{1}{v} + \dfrac{1}{u} = \dfrac{1}{f}$,

$$\therefore \quad \frac{1}{v} + \frac{1}{(+12)} = \frac{1}{(+18)}$$

$$\therefore \quad \frac{1}{v} = \frac{1}{18} - \frac{1}{12} = -\frac{1}{36}$$

$$\therefore \quad v = -36$$

Since v is negative in sign the image is *virtual*, and it is 36 cm from the lens. See Fig. 21.17 (ii).

2. A beam of light, converging to a point 10 cm behind a converging lens, is incident on the lens. Find the position of the point image if the lens has a focal length of 40 cm.

If the incident beam converges to the point O, then O is a *virtual object*, Fig. 21.15. See p. 412. Thus $u = -10$ cm. Also, $f = +40$ cm since the lens is converging. Substituting in $\dfrac{1}{v} + \dfrac{1}{u} = \dfrac{1}{f}$,

$$\frac{1}{v} + \frac{1}{(-10)} = \frac{1}{(+40)}$$

$$\therefore \quad \frac{1}{v} = \frac{1}{40} + \frac{1}{10} = \frac{5}{40}$$

$$\therefore \quad v = \frac{40}{5} = 8$$

Since v is positive in sign the image is *real*, and it is 8 cm from the lens. The image is I in Fig. 21.15.

3. An object is placed 6 cm in front of a diverging lens of focal length 12 cm. Find the image position.

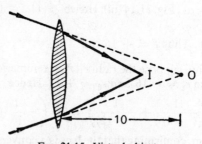

FIG. 21.15. Virtual object.

Since the lens is diverging, $f = -12$ cm. The object is real, and hence $u = +6$ cm. Substituting in $\dfrac{1}{v} + \dfrac{1}{u} = \dfrac{1}{f}$,

$$\therefore \quad \frac{1}{v} + \frac{1}{(+6)} = \frac{1}{(-12)}$$

$$\therefore \quad \frac{1}{v} = -\frac{1}{12} - \frac{1}{6} = -\frac{3}{12}$$

$$\therefore \quad v = -\frac{12}{3} = -4$$

Since v is negative in sign the image is virtual, and it is 4 cm from the lens. See Fig. 21.18 (i).

4. A converging beam of light is incident on a diverging lens of focal length 15 cm. If the beam converges to a point 3 cm behind the lens, find the position of the point image.

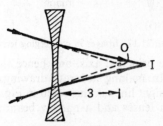

FIG. 21.16. Virtual object.

If the beam converges to the point O, then O is a virtual object, as in example 3, Fig. 21.15. Thus $u = -3$ cm. Since the lens is diverging, $f = -15$ cm. Substituting in $\dfrac{1}{v} + \dfrac{1}{u} = \dfrac{1}{f}$,

$$\therefore \quad \frac{1}{v} + \frac{1}{(-3)} = \frac{1}{(-15)}$$

$$\therefore \quad \frac{1}{v} = -\frac{1}{15} + \frac{1}{3} = \frac{4}{15}$$

$$\therefore \quad v = \frac{15}{4} = 3\tfrac{3}{4}$$

Since v is positive in sign the point image, I, is *real*, and it is $3\tfrac{3}{4}$ cm from the lens, Fig. 21.16.

Images in Lenses

Converging lens. (i) When an object is a very long way from this lens, i.e., at infinity, the rays arriving at the lens from the object are parallel. Thus the image is formed at the focus of the lens, and is real and inverted.

(ii) Suppose an object OP is placed at O perpendicular to the principal axis of a thin converging lens, so that it is farther from the lens than its principal focus, Fig. 21.17 (i). A ray PC incident on the middle, C, of the lens is very slightly displaced by its refraction through the lens, as the opposite surfaces near C are parallel (see Fig. 21.11, which is an exaggerated sketch of the passage of the ray). We therefore consider that PC passes *straight through* the lens, and this is true for any ray incident on the middle of a thin lens.

A ray PL parallel to the principal axis is refracted so that it passes through the focus F. Thus the image, Q, of the top point P of the object

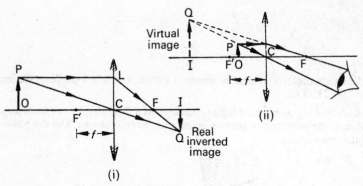

FIG. 21.17. Images in converging lenses.

is formed below the principal axis, and hence the whole image IQ is real and inverted. In making accurate drawings the lens should be represented by a straight line, as illustrated in Fig. 21.17, as we are only concerned with thin lenses and a narrow beam incident close to the principal axis.

(iii) The image formed by a converging lens is always real and inverted until the object is placed nearer the lens than its focal length, Fig. 21.17 (ii). In this case the rays from the top point P *diverge* after refraction through the lens, and hence the image Q is *virtual*. The whole image, IQ, is erect (the same way up as the object) and magnified, besides being virtual, and hence the converging lens can be used as a simple "magnifying glass" (see p. 527).

Diverging lens. In the case of a converging lens, the image is sometimes real and sometimes virtual. In a diverging lens, the image is

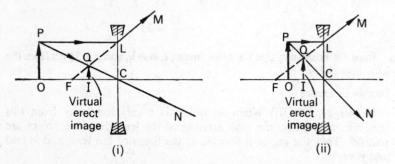

FIG. 21.18. Images in diverging lenses.

always virtual; in addition, the image is always erect and diminished. Fig. 21.18 (i), (ii) illustrate the formation of two images. A ray PL appears to diverge from the focus F after refraction through the lens, a ray PC passes straight through the middle of the lens and emerges along CN, and hence the emergent beam from P appears to diverge from Q on the same side of the lens as the object. The image IQ is thus *virtual*.

The rays entering the eye from a point on an object viewed through a lens can easily be traced. Suppose L is a converging lens, and IQ is the image of the object OP, drawn as already explained, Fig. 21.19. If the eye E observes the top point P of the object through the lens, the cone of rays entering E are those bounded by the image Q of P and the pupil of the eye. If these rays are produced back to meet the lens L, and the

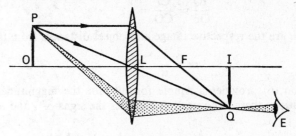

FIG. 21.19. Rays entering the eyes.

points of incidence are joined to P, the rays entering E are shown shaded in the beam. The method can be applied to trace the beam of light entering the eye from any other point on the object; the important thing to remember is to *work back from the eye*.

Another proof of $\frac{1}{v} + \frac{1}{u} = \frac{1}{f}$. We have already shown how the lens equation $\frac{1}{v} + \frac{1}{u} = \frac{1}{f}$ can be derived by considering refraction in turn at the two curved surfaces (p. 480). A proof of the equation can also be obtained from Fig. 21.17 or Fig. 21.18, but it is not as rigid a proof as that already given on page 481.

In Fig. 21.18, triangles CQI, CPO are similar. Hence IQ/PO = CI/CO. Since triangles FQI, FLC are similar, IQ/CL = FI/FC. Now CL = PO. Thus the left sides of the two ratios are equal.

$$\therefore \quad \frac{CI}{CO} = \frac{FI}{FC}$$

But $CI = -v$; $CO = +u$; $FI = FC - IC = -f - (-v) = v - f$; and $FC = -f$.

$$\therefore \quad \frac{-v}{+u} = \frac{v-f}{-f}$$

$$\therefore \quad vf = uv - uf$$

$$\therefore \quad uf + vf = uv$$

Dividing throughout by uvf and simplifying each term,

$$\therefore \quad \frac{1}{v} + \frac{1}{u} = \frac{1}{f}$$

The same result can be derived by considering similar triangles in Fig. 21.17, a useful exercise for the student.

Lateral Magnification

The lateral or transverse or linear magnification, m, produced by a lens is defined by

$$m = \frac{height\ of\ image}{height\ of\ object} \quad . \quad . \quad . \quad . \quad (5)$$

Thus $m = \dfrac{IQ}{OP}$ in Fig. 21.17 or Fig. 21.18. Since triangles QIC, POC are similar in either of the diagrams,

$$\frac{IQ}{OP} = \frac{CI}{CO} = \frac{v}{u},$$

where v, u are the respective image and object distances from the lens.

$$\therefore \qquad \mathbf{m = \frac{v}{u}} \qquad . \qquad . \qquad . \qquad . \qquad . \qquad (6)$$

Equation (6) provides a simple formula for the magnitude of the magnification; there is no need to consider the signs of v and u in this case.

Other formulæ for magnification. Since $\dfrac{1}{v} + \dfrac{1}{u} = \dfrac{1}{f}$, we have, by multiplying throughout by v,

$$1 + \frac{v}{u} = \frac{v}{f}$$

$$\therefore \quad 1 + m = \frac{v}{f}$$

$$\therefore \quad m = \frac{v}{f} - 1 \qquad . \qquad . \qquad . \qquad . \qquad (7)$$

Thus if a real image is formed 25 cm from a converging lens of focal length 10 cm, the magnification, m, $= \dfrac{+25}{+10} - 1 = 1 \cdot 5$.

By multiplying both sides of the lens equation by u, we have

$$\frac{u}{v} + 1 = \frac{u}{f}$$

$$\therefore \quad \frac{1}{m} + 1 = \frac{u}{f}$$

Object at Distance 2f from Converging Lens

When an object is placed at a distance of $2f$ from a convex lens, drawing shows that the real image obtained is the same size as the image

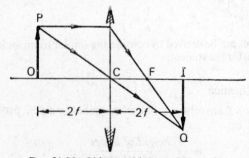

Fig. 21.20. Object and image of same size.

and is also formed at a distance $2f$ from the lens, Fig. 21.20. This result can be accurately checked by using the lens equation $\dfrac{1}{v} + \dfrac{1}{u} = \dfrac{1}{f}.$ Substituting $u = +2f$, and noting that the focal length, f, of a converging lens is positive, we have

$$\frac{1}{v} + \frac{1}{2f} = \frac{1}{f}$$

$$\therefore \quad \frac{1}{v} = \frac{1}{f} - \frac{1}{2f} = \frac{1}{2f}$$

$$\therefore \quad v = 2f = \text{image distance.}$$

$$\therefore \text{ lateral magnification, } m = \frac{v}{u} = \frac{2f}{2f} = 1,$$

showing that the image is the same size as the object.

Least Possible Distance Between Object and Real Image with Converging Lens

It is not always possible to obtain a real image on a screen, although the object and the screen may both be at a greater distance from a converging lens than its focal length. The theory below shows that the distance between an object and a screen must be equal to, or greater than, *four times the focal length* if a real image is required.

Theory. Suppose I is the real image of a point object O in a converging lens. If the image distance $= x$, and the distance OI $= d$, the object distance $=$

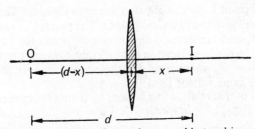

O   I

$|\!\leftarrow\!(d-x)\!\rightarrow\!|\quad|\!\leftarrow x \rightarrow\!|$

$|\!\leftarrow\!\!-\!-\!-\!-\!-\! d \!-\!-\!-\!-\!-\!\rightarrow\!|$

FIG. 21.21. Minimum distance between object and image.

$(d - x)$, Fig. 21.21. Thus $v = +x$, and $u = +(d - x)$. Substituting in the lens equation $\dfrac{1}{v} + \dfrac{1}{u} = \dfrac{1}{f}$, in which f is positive, we have

$$\frac{1}{x} + \frac{1}{d - x} = \frac{1}{f}$$

$$\therefore \quad \frac{d}{x(d - x)} = \frac{1}{f}$$

$$\therefore \quad x^2 - dx + df = 0 \quad . \quad . \quad . \quad . \quad \text{(i)}$$

For a real image, the roots of this quadratic equation for x must be real roots. Applying to (i) the condition $b^2 - 4ac >$ for 0 the general quadratic $ax^2 + bx + c = 0$, then

$$d^2 - 4df > 0$$
$$\therefore \ d^2 > 4df$$
$$\therefore \ d > 4f$$

Thus the distance OI between the object and screen must be greater than $4f$, otherwise no image can be formed on the screen. Hence $4f$ is the minimum distance between object and screen; the latter case is illustrated by Fig. 21.20, in which $u = 2f$ and $v = 2f$. If it is difficult to obtain a real image on a screen when a converging lens is used, possible causes may be (i) the object is nearer to the lens than its focal length, Fig. 21.17 (ii), or (ii) the distance between the screen and object is less than four times the focal length of the lens.

Conjugate Points. Newton's Relation

Suppose that an object at a point O in front of a lens has its image

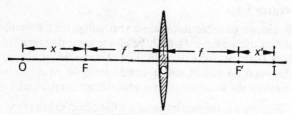

FIG. 21.22. Newton's relation.

formed at a point I. Since light rays are reversible, it follows an object placed at I will give rise to an image at O. The points O, I are thus "interchangeable", and are hence called *conjugate points* (or conjugate foci) with respect to the lens. Newton showed that conjugate points obey the relation $xx' = f^2$, where x, x' are their respective distances from the focus on the same side of the lens.

The proof of this relation can be seen by taking the case of the converging lens in Fig. 21.22, in which $OC = u = x + f$, and $CI = v = x' + f$.

Substituting in the lens equation $\dfrac{1}{v} + \dfrac{1}{u} = \dfrac{1}{f}$,

$$\frac{1}{x' + f} + \frac{1}{x + f} = \frac{1}{f}$$
$$\therefore \ f(x' + x + 2f) = (x' + f)(x + f)$$
$$\therefore \ xx' = f^2 \qquad . \qquad . \qquad . \qquad . \qquad (8)$$

Since $x' = f^2/x$, it follows that x' increases as x decreases. The image I thus recedes from the focus F' away from the lens when the object O approaches the lens.

The property of conjugate points stated above, namely that an object and an image at these points are interchangeable, can also be derived from the lens equation $\dfrac{1}{v} + \dfrac{1}{u} = \dfrac{1}{f}$. Thus if $u = 15$ cm and $v = 10$ cm satisfies this equation, so must $u = 10$ cm and $v = 15$ cm.

Displacement of Lens when Object and Screen are Fixed

Suppose that an object O, in front of a converging lens A, gives rise to an image on a screen at I, Fig. 21.23. Since the image distance AI (v) is

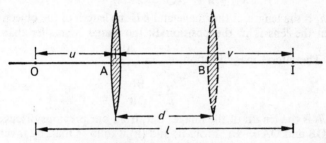

FIG. 21.23. Displacement of lens.

greater than the object distance AO (u), the image is larger than the object. If the object and the screen are kept fixed at O, I respectively, another clear image can be obtained on the screen by moving the lens from A to a position B. This time the image is smaller than the object, as the new image distance BI is less than the new object distance OB.

Since O and I are conjugate points with respect to the lens, it follows that OB = IA and IB = OA. (If this is the case the lens equation will be satisfied by $\dfrac{1}{IB} + \dfrac{1}{OB} = \dfrac{1}{f}$ and by $\dfrac{1}{IA} + \dfrac{1}{OA} = \dfrac{1}{f}$.) If the *displacement*, AB, of the lens $= d$, and the constant distance OI $= l$, then OA + BI $= l - d$. But, from above, OA = IB. Hence OA $= (l - d)/2$. Further, AI = AB + BI = OA + AB $= (l - d)/2 + d = (l + d)/2$.

But $u = $ OA, and $v = $ AI for the lens in the position A. Substituting for OA and AI in $\dfrac{1}{v} + \dfrac{1}{u} = \dfrac{1}{f}$,

$$\frac{1}{(l + d)/2} + \frac{1}{(l - d)/2} = \frac{1}{f}$$

$$\therefore \quad \frac{2}{l + d} + \frac{2}{l - d} = \frac{1}{f}$$

$$\therefore \quad \frac{4l}{l^2 - d^2} = \frac{1}{f}$$

$$\therefore \quad f = \frac{l^2 - d^2}{4l} \qquad . \qquad . \qquad . \qquad . \qquad (9)$$

Thus if the displacement d of the lens, and the distance l between the object and the screen, are measured, the focal length f of the lens can be found from equation (9). This provides a very useful method of measuring the focal length of a lens whose surfaces are inaccessible (for example, when the lens is in a tube), when measurements of v and u cannot be made (see p. 445).

Magnification. When the lens is in the position A, the lateral magnification m_1 of the object $= \dfrac{v}{u} = \dfrac{AI}{OA}$, Fig. 21.23.

$$\therefore \quad \frac{h_1}{h} = \frac{AI}{AO} \quad . \quad . \quad . \quad . \quad . \quad (i)$$

where h_1 is the length of the image and h is the length of the object.

When the lens is in the position B, the image is smaller than the object. The lateral magnification, $m_2 = \dfrac{BI}{OB}$.

$$\therefore \quad \frac{h_2}{h} = \frac{BI}{OB} \quad . \quad . \quad . \quad . \quad (ii)$$

where h_2 is the length of the image. But, from our previous discussion, $AI = OB$ and $OA = BI$. From (i) and (ii) it follows that, by inverting (i),

$$\frac{h}{h_1} = \frac{h_2}{h}$$
$$\therefore \quad h^2 = h_1 h_2$$
$$\therefore \quad h = \sqrt{h_1 h_2} \quad . \quad . \quad . \quad (10)$$

The length, h, of an object can hence be found by measuring the lengths h_1, h_2 of the images for the two positions of the lens. This method of measuring h is most useful when the object is inaccessible, for example, when the width of a slit in a tube is required.

EXAMPLES

1. A converging lens of focal length 30 cm is 20 cm away from a diverging lens of focal length 5 cm. An object is placed 6 metres distant from the former lens (which is the nearer to it) and on the common axis of the system. Determine the position, magnification, and nature of the image formed. (*O. & C.*)

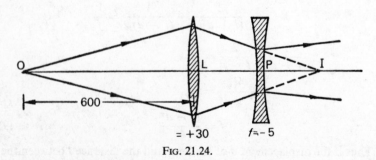

FIG. 21.24.

Suppose O is the object, Fig. 21.24.

For the converging lens,

$$u = + 600 \text{ cm}, \quad f = + 30 \text{ cm}.$$

Substituting in the lens equation,

$$\therefore \quad \frac{1}{v} + \frac{1}{(+600)} = \frac{1}{(+30)}$$

from which

$$v = \frac{600}{19} = 31\tfrac{11}{19} \text{ cm} = \text{LI}$$

$$\therefore \quad \text{PI} = \text{LI} - \text{LP} = 31\tfrac{11}{19} - 20 = 11\tfrac{11}{19} \text{ cm}.$$

For the diverging lens, I is a virtual object. Thus $u = \text{PI} = -11\tfrac{11}{19}$. Also $f = -5$. Substituting in the lens equation, we have

$$\frac{1}{v} + \frac{1}{(-11\tfrac{11}{19})} = \frac{1}{(-5)}$$

from which

$$v = -8 \cdot 8 \text{ cm}.$$

The image is thus virtual, and hence the rays diverge after refraction through P, as shown. The image is 8·8 cm to the left of P.

The magnification, m, is given by $m = m_1 \times m_2$, where m_1, m_2 are the magnifications produced by the converging and diverging lens respectively.

But

$$m_1 = \frac{v}{u} = 31\tfrac{11}{19}/600$$

and

$$m_2 = 8 \cdot 8/11\tfrac{11}{19}$$

$$\therefore \quad m = 31\tfrac{11}{19}/600 \times 8\tfrac{4}{5}/11\tfrac{11}{19} = \frac{1}{25}$$

2. Establish a formula connecting object-distance and image-distance for a simple lens. A small object is placed at a distance of 30 cm from a converging lens of focal length 10 cm. Determine at what distances from this lens a second converging lens of focal length 40 cm must be placed in order to produce (i) an erect image, (ii) an inverted image, in each case of the same size as the object. (*L.*)

First part. See text.

Second part. Suppose O is the object, Fig. 21.25. The image I in the converging lens L is formed at a distance v from L given by

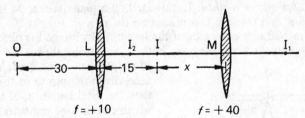

Fig. 21.25.

$$\frac{1}{v} + \frac{1}{(+30)} = \frac{1}{(+10)}$$

from which

$$v = +15.$$

For an erect image. Since $\dfrac{v}{u} = \dfrac{15}{30} = \dfrac{1}{2}$, the image at I is half the object size; also, the image is inverted, since it is real (see Fig. 21.17 (i)). If an erect image is required, the second lens, M, must invert the image at I. Further, if the new image, I_1, say, is to be the same size as the object at O, the magnification produced by M of the image at I must be 2. Suppose $IM = x$ numerically; then, since the magnification $\left(\dfrac{v}{u}\right) = 2$, $MI_1 = 2x$. As I and I_1 are both real, we have, from

$$\frac{1}{v} + \frac{1}{u} = \frac{1}{f},$$

$$\frac{1}{(+2x)} + \frac{1}{(+x)} = \frac{1}{(+40)}$$

$$\therefore \quad \frac{3}{2x} = \frac{1}{40}$$

from which $\qquad\qquad\qquad x = 60 \text{ cm}.$

Thus M must be placed 75 cm from L for an erect image of the same size as O.

For an inverted image. Since the image at I is inverted, the image I_2 of I in M must be erect with respect to I. The lens M must thus act like a magnifying glass which produces a magnification of 2, and the image I_2 is *virtual* in this case. Suppose $IM = x$ numerically; then $I_2M = 2x$ numerically. Substituting in

$$\frac{1}{v} + \frac{1}{u} = \frac{1}{f}$$

$$\therefore \quad \frac{1}{(-2x)} + \frac{1}{(+x)} = \frac{1}{(+40)}$$

$$\therefore \quad x = 20 \text{ cm}.$$

Thus M must be placed 35 cm from L for an inverted image of the same size as O.

SOME METHODS OF MEASURING FOCAL LENGTHS OF LENSES, AND THEIR RADII OF CURVATURE

Converging Lens

(1) *Plane mirror method.* In this method a plane mirror M is placed on a table, and the lens L is placed on the mirror, Fig. 21.26. A pin O is then moved along the axis of the lens until its image I is observed to coincide with O when they are both viewed from above, the method

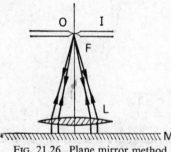

FIG. 21.26. Plane mirror method.

of no parallax being used. The distance from the pin O to the lens is then the focal length, f, of the lens, which can thus be measured.

The explanation of the method is as follows. In general, rays from O pass through the lens, are reflected from the mirror M, and then pass through the lens again to form an image at some place. When O and the image coincide in position, the rays from O incident on M must have returned along their incident path after reflection from the mirror.

This is only possible if the rays are incident *normally* on M. Consequently the rays entering the lens after reflection are all parallel, and hence the point to which they converge must be the focus, F, Fig. 21.26. It will thus be noted that the mirror provides a simple method of obtaining parallel rays incident on the lens.

(2) *Lens formula method.* In this method five or six values of u and v are obtained by using an illuminated object and a screen, or by using two pins and the method of no parallax. The focal length, f, can then be calculated from the equation $\dfrac{1}{v} + \dfrac{1}{u} = \dfrac{1}{f}$, and the average of the values obtained. Alternatively, the values of $\dfrac{1}{u}$ can be plotted against $\dfrac{1}{v}$, and a straight line drawn through the points. When $\dfrac{1}{u} = 0, \dfrac{1}{v} = \text{OA} = \dfrac{1}{f}$, from the lens equation; thus $f = \dfrac{1}{\text{OA}}$, and hence can be calculated, Fig. 21.27. Since $\dfrac{1}{u} = \dfrac{1}{f}$ when $\dfrac{1}{v} = 0$, from the lens equation, $\text{OB} = \dfrac{1}{f}$. Thus f also be evaluated from $\dfrac{1}{\text{OB}}$.

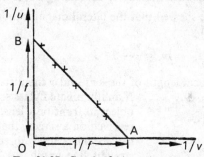

FIG. 21.27. Graph of $1/u$ against $1/v$.

(3) *Displacement method.* In this method, an illuminated object O is placed in front of the lens, A, and an image I is obtained on a screen. Keeping the object and screen fixed, the lens is then moved to a position B so that a clear image is again obtained on the screen, Fig. 21.28. From our discussion on p. 491, it follows that a magnified sharp image is obtained at I when the lens is in the position A, and a diminished sharp image when the lens is in the position B. If the displacement of the lens is d, and the distance between the object and the screen is l, the focal length, f, is given by $f = \dfrac{l^2 - d^2}{4l}$, from p. 491. Thus f can be calculated. The experiment can be repeated by altering the distance between the object and the screen, and the average value of f is then calculated. It should be noted that the screen must be at a distance from the object of at least four times the focal length of the lens, otherwise an image is unobtainable on the screen (p. 490).

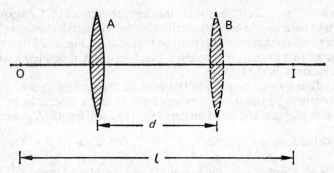

FIG. 21.28. Displacement method for focal length.

Since no measurements need be made to the surfaces of the lens (the "displacement" is simply the distance moved by the holder of the lens), this method can be used for finding the focal length of (i) a thick lens, (ii) an inaccessible lens, such as that fixed inside an eye-piece or telescope tube. Neither of the two methods previously discussed could be used for such a lens.

Lateral Magnification Method of Measuring Focal Length

On p. 488, we showed that the lateral magnification, m, produced by a lens is given by

$$m = \frac{v}{f} - 1 \quad . \quad . \quad . \quad . \quad . \quad \text{(i)}$$

where f is the focal length of the lens and v the distance of the image.

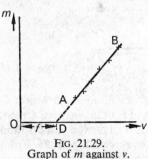

FIG. 21.29.
Graph of m against v.

If an illuminated glass scale is set up as an object in front of a lens, and the image is received on a screen, the magnification, m, can be measured directly. From (i) a straight line graph BA is obtained when m is plotted against the corresponding image distance v, Fig. 21.29. Further, from (i), $v/f - 1 = 0$ when $m = 0$; thus $v = f$ in this case. Hence, by producing BA to cut the axis of v in D, it follows that OD $= f$; the focal length of the lens can thus be found from the graph.

Diverging Lens

(1) *Converging lens method.* By itself, a diverging lens always forms a virtual image of a real object. A real image may be obtained, however, if a *virtual object* is used, and a converging lens can be used to provide such an object, as shown in Fig. 21.30. An object S is placed at a distance from M greater than its focal length, so that a beam converging to a point O is obtained. O is thus a virtual object for the diverging lens L placed as shown in Fig. 21.30, and a real image I can now be obtained.

I is farther away from L and O, since the concave lens makes the incident beam on it diverge more.

The image distance, v, from the diverging lens is CI and can be measured; v is + ve in sign as I is real. The object distance, u, from this

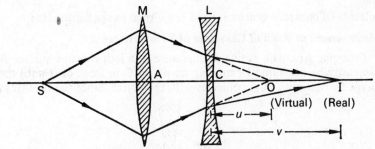

FIG. 21.30. Focal length of diverging lens.

lens = CO = AO − AC, and AC can be measured. The length AO is obtained by removing the lens L, leaving the converging lens, and noting the position of the real image now formed at O by the lens M. Thus u (= CO) can be found; it is a − ve distance, since O is a virtual object for the diverging lens. Substituting for u and v in $\dfrac{1}{v} + \dfrac{1}{u} = \dfrac{1}{f}$, the focal length of the diverging lens can be calculated.

(2) *Concave mirror method.* In this method a real object is placed in front of a diverging lens, and the position of the virtual image is located with the aid of a concave mirror. An object O is placed in front of the lens L, and a concave mirror M is placed behind the lens so that a divergent beam is incident on it, Fig. 21.31. With L and M in the same position, the object O is moved until an image is obtained coincident with it in position, i.e., beside O. The distances CO, CM are then measured.

As the object and image are coincident at O, the rays must be incident *normally* on the mirror M. The rays BA, ED thus pass through the centre of curvature of M, and this is also the position of the virtual image I. The image distance, v, from the lens = IC = IM − CM =

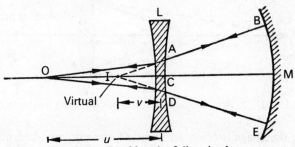

FIG. 21.31. Focal length of diverging lens.

r − CM, where r is the radius of curvature; CM can be measured, while r can be determined by means of a separate experiment, as described on p. 413. The object distance, u, from the lens = OC, and by substituting for u and v in the formula $\frac{1}{v} + \frac{1}{u} = \frac{1}{f}$, the focal length f can be calculated. Of course, v is negative as I is a virtual image for the lens.

Measurement of Radii of Curvature of Lens Surfaces

Diverging lens. The radius of curvature of a lens concave surface A can easily be measured by moving an object O in front of it until the image by reflection at A coincides with the object. Since the rays from

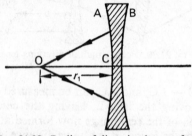

FIG. 21.32. Radius of diverging lens surface.

O are now incident normally on A, its radius of curvature, r_1, = OC, the distance from O to the lens, Fig. 21.32. If the radius of curvature of the surface B is required, the lens is turned round and the experiment is repeated.

Converging lens: Boys' method. Since a convex surface usually gives a virtual image, it is not an easy matter to measure the radius of curvature of such a lens surface. C. V. Boys, however, suggested an ingenious method which is now known by his name, and is illustrated in Fig. 21.33. In Boys' method, an object O is placed in front of a converging lens, and is then moved until an image *by reflection at the back surface* NA is formed beside O. To make the image brighter O should be a well-illuminated object, and the lens can be floated with NA on top of mercury to provide better reflection from this surface.

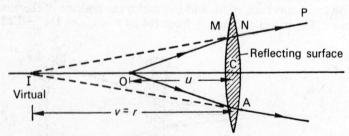

FIG. 21.33. Boys' method for radius of converging lens surface.

Since the image is coincident with O, the rays are incident *normally* on NA. A ray OM from O would thus pass *straight through* the lens along NP after refraction at M. Further, as PN produced passes through I, I is a virtual image of O by *refraction in the lens*. On account of the latter fact, we can apply the lens equation $\frac{1}{v} + \frac{1}{u} = \frac{1}{f}$, where $v = \text{IC} = r$, the radius of curvature of NA, and $u = \text{OC}$. Thus knowing OC and the focal length, f, of the lens, r can be calculated. The same method can be used to measure the radius of curvature of the surface M of the lens, in which case the lens is turned round.

Although reflection from the lens back surface is utilised, the reader should take special pains to note that Boys' method uses the formula $\frac{1}{v} + \frac{1}{u} = \frac{1}{f}$ to calculate the radius of curvature. This is a formula for the *refraction* of light through the lens.

The refractive index of the material of a lens can be found by measuring the radii of curvature, r_1, r_2, of its surfaces and its focal length f. Since $\frac{1}{f} = (n - 1) \left(\frac{1}{r_1} + \frac{1}{r_2} \right)$, where n is the refractive index, the latter can then be calculated.

Combined Focal Length of Two Thin Lenses in Contact

In order to diminish the colouring of the image due to dispersion when an object is viewed through a single lens, the lenses of telescopes and microscopes are made by placing two thin lenses together (see p. 515). The combined focal length, F, of the lenses can be found by considering a point object O placed on the principal axis of two *thin lenses in contact*, which have focal lengths f_1, f_2 respectively, Fig. 21.34. A ray OC from O passes through the middle, C, of both lenses undeviated.

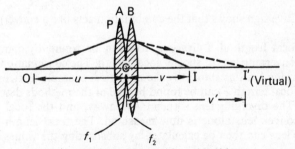

Fig. 21.34. Focal length of combined lenses.

A ray OP from O is refracted through the first lens A to intersect OC at I′, which is therefore the image of O in A. If OC = u, CI′ = v',

$$\therefore \quad \frac{1}{v'} + \frac{1}{u} = \frac{1}{f_1} \qquad . \qquad . \qquad . \qquad . \qquad . \qquad \text{(i)}$$

The beam of light incident on the second lens B converges to I′, which is therefore a *virtual* object for this lens. The image is formed at I at a distance CI, or v, from the lens. Thus since the object distance

CI' is virtual, $u = -v'$ for refraction in this case. For lens B, therefore,

we have
$$\frac{1}{v} + \frac{1}{(-v')} = \frac{1}{f_2},$$

or
$$\frac{1}{v} - \frac{1}{v'} = \frac{1}{f_2} \quad . \quad . \quad . \quad . \quad . \quad \text{(ii)}$$

Adding (i) and (ii) to eliminate v',

$$\therefore \quad \frac{1}{v} + \frac{1}{u} = \frac{1}{f_1} + \frac{1}{f_2}.$$

Since I is the image of O by refraction through both lenses,

$$\frac{1}{v} + \frac{1}{u} = \frac{1}{F},$$

where F is the *focal length of the combined lenses*. Hence

$$\frac{1}{F} = \frac{1}{f_1} + \frac{1}{f_2} \quad . \quad . \quad . \quad . \quad \text{(11)}$$

This formula for F applies to any two thin lenses in contact, such as two diverging lenses, or a converging and diverging lens. When the formula is used, the signs of the focal lengths must be inserted. As an illustration, suppose that a thin converging lens of 8 cm focal length is placed in contact with a diverging lens of 12 cm focal length. Then $f_1 = +8$, and $f_2 = -12$. The combined focal length, F, is thus given by

$$\frac{1}{F} = \frac{1}{(+8)} + \frac{1}{(-12)} = \frac{1}{8} - \frac{1}{12} = +\frac{1}{24}.$$

$$\therefore \quad F = +24 \text{ cm}.$$

The positive sign shows that the combination acts like a converging lens.

The focal length of a diverging lens can be found by combining it with a converging lens of shorter focal length. The combination acts like a converging lens, as shown by the numerical example just considered, and its focal length F can be found by one of the methods described on p. 494. The diverging lens is then taken away, and the focal length f_1 of the convex lens alone is now measured. The focal length f_2 of the concave lens can then be calculated by substituting the values of F and f_1 in the formula $\frac{1}{F} = \frac{1}{f_1} + \frac{1}{f_2}$.

Refractive Index of a small Quantity of Liquid

Besides the method given on p. 425, the refractive index of a small amount of liquid can be found by smearing it over a plane mirror and placing a converging lens on top, as shown in the exaggerated sketch of Fig. 21.35. An object O is then moved along the principal axis until the inverted image I seen looking down into the mirror is coincident with O in position. In this case the rays which pass through the lens and liquid are incident *normally* on the mirror, and the distance from O to the lens is now the focal length, F, of the *lens and liquid* combination (see p. 494). If the experiment is repeated with the convex lens alone on the mirror, the focal length f_1 of the lens can be

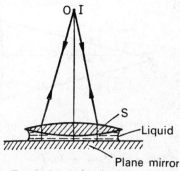

FIG. 21.35. Refractive index liquid.

measured. But $\dfrac{1}{F} = \dfrac{1}{f_1} + \dfrac{1}{f_2}$, where

f_2 is the focal length of the liquid lens.
Thus, knowing f_1 and F, f_2 can be calculated.

From Fig. 21.35, it can be seen that the liquid lens is a plano-concave type; its lower surface corresponds to the plane surface of the mirror, while the upper surface corresponds to the Surface S of the converging lens. If the latter has a radius of curvature r, then, from equation (4) on p. 483.

$$\frac{1}{f_2} = (n - 1)\left(\frac{1}{r} + \frac{1}{\infty}\right).$$

$$\therefore \quad \frac{1}{f_2} = (n - 1)\frac{1}{r}$$

$$\therefore \quad n - 1 = \frac{r}{f_2}$$

$$\therefore \quad n = 1 + \frac{r}{f_2} \quad . \quad . \quad . \quad . \quad . \quad \text{(i)}$$

The radius of curvature r of the surface S of the lens can be measured by Boys' method (p. 498). Since f_2 has already been found, the refractive index n of the liquid can be calculated from (i). This method of measuring n is useful when only a small quantity of the liquid is available.

EXAMPLES

1. Draw a diagram to illustrate the principle of the convex driving mirror on a motor-car. A converging lens of focal length 24 cm is placed 12 cm in front of a convex mirror. It is found that when a pin is placed 36 cm in front of the lens it coincides with its own inverted image formed by the lens and mirror. Find the focal length of the mirror.

First part. See text.

Second part. Suppose O is the position of the pin, Fig. 21.36. Since an inverted image of the pin is formed at O, the rays from O strike the convex mirror normally. Thus the image, I, in the lens of O is at the centre of curvature of the mirror.

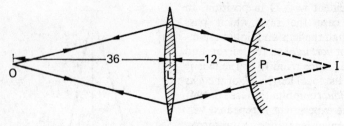

FIG. 21.36. Example.

Since $u = OL = +36$, $f = +24$, it follows from the lens equation that

$$\frac{1}{v} + \frac{1}{36} = \frac{1}{24}$$

from which $v = 72 \text{ cm} = LI$

$$\therefore \quad PI = LI - LP = 72 - 12 = 60$$

$\therefore$ radius of curvature, r, of mirror = 60 cm.

$$\therefore \quad \text{focal length of mirror} = \frac{r}{2} = 30 \text{ cm}.$$

2. Give an account of a method of measuring the focal length of a diverging lens, preferably without the aid of an auxiliary converging lens. A luminous object and a screen are placed on an optical bench and a converging lens is placed between them to throw a sharp image of the object on the screen; the linear magnification of the image is found to be 2·5. The lens is now moved 30 cm nearer the screen and a sharp image again formed. Calculate the focal length of the lens. (*N.*)

First part. A concave mirror can be used, p. 497.

Second part. If O, I are the object and screen positions respectively, and L_1, L_2 are the two positions of the lens, then $OL_1 = IL_2$, Fig. 112. See *Displacement method for focal length*, p. 495. Suppose $OL_1 = x = L_2I$.

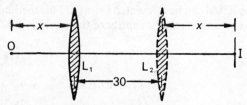

FIG. 21.37. Example.

For the lens in the position L_1, $u = OL_1 = x$, $v = L_1I = 30 + x$.

But magnification, $m = \frac{v}{u} = 2 \cdot 5$

$$\frac{30 + x}{x} = 2 \cdot 5$$

$$\therefore \quad x = 20 \text{ cm}.$$

$$\therefore \quad u = OL_1 = 20 \text{ cm}.$$

$$v = L_1I = 30 + x = 50 \text{ cm}.$$

Substituting in $\dfrac{1}{v} + \dfrac{1}{u} = \dfrac{1}{f}$,

$$\therefore \quad \frac{1}{20} + \frac{1}{50} = \frac{1}{f}$$

from which $\qquad f = 14\frac{2}{7}$ cm.

3. Describe two methods for the determination of the focal length of a diverging lens. A thin equiconvex lens is placed on a horizontal plane mirror, and a pin held 20 cm vertically above the lens coincides in position with its own image. The space between the under surface of the lens and the mirror is filled with water (refractive index 1·33) and then, to coincide with its image as before, the pin has to be raised until its distance from the lens is 27·5 cm. Find the radius of curvature of the surfaces of the lens. (*N.*)

First part. See text.

Second part. The focal length, f_1, of the lens = 20 cm, and the focal length, F, of the water and glass lens combination = 27·5 cm. See p. 501 and Fig. 21.35. The focal length, f, of the water lens is given by

$$\frac{1}{F} = \frac{1}{f} + \frac{1}{f_1}$$

$$\therefore \quad \frac{1}{(+27\cdot5)} = \frac{1}{f} + \frac{1}{(+20)}$$

Solving, $\qquad \therefore \quad \dfrac{1}{f} = \dfrac{1}{27\cdot5} - \dfrac{1}{20} = -\dfrac{3}{220}$

the minus showing that the water lens is a diverging lens.

But $\qquad \dfrac{1}{f} = (n-1)\dfrac{1}{r}$,

where $n = 1\cdot33$, and $r =$ radius of the curved face of the lens.

$$\therefore \quad -\frac{3}{220} = (1\cdot33 - 1)\frac{1}{r}$$

$$\therefore \quad r = -24\cdot2 \text{ cm.}$$

The glass lens is equiconvex, and hence the radii of its surfaces are the same.

4. Derive an expression for the equivalent focal length of a system of two thin lenses of focal lengths f_1 and f_2 in contact. Two equiconvex lenses of focal length 20 cm are placed in contact and the space between them filled with water. Find the focal length of the combination ($_a n_g = 3/2$, $_a n_w = 4/3$). (*L.*)

First part. See text.

Second part. Since the lenses are equiconvex, the radii of curvature, r, of their surfaces are equal. Now

$$\frac{1}{f} = (n-1)\left(\frac{1}{r_1} + \frac{1}{r_2}\right)$$

$$\therefore \quad \frac{1}{20} = \left(\frac{3}{2} - 1\right)\left(\frac{1}{r} + \frac{1}{r}\right)$$

$$\therefore \quad r = 20 \text{ cm.}$$

The water between the lenses forms an equiconcave lens of refractive index 4/3 and radii 20 cm. Its focal length f_1 is thus given by

$$\frac{1}{f_1} = (n - 1)\left(\frac{1}{r_1} + \frac{1}{r_2}\right)$$

$$= \left(\frac{4}{3} - 1\right)\left(\frac{1}{-20} + \frac{1}{-20}\right)$$

$$\therefore \quad f_1 = -30 \text{ cm.}$$

The focal length, F, of the combination is given by

$$\frac{1}{F} = \frac{1}{f} + \frac{1}{f_1} + \frac{1}{f} = \frac{2}{f} + \frac{1}{f_1}$$

where f is the focal length of a glass lens.

$$\therefore \quad \frac{1}{F} = \frac{2}{(+20)} + \frac{1}{(-30)} = \frac{2}{30}$$

$$\therefore \quad F = \frac{30}{2} = 15 \text{ cm.}$$

EXERCISES 21

Refraction at a Single Curved Surface

1. A solid glass sphere has a radius of 10 cm and a refractive index of 1·5. Find the position from the centre, and nature, of the image of an object (i) 20 cm, (ii) 40 cm from the centre due to refraction at the nearest part of the sphere.

2. Obtain a formula connecting the distances of the object and the image from a spherical refracting surface. A transparent sphere of refractive index 4/3 has a radius of 12 cm. Find the positions of the image of a small object inside it 4 cm from the centre, when it is viewed first on one side and then on the other side of the sphere, in the direction of the line joining the centre to the object.

3. Viewed normally through its flat surface, the greatest thickness of a plano-convex lens appears to be 2·435 cm, and through its curved surface 2·910 cm. Actually it is 3·665 cm. Find (a) the refractive index of the glass, (b) the radius of curvature of the convex surface. Do you consider this is a satisfactory method of finding the radius of curvature? (N.)

4. A large glass sphere is placed immediately behind a small hole in an opaque screen, and a small filament lamp is placed at such a distance u in front of the hole that its image falls within the sphere, and at a distance v behind the hole. (a) Sketch the course taken by the light rays in the formation of this image. (b) Derive a formula connecting the quantities u and v with the refractive index n of the glass. (c) If $n = 1·5$ and $u = 4r$, where r is the sphere's radius, find the image position due to refraction at the nearest part of the glass surface only.

Refraction Through Lenses

5. An object is placed (i) 12 cm, (ii) 4 cm from a converging (convex) lens of focal length 6 cm. Calculate the image position and the magnification in each case, and draw sketches illustrating the formation of the image.

6. What do you know about the image obtained with a diverging (concave) lens? The image of a real object in a diverging lens of focal length 10 cm is formed 4 cm from the lens. Find the object distance and the magnification. Draw a sketch to illustrate the formation of the image.

7. The image obtained with a converging lens is erect and three times the length of the object. The focal length of the lens is 20 cm. Calculate the object and image distances.

8. A beam of light converges to a point 9 cm behind (i) a converging lens of focal length 12 cm, (ii) a diverging lens of focal length 15 cm. Find the image position in each case, and draw sketches illustrating them.

9. (i) The surfaces of a biconvex lens are 8 cm and 12 cm radius of curvature. If the refractive index of the glass is 1·5, calculate the focal length. (ii) The curved surface of a plano-concave lens is 10 cm radius of curvature, and the refractive index of the glass is 1·6. Calculate the focal length.

10. Describe an experiment to obtain an accurate value for the focal length of a thin converging lens.

A thin converging lens is fixed inside a tube *AB*. A sharp image of an illuminated object is formed on a screen when the end *A* of the tube is 90·0 cm from the screen and again when it is 140·0 cm from the screen. If the distance between object and screen is 250 cm in each case, how far is the lens from *A*? (*L*.)

11. Why is a sign convention used in geometrical optics?

A thin equiconvex lens of glass of refractive index 1·50 whose surfaces have a radius of curvature of 24·0 cm is placed on a horizontal plane mirror. When the space between the lens and mirror is filled with a liquid, a pin held 40·0 cm vertically above the lens is found to coincide with its own image. Calculate the refractive index of the liquid. (*N*.)

13. State clearly the sign convention you employ in optics. Light from a point *O* in air on the axis of a simple spherical interface between air and glass is refracted so as to form an image at *I*. Derive a formula connecting the distances of *O* and *I* from the pole when *I* is (*a*) real, (*b*) virtual. Show that the two formulae can be reduced to a single formula by use of your sign convention.

Calculate the focal length in air of a thin converging meniscus lens with surfaces having radii of curvature 16·0 cm and 24·0 cm, the refractive index of the glass being 1·60.

Indicate briefly a method for measuring this focal length. (*L*.)

12. What do you understand by *a virtual object* in optics? Describe a direct method, not involving any calculation, of finding (*a*) the focal length of a double concave lens, and (*b*) the radius of curvature of one of its faces. You may use other lenses and mirrors if you wish.

A converging lens of 20 cm focal length is arranged coaxially with a diverging lens of focal length 8·0 cm. A point object lies on the same side as the converging lens and very far away on the axis. What is the smallest possible distance between the lenses if the combination is to form a real image of the object?

If the lenses are placed 6·0 cm apart, what is the position and nature of the final image of this distant object? Draw a diagram showing the passage of a wide beam of light through the system in this case. (*C*.)

14. Explain in detail how, with the aid of a pin and a plane mirror, you would determine the focal length of a thin biconvex lens.

Having found the focal length of this lens, explain how you would find the radius of curvature of one of its faces by Boys' method. Discuss whether or not this method can be used to find the radii of curvature of the faces of a thin converging meniscus lens.

The radii of curvature of the faces of a thin converging meniscus lens of material of refractive index 3/2 are 10 cm and 20 cm. What is the focal length of the lens (a) in air, (b) when completely immersed in water of refractive index 4/3? (N.)

15. A thin planoconvex lens is made of glass of refractive index 1·5. When an object is set up 10 cm from the lens, a virtual image ten times its size is formed. What is (a) the focal length of the lens, (b) the radius of curvature of the lens surface?

If the lens is floated on mercury with the curved side downwards and a luminous object placed vertically above it, how far must the object be from the lens in order that it may coincide with the image produced by reflection in the curved surface? (L.)

16. Deduce an expression for the focal length of a lens in terms of u and v, the object and image distance from the lens.

A lens is set up and produces an image of a luminous point source on a screen 25 cm away. *If the aperture of the lens is small,* where must the screen be placed to receive the image when a parallel slab of glass 6 cm thick is placed at right angles to the axis of the lens and between the lens and the screen, if the refractive index of the glass is 1·6? Deduce any formula you use. (L.)

17. Derive an expression for the focal length of a lens in terms of the radii of curvature of its faces and its refractive index.

Find the condition that the distance between the object and image is a minimum, and explain how you would verify your result experimentally. (C.)

18. Give an account of a method of finding the focal length of a thin concave lens using an auxiliary convex lens which is not placed in contact with it.

A thin equiconvex lens of refractive index 1·50 is placed on a horizontal plane mirror, and a pin fixed 15·0 cm above the lens is found to coincide in position with its own image. The space between the lens and the mirror is now filled with a liquid and the distance of the pin above the lens when the image and object coincide is increased to 27·0 cm. Find the refractive index of the liquid. (N.)

19. A thin converging lens is mounted coaxially inside a short cylindrical tube whose ends are closed by means of thin windows of parallel-sided glass. Explain the *principles* of two methods by which the focal length of the lens could be determined.

When a thin biconvex lens is placed on a horizontal mirror, a pin placed 14·0 cm above the lens on the axis is found to coincide with its own image. When a little water of refractive index 1·33 is inserted between lens and mirror the self-conjugate positions of the pin are respectively 17·2 cm and 26·2 cm above the lens, with first one and then the other face of the lens in contact with the water.

Deduce what information you can about the lens from these data. (L.)

20. Describe in detail how you would determine the focal length of a diverging lens with the help of (a) a converging lens, (b) a concave mirror.

A converging lens of 6 cm focal length is mounted at a distance of 10 cm from a screen placed at right angles to the axis of the lens. A diverging lens of 12 cm focal length is then placed coaxially between the converging lens and the screen so that an image of an object 24 cm from the converging lens is focused on the screen. What is the distance between the two lenses? Before commencing the calculation state the sign convention you will employ. (N.)

21. (a) Find an expression for the focal length of a combination of two thin lenses in contact. (b) A symmetrical convex glass lens, the radii of curvature of which are 3 cm, is situated just below the surface of a tank of water which is 40 cm deep. An illuminated scratch on the bottom of the tank is viewed vertically downwards through the lens and the water. Where is the image, and where should the eye of the observer be placed in order to see it? The refractive indices of glass and water may be taken as 3/2 and 4/3 respectively. (O. & C).

22. Find the relation between the focal lengths of two thin lenses in contact and the focal length of the combination.

The curved face of a planoconvex lens ($n = 1.5$) is placed in contact with a plane mirror. An object at 20 cm distance coincides with the image produced by the lens and reflection by the mirror. A film of liquid is now placed between the lens and the mirror and the coincident object and image are at 100 cm distance. What is the index of refraction of the liquid? (L.)

23. Describe an optical method of finding the radius of curvature of a surface of a thin convex lens.

An object is placed on the axis of a thin planoconvex lens, and is adjusted so that it coincides with its own image formed by light which has been refracted into the lens at its first surface, internally reflected at the second surface, and refracted out again at the first surface. It is found that the distance of the object from the lens is 20·5 cm when the convex surface faces the object, and 7·9 cm when the plane surface faces the object. Calculate (a) the focal length of the lens, (b) the radius of curvature of the surface, (c) the refractive index of the glass. (C).

24. Define *focal length, conjugate foci, real image*. Obtain an expression for the transverse magnification produced by a thin converging lens.

Light from an object passes through a thin converging lens, focal length 20 cm, placed 24 cm from the object and then through a thin diverging lens, focal length 50 cm, forming a real image 62·5 cm from the diverging lens. Find (a) the position of the image due to the first lens, (b) the distance between the lenses, (c) the magnification of the final image. (L.)

25. Describe how you would determine the focal length of a diverging lens if you were provided with a converging lens (a) of shorter focal length, (b) of longer focal length.

An illuminated object is placed at right angles to the axis of a converging lens, of focal length 15 cm, and 22·5 cm from it. On the other side of the converging lens, and coaxial with it, is placed a diverging lens of focal length 30 cm. Find the position of the final image (a) when the lenses are 15 cm apart and a plane mirror is placed perpendicular to the axis 40 cm beyond the diverging lens, (b) when the mirror is removed and the lenses are 35 cm apart. (N.)

chapter twenty-two

Defects of vision
Defects of lenses

DEFECTS OF VISION

THERE are numerous defects of vision, each necessitating the use of a different kind of spectacles. The use of convex lenses in spectacles was fairly widespread by 1300, but concave lenses were not in common use until about 1550, and were then highly valued. We propose to discuss briefly the essential optical principles of some of the main defects of vision and their "correction", and as a necessary preliminary we must mention certain topics connected with the eye itself.

Far and Near Points of Eye. Accommodation

An account of the essential features of the eye was given in Chapter 15 on p. 389, and it was mentioned there that the image formed by the eye-lens L must appear on the retina R, the light-sensitive screen of the eye, in order to be clearly seen, Fig. 22.1. The ciliary muscles enable the eye to focus objects at different distances from it, a property of the eye known as its power of *accommodation*. The most distant point it can focus (the "*far point*") is at infinity for a normal eye; and as the ciliary muscles are then completely relaxed, the eye is said to be "unaccommodated", or "at rest". In this case parallel rays entering the eye are focused on the retina, Fig. 22.1 (i).

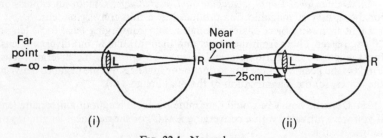

(i)   (ii)

FIG. 22.1. Normal eye.

On the other hand, an object is seen in great detail when it is placed as near the eye as it can be while remaining focused; this distance from the eye known as its *least distance of distinct vision*. The point at this distance from the eye is called its *near point*, and the distance is about 25 cm for a normal eye, Fig. 22.1 (ii). The eye is said to be "fully accommodated" when viewing an object at its near point, as the ciliary muscles are then fully strained.

Short Sight

If the focal length of the eye is too short, owing to the eye-ball being too long, parallel rays will be brought to a focus at a point D in front of the retina, Fig. 22.2 (i). In this case the far point of the eye is not at infinity, but at a point P nearer to the eye, and the defect of vision is known as *short sight*, or *myopia*.

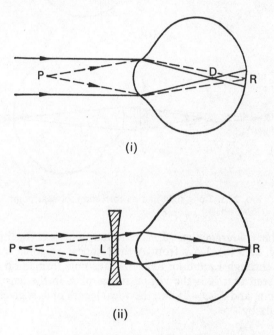

(i)

(ii)

FIG. 22.2. Short sight and its correction.

A suitable *diverging* lens, L, is required to correct for short sight, Fig. 22.2 (ii). Parallel rays refracted through L are now made divergent, and if they appear to come from the far point P of the eye, the rays are brought to a focus on the retina R. From Fig. 22.2 (ii), it can be seen that the focal length of the required lens is equal to PL, which is practically equal to the distance of the far point from the eye.

Long sight

If a person's far point is normal, i.e. at infinity, but his near point is farther from the eye than the normal least distance of distinct vision, 25 cm, the person is said to be "long-sighted". In Fig. 22.3 (i), X is the near point of a person suffering from long sight, due to a short eyeball, for example. Rays from X are brought to a focus on the retina R; whereas rays from the normal near point A, 25 cm from the eye, are brought to a focus at B behind the retina.

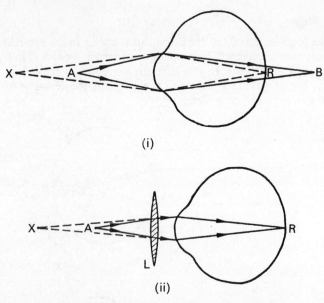

(i)

(ii)

FIG. 22.3. Long sight and its correction for near point.

A suitable *converging* lens, L, is required to correct for this defect of vision, Fig. 22.3 (ii). Rays from A then appear to come from X after refraction through L, and an image is thus now formed on the retina. It can be seen that X is the virtual image of A in the lens L. Thus if XL = 50 cm, and AL = 25 cm, the focal length of L is given from the lens formula by

$$\frac{1}{(-50)} + \frac{1}{(+25)} = \frac{1}{f}$$

$$\therefore \quad \frac{1}{50} = \frac{1}{f}$$

$$\therefore \quad f = 50 \text{ cm.}$$

Correction for far point. Presbyopia

Some people are able to focus only those beams of light which converge to a point behind the retina, in which case the far point is virtual. A parallel beam of light is then brought to a focus behind the retina, R, Fig. 22.4 (i). This defect of vision is corrected by using a suitable converging lens L, Fig. 22.4 (ii).

Short sight and long sight are defects of refraction. A person with these defects of vision still has the power of accommodation. *Presbyopia*, however, is a defect of vision where the eyelens becomes inelastic through old age. Thus the eye is unable to accommodate. *Bifocals*, to

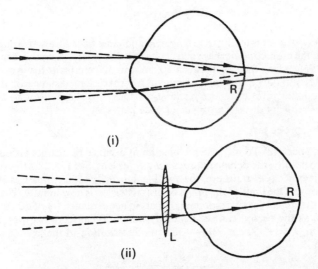

(i)

(ii)

FIG. 22.4. Long sight, and its correction for far point.

correct for both near and far points, may be required. Fig. 22.5 illustrates a bifocal lens, with a positive reading section below and a negative section above for far vision.

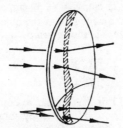

FIG. 22.5. Bifocal lens. Correction for near and far vision.

Astigmatism

If the cornea, the refracting surface in front of the eye (p. 389), has widely varying curvatures in different planes, the rays from an object in one plane will be brought to a focus by the eye at a different place from rays in another plane. In this case the lines in one direction in Fig. 22.6 will be sharply focused compared with the other lines. A strain is thus imposed on the eye when viewing an object, and this defect of vision is called *astigmatism*. It is corrected by using a *cylindrical lens*, whose curvature compensates for the curvature of the cornea in the particular astigmatic plane.

FIG. 22.6.
Test of astigmatism.

EXAMPLES

1. Draw diagrams to illustrate *long sight* and *short sight*. Draw also diagrams showing the correction of these defects by suitable lenses. A person can focus objects only when they lie between 50 cm and 300 cm from his eyes. What spectacles should he use (*a*) to increase his maximum distance of distinct vision to infinity, (*b*) to reduce his least distance of distinct vision to 25 cm? Find this range of distinct vision using each pair. (*N*)

First part. See text.

Second part. (*a*) To increase the maximum distance of distinct vision from 300 cm to infinity, the person requires a *diverging* lens. See Fig. 22.2 (ii). Assuming the lens is close to his eye, the focal length PL = 300 cm as P is 300 cm from the eye. One limit of the range of distinct vision is now infinity. The other limit is the object (*u*) corresponding to an image distance (*v*) of 50 cm from the lens, as the person can see distinctly things 50 cm from his eyes. In this case, then, $v = -50$ cm, $f = -300$ cm. Substituting in the lens equation, we have

$$\frac{1}{(-50)} + \frac{1}{u} = \frac{1}{(-300)}$$

from which $u = 60$ cm.

The range of distinct vision is thus from 60 cm to infinity.

(*b*) To reduce his least distance of distinct vision from 50 cm to 25 cm, the person requires a *converging* lens. See Fig. 22.3 (ii). In this case, assuming the lens is close to the eye, $u = +25$ cm, $v = -50$ cm, as the image must be formed 50 cm from the eye on the same side as the object, making the image virtual. The focal length of the lens is thus given by

$$\frac{1}{(-50)} + \frac{1}{(+25)} = \frac{1}{f}$$

from which $f = +50$ cm.

Objects placed at the focus of this lens appear to come from infinity. The maximum distance of distinct vision, *u*, is given by substituting $v = -300$ cm and $f = +50$ cm in the lens formula.

Thus $$\frac{1}{(-300)} + \frac{1}{u} = \frac{1}{(+50)}$$

from which $$u = \frac{300}{7} = 42\tfrac{6}{7} \text{ cm}.$$

The range of distinct vision is thus from 25 to $42\tfrac{6}{7}$ cm.

2. Explain what is meant by the magnifying power of an optical instrument, considering the cases of microscope and telescope. A thin converging lens of focal length 5 cm is laid on a map situated 60·5 cm below the eye of an observer whose least distance of distinct vision is 24·5 cm. Describe what is seen (*a*) then, and when the lens is raised (*b*) 5 cm, (*c*) $5\tfrac{1}{2}$ cm, (*d*) 6 cm above the map. (*L*.)

First part. See later text, p. 526.

Second part. (*a*) When the lens is on top of the map, it acts as a thin piece of glass which very slightly raises the map to the observer (p. 427). The map appears unaltered in size and is the same way up.

(*b*) When the map is 5 cm from the lens, i.e., at its focal plane, the map is the same way up but now appears bigger. In this case the image is at infinity.

(*c*) When the map is $5\frac{1}{2}$ cm from the lens, the image is inverted. Its distance v is given by $\dfrac{1}{v} + \dfrac{1}{u} = \dfrac{1}{f}$, from which

$$\frac{1}{v} + \frac{1}{(+5\frac{1}{2})} = \frac{1}{(+5)}$$

or $\qquad v = 55$ cm.

The image is thus $60\cdot5 - (55 + 5\frac{1}{2})$ cm from the observer, i.e., the image is formed at the eye. A blurred image is seen.

(*d*) When the map is 6 cm from the lens, the image is inverted and its distance given by

$$\frac{1}{v} + \frac{1}{(+6)} = \frac{1}{(+5)}$$

$$\therefore \quad v = 30 \text{ cm.}$$

The image is thus $60\cdot5 - (30 + 5\cdot5)$ cm, or 25 cm, from the eye of the observer. The inverted map is thus seen clearly. Since $v/u = 30/6 = 5$, the inverted map is magnified five times.

DEFECTS OF LENSES

We have just considered defects inherent in the eye; we have now to consider defects of an image produced by a lens, which is quite a different matter.

Chromatic Aberration

The image of an object formed by a single lens is distorted from a variety of causes. The main defect of the lens is the colouring of the image it produces, which is known as *chromatic aberration*.

Experiment shows that if a parallel beam of white light is incident on a converging lens, the red rays in the light are brought to a focus R, and the blue rays are brought to slightly nearer focus at B, Fig. 22.7.

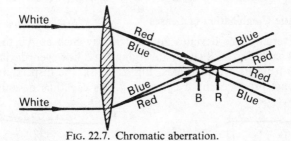

Fig. 22.7. Chromatic aberration.

Thus a single lens produces coloured images of an object which are at slightly different positions. Because he did not know how to eliminate the chromatic aberration Newton decided to abandon the use of lenses for large telescopes (p. 533).

It has already been noted that the refractive index, n, of a material varies with the colour of the light (p. 458). Thus since $1/f = (n - 1) \times \left(\dfrac{1}{r_1} + \dfrac{1}{r_2}\right)$, where f is the focal length of a lens, n is the refractive index of the material, and r_1, r_2 are the radii of curvature of the surfaces, it follows that f has different values for different colours. For example, the focal length of a lens for blue light, f_b, is given by $\dfrac{1}{f_b} = (n_b - 1) \times \left(\dfrac{1}{r_1} + \dfrac{1}{r_2}\right)$, where n_b is the refractive index for blue light. Suppose the lens in Fig. 22.7 is made of crown glass, for which $n_b = 1\cdot523$, and suppose r_1, r_2 are 15 and 12 cm respectively. Then

$$\frac{1}{f_b} = (1\cdot523 - 1) \left(\frac{1}{(+15)} + \frac{1}{(+12)} \right)$$

from which $f_b = 12\cdot86$ cm.

For the same glass, the refractive index n_r for red light $= 1\cdot513$. The focal length f_r for red light is hence given by

$$\frac{1}{f_r} = (n_r - 1) \left(\frac{1}{r_1} + \frac{1}{r_2} \right)$$

$$= (1\cdot513 - 1)\left(\frac{1}{(+15)} + \frac{1}{(+12)} \right),$$

from which $f_r = 13\cdot00$ cm.

The separation BR of the two foci is thus given by

$$f_r - f_b = 13\cdot00 - 12\cdot86 = 0\cdot14 \text{ cm.}$$

The *ratio* of the focal lengths for the two colours $= \dfrac{f_r}{f_b} = \dfrac{n_b - 1}{n_r - 1}$,

since $\dfrac{1}{f_b} = (n_b - 1) \left(\dfrac{1}{r_1} + \dfrac{1}{r_2} \right)$ and $\dfrac{1}{f_r} = (n_r - 1) \left(\dfrac{1}{r_1} + \dfrac{1}{r_2} \right)$.

Achromatic Combination of Lenses

A converging lens deviates an incident ray such as AB towards its principal axis, Fig. 22.8 (i). A diverging lens, however, deviates a ray PQ away from its principal axis, Fig. 22.8 (ii). The dispersion between two colours produced by a converging lens can thus be annulled by

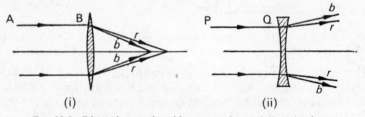

FIG. 22.8. Dispersion produced by converging and diverging lens.

placing a suitable diverging lens beside it. Except for the mathematics, the making of *achromatic lenses* is analogous to the case of achromatic prisms, discussed on p. 459. There it was shown that two prisms of different material, with angles pointing in opposite directions, can act as an achromatic combination. Fig. 22.9 illustrates an achromatic lens combination, known as an *achromatic doublet*. The biconvex lens is made of crown glass, while the concave lens is made of flint glass and is a plano-concave lens. So that the lenses can be cemented together with Canada balsam, the radius of curvature of the curved surface of the plano-concave lens is made numerically the same as that of one surface of the convex lens. The achromatic combination acts as a convex lens when used as the objective lens in a telescope or microscope.

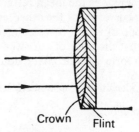

Crown   Flint

FIG. 22.9.
Achromatic doublet in
telescopic objective.

It should be noted that chromatic aberration would occur if the diverging and converging lenses were made of the *same* material, as the two lenses together would then constitute a single thick lens of one material.

Condition for an Achromatic Combination

Achromatic lenses were first made about 1729, years after Newton had considered they were impossible to construct. The necessary condition for an achromatic combination of lenses is derived on p. 516. Here we shall accept the result, which states: *Two lenses form an achromatic doublet for two colours, if the ratio of their focal lengths is numerically equal to the ratio of the corresponding dispersive powers* (p. 459) *of their materials.* Hence, since f_1, f_2 are of opposite signs because one lens must be converging and the other diverging,

$$\frac{\mathbf{f_1}}{\mathbf{f_2}} = -\frac{\omega_1}{\omega_2} \qquad . \qquad . \qquad . \qquad . \qquad (1)$$

where f_1, f_2 are the mean focal lengths and ω_1, ω_2 are the respective dispersive powers.

If a lens combination of focal length F is required, then f_1, f_2 must satisfy in addition the relation

$$\frac{1}{f_1} + \frac{1}{f_2} = \frac{1}{F} \qquad . \qquad . \qquad . \qquad . \qquad (i)$$

Knowing ω_1, ω_2, and F, the magnitudes of f_1, f_2 can be found by solving the equations (i) and (1).

There still remains, however, the practical matter of fitting the surfaces of the two lenses together, as the lenses must be in good optical contact to function efficiently. Suppose the flint glass lens has one surface plane, as shown in Fig. 22.9, and has the focal length f_1. Then

$$\frac{1}{f_1} = (n-1)\left(\frac{1}{r_1} + \frac{1}{\infty}\right) = (n-1)\frac{1}{r_1},$$

where n is the mean refractive index of flint glass and r_1 is the radius of curvature of the lens. Since f_1 and n are known, r_1 can be calculated. The focal length f_2 of the crown glass converging lens is given by

$$\frac{1}{f_2} = (n' - 1) \left(\frac{1}{r_1} + \frac{1}{r_2}\right)$$

where n' is the mean refractive index of crown glass and r_2 is the radius of curvature of the other lens surface. Knowing f_2, n', and r_1, the magnitude of r_2 can be calculated. In this way the lens manufacturer knows what the radii of curvature of the crown and flint glass lenses must be to form an achromatic doublet of a specified focal length F.

Condition for achromatic lenses. Since $\frac{1}{f} = (n - 1)\left(\frac{1}{r_1} + \frac{1}{r_2}\right)$ with the usual notation,

$$\frac{1}{f_b} = (n_b - 1)\left(\frac{1}{r_1} + \frac{1}{r_2}\right)$$

and

$$\frac{1}{f_r} = (n_r - 1)\left(\frac{1}{r_1} + \frac{1}{r_2}\right)$$

$$\therefore \quad \frac{1}{f_b} - \frac{1}{f_r} = (n_b - n_r)\left(\frac{1}{r_1} + \frac{1}{r_2}\right) \quad . \quad . \quad . \quad \text{(i)}$$

Now the magnitudes of $\frac{1}{f_b}$ and $\frac{1}{f_r}$ are very close to each other since f_b is nearly equal to f_r. Thus, using the calculus notation, $\frac{1}{f_b} - \frac{1}{f_r} = \delta\left(\frac{1}{f}\right)$; the latter represents the small change in $\frac{1}{f}$ when blue, and then red, rays are incident on the lens. From (i),

$$\delta\left(\frac{1}{f}\right) = (n_b - n_r)\left(\frac{1}{r_1} + \frac{1}{r_2}\right) \quad . \quad . \quad . \quad \text{(ii)}$$

But

$$\frac{1}{f} = (n - 1)\left(\frac{1}{r_1} + \frac{1}{r_2}\right) \quad . \quad . \quad . \quad \text{(iii)}$$

where f is the focal length of the lens when yellow light is incident on the lens, and n is the refractive index for yellow light. Dividing (ii) by (iii) and simplifying for $\delta\left(\frac{1}{f}\right)$, we obtain

$$\delta\left(\frac{1}{f}\right) = \frac{n_b - n_r}{n - 1} \cdot \frac{1}{f}$$

Now the dispersive power, ω, of a material is defined by the relation $\omega = \frac{n_b - n_r}{n - 1}$ (see p. 459.)

$$\therefore \quad \delta\left(\frac{1}{f}\right) = \frac{\omega}{f} \quad . \quad . \quad . \quad . \quad . \quad \text{(iv)}$$

Combined lenses. Suppose f_1, f_2 are the respective focal lengths of two thin lenses in contact, ω_1, ω_2 are the corresponding dispersive powers of their materials, and F is the combined focal length. If the combination is achromatic for blue and red light, the focal length F_b for blue light is the same as the focal length F_r for red light, i.e., $F_b = F_r$.

$$\therefore \qquad \frac{1}{F_b} = \frac{1}{F_r}$$

$$\therefore \quad \frac{1}{F_b} - \frac{1}{F_r} = 0$$

$$\therefore \quad \delta\left(\frac{1}{F}\right) = 0 \quad . \quad . \quad . \quad . \quad . \quad \text{(v)}$$

But
$$\frac{1}{F} = \frac{1}{f_1} + \frac{1}{f_2}$$

$$\therefore \quad \delta\left(\frac{1}{F}\right) = \delta\left(\frac{1}{f_1}\right) + \delta\left(\frac{1}{f_2}\right)$$

From (v),
$$0 = \delta\left(\frac{1}{f_1}\right) + \delta\left(\frac{1}{f_2}\right)$$

$$\therefore \quad \delta\left(\frac{1}{f_1}\right) = -\delta\left(\frac{1}{f_2}\right)$$

Now
$$\delta\left(\frac{1}{f_1}\right) = \frac{\omega_1}{f_1} \text{ from equation (iv)}$$

and similarly
$$\delta\left(\frac{1}{f_2}\right) = \frac{\omega_2}{f_2}$$

$$\therefore \quad \frac{\omega_1}{f_1} = -\frac{\omega_2}{f_2} \text{ from above}$$

$$\therefore \quad \frac{f_1}{f_2} = -\frac{\omega_1}{\omega_2} \quad . \quad . \quad . \quad . \quad \text{(2)}$$

Thus the ratio of the focal lengths is equal to the ratio of the dispersive powers of the corresponding lens materials, as stated on p. 515. Since ω_1, ω_2 are positive numbers, it follows from (2) that f_1 and f_2 must have opposite signs. Thus a concave lens must be combined with a convex lens to form an achromatic combination (see Fig. 22.9).

Achromatic separated lenses. An achromatic combination can be made with two separated lenses of the *same* material, as we shall now show.

Suppose two lenses of focal lengths f_1, f_2 respectively are situated at a distance d apart. Their combined focal length F is then given by

$$\frac{1}{F} = \frac{1}{f_1} + \frac{1}{f_2} - \frac{d}{f_1 f_2}$$

$$\therefore \quad \delta\left(\frac{1}{F}\right) = \delta\left(\frac{1}{f_1}\right) + \delta\left(\frac{1}{f_2}\right) - \delta\left(\frac{d}{f_1 f_2}\right)$$

$$= \delta\left(\frac{1}{f_1}\right) + \delta\left(\frac{1}{f_2}\right) - \frac{d}{f_1}\delta\left(\frac{1}{f_2}\right) - \frac{d}{f_2}\delta\left(\frac{1}{f_1}\right)$$

For an achromatic combination of two colours, $\delta\left(\frac{1}{F}\right) = 0$, as previously

shown. Further, $\delta\left(\frac{1}{f_1}\right) = \frac{\omega}{f_1}$, $\delta\left(\frac{1}{f_2}\right) = \frac{\omega}{f_2}$, where ω is the dispersive power of the material of the lenses. From above,

$$0 = \frac{\omega}{f_1} + \frac{\omega}{f_2} - \frac{\omega d}{f_1 f_2} - \frac{\omega d}{f_1 f_2}$$

$$\therefore \quad 2d = f_1 + f_2$$

$$\therefore \quad d = \frac{f_1 + f_2}{2} \qquad . \qquad . \qquad . \qquad . \qquad . \qquad (3)$$

Thus the distance between the lenses must be equal to the average of the focal lengths. This condition is utilised in the design of an efficient telescope eye-piece (p. 519).

Spherical Aberration

We have now to consider another defect of an image due to a single lens, known as *spherical aberration*.

The lens formula $\frac{1}{v} + \frac{1}{u} = \frac{1}{f}$ has been obtained by considering a narrow beam of rays incident on the central portion of a lens. In this case the angles of incidence and refraction at the surfaces of the lens are small, and sin i and sin r can then be replaced respectively by i and r in radians, as shown on p. 472. This leads to the lens formula and a unique focus, F. If a *wide* parallel beam of light is incident on the lens, however, experiment shows that the rays are not all brought to the same focus, Fig. 22.10. It therefore follows that the image of an

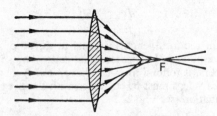

FIG. 22.10. Spherical aberration.

object is distorted if a wide beam of light falls on the lens, and this is known as *spherical aberration*. The aberration may be reduced by surrounding the lens with an opaque disc having a hole in the middle, so that light is incident only on the middle of the lens, but this method reduces the brightness of the image since it reduces the amount of light energy passing through the lens.

As rays converge to a single focus for small angles of incidence, spherical aberration can be diminished if the angles of incidence on the lens' surfaces are diminished. In general, then, the *deviation* of the light by a lens should be shared as equally as possible by its surfaces, as each angle of incidence would then be as small as possible. A practical method of reducing spherical aberration is to utilise *two* lenses, when four surfaces are obtained, and to share the deviation equally between the lenses. The lenses are usually plano-convex.

Eye-pieces. Huygens' Eye-piece

The eye-piece of a telescope should be designed with a view to re-ducing chromatic and spherical aberration; in practice this can most conveniently be done by using *two* lenses as the eye-piece. We have already shown that such lenses, made of the same material, are achro-matic if their distance apart is equal to the average of their focal lengths. Further, the lenses reduce spherical aberration if their distance apart is equal to the difference between their focal lengths.

Huygens designed an eye-piece consisting of two plano-convex lenses; one lens, F, had three times the focal length of the other, E, Fig. 22.11 (i). The lens F pointing to the telescope objective is known as

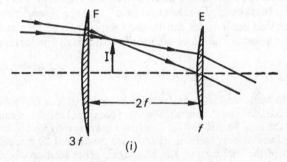

FIG. 22.11 (i). Huygens' eye-piece.

the *field lens*, while the lens E close to the eye is known as the eye lens, and F and E are at a distance $2f$ apart, where f is the focal length of E. Since $3f$ is the focal length of F, it follows from above that the eye-piece eliminates chromatic aberration and reduces spherical aberration.

Since the image formed by the objective of a telescope is at a distance equal to, or less than, the focal length of the eye lens E (p. 535), the image I formed by the objective must be situated between F and E. This, then, is the place where cross-wires must be placed if measurement of the final image is required. But the cross-wires are viewed through one lens, E, while the distant object is viewed by rays refracted through both lenses, F, E. The relative lengths of the cross-wires and image are thus rendered disproportionate, and hence cross-wires cannot be used with Huygens' eye-piece, which is a disadvantage.

Ramsden's Eye-piece

Ramsden's eye-piece is more commonly used than Huygens' eye-piece. It consists of two plano-convex lenses of equal focal length f, the distance between them being $2f/3$, Fig. 22.11 (ii). The achromatic con-dition requires that the distance between the lenses should be f, the average of the focal lengths. If the field lens F were at the focus of the eye lens E, however, E would magnify any dust on F, and vision would then be obscured. F is placed at a distance $f/4$ from the focus of the objective of the telescope, where the real image is formed, in which case the image in F is formed at a distance from E equal to f, its focal length, and parallel rays emerge from E.

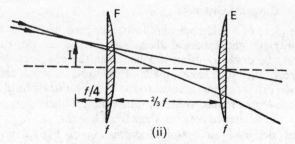

FIG. 22.11 (ii). Ramsden's eye-piece.

The chromatic aberration of Ramsden's eye-piece is small, as is the spherical aberration. The advantage of the eye-piece, however, lies in the fact that cross-wires can be used with it; they are placed outside the combination at the place where the real image I is formed.

EXAMPLES

1. A thin biconvex lens is placed with its principal axis first along a beam of parallel red light and then along a beam of parallel blue light. If the refractive indices of the lens for red and for blue light are respectively 1·514 and 1·524, and if the radii of curvature of its faces are 30 cm and 20 cm, calculate the separation of the foci for red and blue light. What relation does the result bear to the dispersive power of the lens for the two kinds of light? (*N.*)

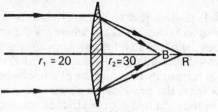

FIG. 22.12.

(a) $\qquad \dfrac{1}{f} = (n-1)\left(\dfrac{1}{r_1} + \dfrac{1}{r_2}\right)$, with the usual notation.

$$\therefore \frac{1}{f_r} = (1\cdot514 - 1)\left(\frac{1}{+20} + \frac{1}{+30}\right)$$

$$\therefore \frac{1}{f_r} = 0\cdot514\left(\frac{1}{20} + \frac{1}{30}\right) = \frac{0\cdot514}{12}$$

$$\therefore f_r = 23\cdot33 \text{ cm.}$$

Also, $\dfrac{1}{f_b} = (1\cdot524 - 1)\left(\dfrac{1}{+20} + \dfrac{1}{+30}\right) = \dfrac{0\cdot524}{12}$

$$\therefore f_b = 22\cdot90 \text{ cm.}$$

$$\therefore \text{separation} = 23\cdot33 - 22\cdot90 = 0\cdot43 \text{ cm.}$$

(b) We know that $\dfrac{1}{f_b} - \dfrac{1}{f_r} = \dfrac{n_b - n_r}{n - 1} \cdot \dfrac{1}{f}$ (See p. 516.)

i.e., $\qquad \dfrac{1}{f_b} - \dfrac{1}{f_r} = \dfrac{\omega}{f}$

$$\therefore \quad f_{\mathrm{r}} - f_{\mathrm{b}} = \frac{\omega}{f} \cdot f_{\mathrm{b}} f_{\mathrm{r}}$$

Now $f_{\mathrm{b}} f_{\mathrm{r}}$ is approximately equal to f^2,

$$\therefore \quad f_{\mathrm{r}} - f_{\mathrm{b}} = \frac{\omega}{f} \cdot f^2 = \omega f$$

$$\therefore \quad \frac{f_{\mathrm{r}} - f_{\mathrm{b}}}{\omega} = f$$

$$\therefore \quad \frac{\text{separation}}{\text{dispersive power}} = \text{focal length,}$$

and is the relation required. The focal length, f, is that for the mean (yellow) light.

2. Define the dispersive power of a medium, and describe, giving the necessary theory, how two thin lenses in contact can be used to form an achromatic combination. A lens of crown glass of dispersive power 0·018 has a focal length of 50 cm. What is the focal length of a flint glass of dispersive power 0·045, which will form an achromatic combination with it? Calculate the focal length of the combination. (*C*)

First part. See text.

Second part. The focal lengths of the two lenses must be proportional to their respective dispersive powers to form an achromatic combination (p. 517). Thus if f is the focal length of the flint glass lens,

$$\frac{f}{50} = -\frac{0\cdot045}{0\cdot018}$$

$$\therefore \quad f = -125 \text{ cm.}$$

The flint glass lens must be a diverging lens, and the crown glass lens must be a converging lens; otherwise the combination would not act as an achromatic converging lens (p. 515). The focal length, F, of the combination is hence given by

$$\frac{1}{F} = \frac{1}{(+50)} + \frac{1}{(-125)}$$

from which $\qquad F = 83\frac{1}{3}$ cm.

3. Explain how it is possible to construct achromatic lenses. Why did Newton consider it impossible? An achromatic objective of 100 cm focal length is to be made, using two lenses of the glasses shown below. Find the focal length of each lens, stating whether it is convergent or divergent.

	Glass A	Glass B
n red	1·5155	1·641
n blue	1·5245	1·659

(N.)

First part. See text. Newton considered it impossible because there was only a small range of glasses in his time, and the dispersive powers of the different glass materials were about the same.

Second part. Let $f_1, f_2 = $ the focal lengths of the lenses.

$$\therefore \quad \frac{1}{f_1} + \frac{1}{f_2} = \frac{1}{100} \quad . \quad . \quad . \quad . \quad \text{(i)}$$

Now the mean refractive index, n, for glass A $= \dfrac{1\cdot5245 + 1\cdot5155}{2} = 1\cdot52$

$\therefore$ dispersive power, ω_2, for glass A $= \dfrac{n_b - n_r}{n - 1} = \dfrac{1\cdot5245 - 1\cdot5155}{1\cdot52 - 1} = \dfrac{9}{520}$

and mean refractive index, n, for glass B $= \dfrac{1\cdot659 + 1\cdot641}{2} = 1\cdot65$

$\therefore$ dispersive power, ω_1, for glass B $= \dfrac{1\cdot659 - 1\cdot641}{1\cdot65 - 1} = \dfrac{18}{650}$

The condition for achromatism is $\dfrac{f_1}{f_2} = - \dfrac{\omega_1}{\omega_2}$

$$\therefore \quad \frac{f_1}{f_2} = - \frac{18}{650} \Big/ \frac{9}{520} \qquad . \qquad . \qquad . \qquad . \qquad \text{(ii)}$$

From (ii), $\qquad\qquad f_1 = - \dfrac{18}{650} \times \dfrac{520}{9} \cdot f_2 = - \dfrac{8}{5} f_2$

Substituting in (i),

$$\therefore \quad - \frac{5}{8 f_2} + \frac{1}{f_2} = \frac{1}{100}$$

$$\therefore \quad f_2 = + 37\cdot5$$

$$\therefore \quad f_1 = - \frac{8}{5} \times + 37\cdot5 = - 60$$

Thus a converging lens of 37·5 cm focal length of glass A should be combined with a diverging lens of 60 cm focal length of glass B.

EXERCISES 22

Defects of Vision

1. Explain how the eye is focused for viewing objects at different distances. Describe and explain the defects of vision known as *long sight* and *short sight*, and their correction by the use of spectacles.

Explain the advantages we gain by the use of two eyes instead of one.

A certain person can see clearly objects at distances between 20 cm and 200 cm from his eye. What spectacles are required to enable him to see distant objects clearly, and what will be his least distance of distinct vision when he is wearing them? (*L.*)

2. Give an account of the common optical defects of the human eye and explain how their effects may be corrected.

An elderly person cannot see clearly, without the use of spectacles, objects nearer than 200 cm. What spectacles will he need to reduce this distance to 25 cm? If his eyes can focus rays which are converging to points not less than 150 cm behind them, calculate his range of distinct vision when using the spectacles. (*N.*)

3. Describe the optical functions of the cornea and lens in the human eye and explain how the corresponding purposes are served in a camera.

In order to correct his near point to 25 cm a man is given spectacles with converging lenses of 50 cm focal length, and to correct his far point to infinity he is given diverging lenses of 200 cm focal length. Ignoring the separation of lens and eye determine the distances of his near and far points when not wearing spectacles and suggest reasons for his defects of vision.

The glass used for the lenses has a refractive index of 1·50 and the back surface of each lens is concave to the eye with a radius of curvature of 50 cm.

Calculate the radii of the front surfaces and state whether the surfaces are convex or concave outwards. (*N.*)

4. Describe a method of determining the focal length of a thin diverging lens of power numerically about 6 dioptres. Give the theory of the method, assuming any necessary relation between object and image distances, but showing clearly the signs to be attached to the numerical data.

Explain fully (*a*) the use of diverging lenses to correct a defect of vision, (*b*) the optical arrangement of a telescope with a diverging lens as eyepiece. (*L.*)

5. State the sign convention you employ in solving optical problems. Discuss its application to the radii of curvature of mirrors. Write down the formula relating the object distance with the image distance and the radius of curvature for a spherical mirror of small aperture. Using the same convention and notation, write down the formula connecting object and image distance for a plane mirror.

In order to read a book held as close as 20 cm away from his eyes, a man requires spectacles with converging lenses of focal length 22 cm. How far away from him is the closest object he can clearly see without the spectacles? Assume that the distance between the eye and the spectacle lens is negligible. Outline the physics of two possible causes of this defect of vision. (*C.*)

6. Describe the optical system of the eye and explain the meaning of *long sight*, *short sight*, and *least distance of distinct vision*. Illustrate with clear diagrams the two defects of vision mentioned above and show how they are corrected by lenses.

If the range of vision of a short-sighted man is from 10 to 20 cm from the eye, what lens should be used in order to enable him to see distant objects clearly? What would be the range of accommodation when using this lens? (*L.*)

7. Describe the optical arrangement of the eye, illustrating the description with a labelled diagram.

A person wears bifocal converging spectacles, one surface of each lens being spherical and the other cylindrical. Describe the defects in his vision and explain how the spectacles correct them. (*N.*)

8. Give clear diagrams to illustrate the common optical defects of the human eye.

In a certain case the range of distinct vision is found to be limited to objects distant 15 cm to 30 cm from the eye. What lens would be suitable for the distant vision of distant objects, and what would be the nearer limit of distinct vision when this lens is in use? (*W.*)

Defects of Lenses

9. Define *dispersive power* of a transparent medium.

A crosswire illuminated from behind by a white source of light is placed axially and 30·0 cm from a thin converging lens whose focal length for yellow sodium light is 25·0 cm. If the dispersive power of the glass of the lens for the extreme wavelengths in the light source is 0·0170, find the axial separation of the images formed by light of these wavelengths.

Find also the focal length and the nature of a thin lens, made of glass of dispersive power 0·0270, which in contact with the first lens will form an achromatic doublet. (*L.*)

10. What is meant by *chromatic aberration, dispersion, dispersive power*? Derive the necessary relation between the focal lengths of the component lenses and the dispersive powers of the glasses of which they are made, for a combination of two thin lenses in contact to be achromatic.

A biconvex lens A is combined with a planoconcave lens B to form an achromatic pair. The adjacent faces are in contact and have a common radius of 15·34 cm. Find the focal length of the combination, given the following refractive indices:

	Blue	Red	Yellow
Lens A	1·5235	1·5149	1·5192
Lens B	1·6635	1·6463	1·6549 (L.)

11. Explain what is meant by (a) spherical aberration, (b) chromatic aberration. Give an account of ways in which these defects are minimised in optical instruments. (N.)

12. What is the condition for two thin lenses to form an achromatic doublet? Draw a diagram of such a combination.

A convex lens of crown glass has a focal length of 40 cm and a dispersive power of 0·025. Find the focal length of a flint glass lens of dispersive power 0·04 which will form an achromatic convex doublet with it. Calculate the focal length of the combination.

13. Explain the dispersion produced by a simple lens, and show how the defect may be corrected.

Why is such correction unnecessary in the case of a simple converging lens used as a magnifying glass held close to the eye? (O. & C.)

14. Derive the expression for the focal length of a thin lens in air in terms of the radii of curvature of the faces and the refractive index of the material of the lens.

A white disc, 1 cm in diameter, is placed 100 cm in front of a thin converging lens of 50 cm mean focal length. The refractive index of the material of the lens is 1·524 for red light and 1·534 for violet light. Calculate the diameters of the images for red and violet light and their distance apart. (W.)

15. Discuss in general terms the defects in the image formed by a single converging lens in a camera and indicate how they may be remedied.

A camera lens forms an image of the same size as the object when the screen is in a certain position. When the screen is moved 10 cm further from the lens and the object is moved until the image is again in focus, the magnification is found to be 2. What is the focal length of the lens? (C.)

16. Describe the colour effects which you would expect to see in the image of a small source of white light formed on a screen by a lens.

Derive the condition that a combination of two thin lenses in contact shall be nearly free from this defect. (L.)

17. A thin spherical lens is made from glass with a refractive index n, and its surface have radii of curvature r_1 and r_2. Find an expression for the focal length f of the lens when it is surrounded by a medium of refractive index unity.

If R_1, R_2 and F are the corresponding numerical values of r_1, r_2 and f (i.e. ignoring signs) and $R_1 < R_2$, write down the expressions for F for (a) the two possible types of converging lens, and (b) the two possible types of diverging lens.

A single thin spherical lens is used to form the image of a star on the axis of the lens. State and explain the effect on the image of (i) spherical aberration and (ii) chromatic aberration, and describe how one of these aberrations may be reduced. (O. & C.)

chapter twenty-three

Optical instruments

WHEN a telescope or a microscope is used to view an object, the appearance of the final image is determined by the cone of rays entering the eye. A discussion of optical instruments and their behaviour must therefore be preceded by a consideration of the image formed by the eye, and we must now recapitulate some of the points about the eye mentioned in previous pages.

Firstly, the image formed by the eye lens L must appear on the retina R at the back of the eye if the object is to be clearly seen, Fig. 23.1. Secondly, the normal eye can focus an object at infinity (the "far point" of the normal eye), in which case the eye is said to be "unaccommodated". Thirdly, the eye can see an object in greatest detail when it is placed at a certain distance D from the eye, known as the *least distance of distinct vision*, which is about 25 cm, for a normal eye (p. 508). The point at a distance D from the eye is known as its "near point".

Visual Angle

Consider an object O placed some distance from the eye, and suppose θ is the angle in radians subtended by it at the eye, Fig. 23.1. Since

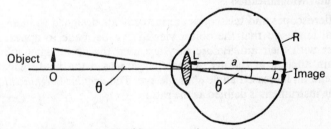

FIG. 23.1. Length of image on retina, and visual angle.

vertically opposite angles are equal, it follows that the length b of the image on the retina is given by $b = a\theta$, where a is the distance from R to L. But a is a constant; hence $b \propto \theta$. We thus arrive at the important conclusion that *the length of the image formed by the eye is proportional to the angle subtended at the eye by the object*. This angle is known as the *visual angle*; the greater the visual angle, the greater is the apparent size of the object.

Fig. 23.2 (i) illustrates the case of an object moved from A to B, and viewed by the eye in both positions. At B the angle β subtended at the eye is greater than the visual angle α subtended at A. Hence the object

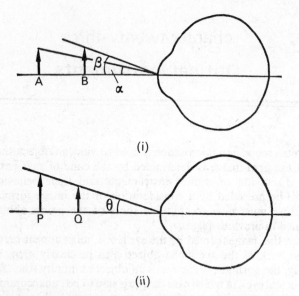

FIG. 23.2. Relation between visual angle and length of image.

appears larger at B than at A, although its physical size is the same. Fig. 23.2 (ii) illustrates the case of two objects, at P, Q respectively, which subtend the same visual angle θ at the eye. The objects thus appear to be of equal size, although the object at P is physically bigger than that at Q. It should be remembered that an object is not clearly seen if it is brought closer to the eye than the near point.

Angular Magnification

Microscopes and telescopes are instruments designed to increase the visual angle, so that the object viewed can be made to appear much larger with their aid. Before they are used the object may subtend a certain angle α at the eye; when they are used the final images may subtend an increased angle α' at the eye. The *angular magnification*, *M*, of the instrument is defined as the ratio

$$\mathbf{M} = \frac{\alpha'}{\alpha} \qquad \cdot \qquad \cdot \qquad \cdot \qquad \cdot \qquad \cdot \qquad \cdot \qquad (1)$$

and this is also popularly known as the *magnifying power* of the instrument. It should be carefully noted that we are concerned with visual angles in the theory of optical instruments, and not with the physical sizes of the object and the image obtained.

Microscopes

At the beginning of the seventeenth century single lenses were developed as powerful magnifying glasses, and many important discoveries in human and animal biology were made with their aid. Shortly afterwards two or more convex lenses were combined to form powerful microscopes, and with their aid HOOKE, in 1648, discovered the existence of "cells" in animal and vegetable tissue.

A microscope is an instrument used for viewing *near* objects. When it is in normal use, therefore, the image formed by the microscope is usually at the least distance of distinct vision, D, from the eye, i.e., at the near point of the eye. With the unaided eye (i.e., without the instrument), the object is seen clearest when it is placed at the near point. Consequently the angular magnification of a microscope in *normal* use is given by

$$M = \frac{\alpha'}{\alpha},$$

where α' is the angle subtended at the eye by the image at the near point, and α is the angle subtended at the unaided eye by the object at the near point.

Simple Microscope or Magnifying Glass

Suppose that an object of length h is viewed at the near point, A, by the unaided eye, Fig. 23.3 (i). The visual angle, α, is then h/D in radian

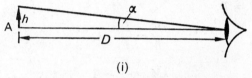

(i)

FIG. 23.3 (i). Visual angle with unaided eye.

measure. Now suppose that a convex lens L is used as a magnifying glass to view the same object. An erect, magnified image is obtained when the object O is nearer to L than its focal length (p. 486), and the observer moves the lens until the image at I is situated at his near point. If the observer's eye is close to the lens at C, the distance IC is then equal to D, the least distance of distinct vision, Fig. 23.3 (ii). Thus the new visual angle α' is given by h'/D, where h' is the length of the virtual image, and it can be seen that α' is greater than α by comparing Fig. 23.3 (i) with Fig. 23.3 (ii).

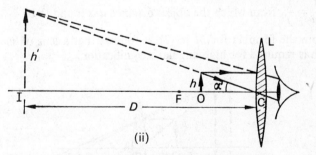

(ii)

FIG. 23.3 (ii). Simple microscope, or magnifying glass.

The angular magnification, M, of this simple microscope can be evaluated in terms of D and the focal length f of the lens. From the definition of M (p. 526), $M = \alpha'/\alpha$.

But
$$a' = \frac{h'}{D}, a = \frac{h}{D}$$

$$\therefore \quad M = \frac{h'}{D} \bigg/ \frac{h}{D} = h'/h \qquad . \qquad . \qquad . \qquad \text{(i)}$$

Now h'/h is the "linear magnification" produced by the lens, and is given by $h'/h = v/u$, where v is the image distance CI and u is the object distance CO (see p. 488). Since $\frac{1}{v} + \frac{1}{u} = \frac{1}{f}$, with the usual notation, we have

$$1 + \frac{v}{u} = \frac{v}{f},$$

by multiplying throughout by v.

$$\therefore \quad \frac{v}{u} = \frac{v}{f} - 1 = \frac{D}{f} - 1$$

since $v = CI = D$.

$$\therefore \quad \frac{h'}{h} = \frac{D}{f} - 1.$$

$$\therefore \quad M = \frac{D}{f} - 1 \qquad . \qquad . \qquad . \qquad . \qquad \text{(2)}$$

from (i) above.

If the magnifying glass has a focal length of 2 cm, $f = +2$ as it is converging; also, if the least distance of distinct vision is 25 cm, $D = -25$ as the image is virtual, see Fig. 23.3 (ii). Substituting in (2),

$$M = \frac{-25}{+2} - 1 = -13\tfrac{1}{2}.$$

Thus the angular magnification is $13\tfrac{1}{2}$. The position of the object O is given by substituting $v = -25$ and $f = +2$ in the lens equation $\frac{1}{v} + \frac{1}{u} = \frac{1}{f}$, from which the object distance u is found to be $+1.86$ cm.

From the formula for M in (2), it follows that a lens of *short* focal length is required for high angular magnification.

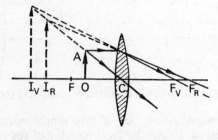

FIG. 23.4. Dispersion with magnifying glass.

When an object OA is viewed through a convex lens acting as a *magnifying glass*, various coloured virtual images, corresponding to I_R, I_V for red and violet rays for example, are formed, Fig. 23.4. The top point of each image lies on the line CA. Hence each image subtends the same angle at the eye close to the lens, so that the colours received by the eye will practically overlap. Thus the virtual image seen in a magnifying glass is almost free of chromatic aberration. A little colour is observed at the edges as a result of spherical aberration. A *real* image formed by a lens has chromatic aberration, as explained on p. 513.

Magnifying Glass with Image at Infinity

We have just considered the normal use of the simple microscope, in which case the image formed is at the near point of the eye and the eye is accommodated (p. 527). When the image is formed at infinity, however, which is not a normal use of the microscope, the eye is undergoing the least strain and is then unaccommodated (p. 508). In this case the object must be placed at the focus, F, of the lens. Fig. 23.5.

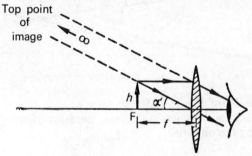

FIG. 23.5. Final image at infinity.

Suppose that the focal length of the lens is f. The visual angle a' now subtended at the eye is then h/f if the eye is close to the lens, and hence the angular magnification, M, is given by

$$M = \frac{a'}{a} = \frac{h/f}{h/D},$$

as $a = h/D$, Fig. 23.3 (i).

$$\therefore \quad M = \frac{D}{f} \qquad . \qquad . \qquad . \qquad . \qquad . \qquad (3)$$

When $f = +2$ cm and $D = -25$ cm, $M = -12\frac{1}{2}$. The angular magnification was $-13\frac{1}{2}$ when the image was formed at the near point (p. 528). It can easily be verified that the angular magnification varies between $12\frac{1}{2}$ and $13\frac{1}{2}$ when the image is formed between infinity and the near point, and the maximum angular magnification is thus obtained when the image is at the near point.

Compound Microscope

From the formula $M = D/f - 1$, M is greater the smaller the focal length of the lens. As it is impracticable to decrease f beyond a certain limit, owing to the mechanical difficulties of grinding a lens of short focal length (great curvature), *two* separated lenses are used to obtain a high angular magnification, and constitute a *compound* microscope. The lens nearer to the object is called the *objective*; the lens through which the final image is viewed is called the *eye-piece*. The objective and the eye-piece are both converging, and both have small focal lengths for a reason explained later (p. 531).

When the microscope is used, the object O is placed at a slightly *greater* distance from the objective than its focal length. In Fig. 23.6, F_o is the focus of this lens. An inverted real image is then formed at I_1 in the microscope tube, and the eye-piece is adjusted so that a large virtual image is formed by it at I_2. Thus I_1 is *nearer* to the eye-piece

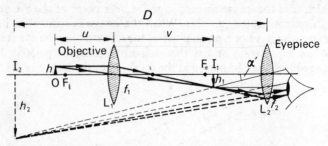

FIG. 23.6. Compound microscope in *normal* use.

than the focus F_e of this lens. It can now be seen that the eye-piece functions as a simple magnifying glass, used for viewing the image formed at I_1 by the objective.

Angular Magnification with Microscope in Normal Use

When the microscope is in normal use the image at I_2 is formed at the least distance of distinct vision, D, from the eye (p. 527). Suppose that the eye is close to the eye-piece, as shown in Fig. 23.6. The visual angle a' subtended by the image at I_2 is then given by $a' = h_2/D$, where h_2 is the height of the image. With the unaided eye, the object subtends a visual angle given by $a = h/D$, where h is the height of the object, see Fig. 23.3 (i).

$$\therefore \quad \text{angular magnification, } M = \frac{a'}{a}$$

$$= \frac{h_2/D}{h/D} = \frac{h_2}{h}.$$

Now $\dfrac{h_2}{h}$ can be written as $\dfrac{h_2}{h_1} \times \dfrac{h_1}{h}$, where h_1 is the length of the intermediate image formed at I_1.

$$\therefore \quad M = \frac{h_2}{h_1} \cdot \frac{h_1}{h} \qquad . \qquad . \qquad . \qquad . \qquad \text{(i)}$$

The ratio h_2/h_1 is the linear magnification of the "object" at I_1 produced by the *eye-piece*, and we have shown on p. 528 that the linear magnification is also given by $\dfrac{v}{f_2} - 1$, where v is the image distance

from the lens and f_2 is the focal length. Since $v = D$ numerically in this case (the image at I_2 is at a distance – D from the eye-piece), it follows that

$$\frac{h_2}{h_1} = \frac{D}{f_2} - 1 \qquad . \qquad . \qquad . \qquad . \qquad \text{(ii)}$$

Also, the ratio h_1/h is the linear magnification of the object at O produced by the *objective* lens. Thus if the distance of the image I_1 from this lens is denoted by v, we have

$$\frac{h_1}{h} = \frac{v}{f_1} - 1 \qquad . \qquad . \qquad . \qquad . \qquad \text{(iii)}$$

$$\therefore \quad M = \frac{h_2}{h_1} \cdot \frac{h_1}{h} = \left(\frac{D}{f_2} - 1\right)\left(\frac{v}{f_1} - 1\right) \qquad . \qquad . \qquad \text{(4)}$$

It can be seen that if f_1 and f_2 are small, M is large. Thus the angular magnification is high if the focal lengths of the objective and the eye-piece are small.

Microscope with Image at Infinity

The compound microscope can also be used with the final image formed at infinity, which is not the normal use of the instrument. In this case the eye is unaccommodated, or "at rest". The image of the object in the objective must now be formed at the focus, F_e, of the eye-piece, as shown in Fig. 23.4, and the visual angle α' subtended at the eye by the final image at infinity is then given by $\alpha' = h_1/f_2$, where h_1 is the length of the image at I_1 and f_2 is the focal length of the eye-piece.

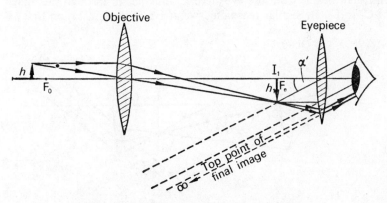

FIG. 23.7. Microscope with image at infinity.

The angular magnification, M, is given by $M = \alpha'/a$, where $a = h/D$ (p. 527).

$$\therefore \quad M = \frac{a'}{a} = \frac{h_1/f_2}{h/D} = \frac{h_1}{h} \cdot \frac{D}{f_2}$$

But, from above, $\dfrac{h_1}{h} = \dfrac{v}{f_1} - 1$.

$$\therefore \quad M = \left(\frac{v}{f_1} - 1\right) \frac{D}{f_2} \qquad . \quad . \quad . \quad (5)$$

Comparing equations (4) and (5), it can be seen, since D is a negative (virtual) quantity, that the angular magnification is greater when the final image is formed at the near point. Further, it will be noted that the eye-piece is nearer to the image at I_1 in the latter case.

The Best Position of the Eye. The Eye-Ring

When an object is viewed by an optical instrument, only those rays from the object which are bounded by the perimeter of the objective lens enter the instrument. The lens thus acts as a *stop* to the light from the object. Similarly, the only rays from the image causing the sensation of vision are those which enter the pupil of the eye. The pupil thus acts as a natural stop to the light from the image; and with a given objective, the best position of the eye is one where it collects as much light as possible from that passing through the objective.

Fig. 23.8 illustrates three of the rays from a point X on an object at O placed in front of a compound microscope. Two of the rays are refracted at the boundary of the objective L_1 to pass through Y on the real image at I_1, while the ray OC_1 through the middle C_1 of the objective passes straight through to Y. The cone of light is then incident on the eye-piece lens L_2, where it is refracted and forms the point T on the final image, corresponding to X on the object. Now the central ray of the beams of light incident on L_1 from every point on the object passes through C_1, the centre of the objective lens. The central ray of the emergent beams from the eye-piece L_2 thus passes through the image of C_1 in L_2. By similar reasoning, we arrive at the conclusion that *all*

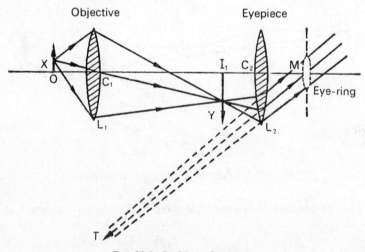

Fig. 23.8. Position of eye-ring.

the emergent rays pass through the image M of the objective in the eye-piece. This image is known as the *eye-ring*, and the best position of the eye is thus at M.

Suppose the objective is 16 cm from L_2, which has a focal length of 2 cm. The image distance, v, in L_2 is given by $\dfrac{1}{v} + \dfrac{1}{(+16)} = \dfrac{1}{(+2)}$, from which $v = 2.3$ cm. Thus M is a short distance from the eye-piece, and in practice the eye should be farther from the eye-piece than in Fig. 23.8. This is arranged in commercial microscopes by having a circular opening fixed at the eye-ring distance from the eye-piece, so that the observer's eye has automatically the best position when it is placed close to the opening.

Angular Magnification of Telescopes

Telescopes are instruments used for viewing distant objects, and they are used extensively at sea and at astronomical observatories. The first telescope is reputed to have been made about 1608, and in 1609 Galileo made a telescope through which he observed the satellites of Jupiter and the rings of Saturn. The telescope thus paved the way for great astronomical discoveries, particularly in the hands of KEPLER. Newton also designed telescopes, and was the first person to suggest the use of curved mirrors for telescopes free from chromatic aberration (see p. 513).

If α is the angle subtended at the unaided eye by a distant object, and α' is the angle subtended at the eye by its image when a telescope is used, the angular magnification M of the instrument is given by

$$M = \frac{\alpha'}{\alpha}.$$

It should be carefully noted that α is *not* the angle subtended at the unaided eye by the object at the near point, as was the case with the microscope, because the telescope is used for viewing distant objects.

Astronomical Telescope in Normal Adjustment

An astronomical telescope made from lenses consists of an objective of long focal length and an eye-piece of short focal length, for a reason given on p. 534. Both lenses are converging. *The telescope is in normal adjustment when the final image is formed at infinity,* and the eye is then unaccommodated when viewing the image. The unaided eye is also unaccommodated when the distant object is viewed, as the latter may be considered to be at infinity.

Fig. 23.9 illustrates the formation of the final image when the telescope is used normally. The image I of the distant object is formed at the focus, F_o, of the objective since the rays incident on the latter are parallel; and since the final image is formed at infinity, the focus F_e of the eye-piece must also be at F_o. Fig. 23.9 shows three of the many rays from the *top* point of the object, marked a, and three of the many rays from the foot of the object, marked b. These rays pass respectively through the top and foot of the image I, as shown.

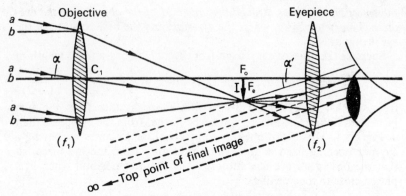

FIG. 23.9. Telescope in *normal* use.

We can now obtain an expression for the angular magnification, M, of the telescope; in so doing we shall assume that the eye is close to the eye-piece. Since the length between the objective and the eye-piece is very small compared with the distance of the object from either lens, we can take the angle α subtended at the unaided eye by the object as that subtended at the objective lens, Fig. 23.9. The angle α' subtended at the eye when the telescope is used is given by $\alpha' = h/f_2$, where h is the length of the image I and f_2 is the focal length of the eye-piece.

But $\alpha = h/f_1,$

where f_1 is the focal length of the objective, since I is at a distance f_1 from C_1.

$$\therefore \quad M = \frac{\alpha'}{\alpha} = \frac{h/f_2}{h/f_1}$$

$$\therefore \quad \mathbf{M} = \frac{\mathbf{f_1}}{\mathbf{f_2}} \quad . \quad . \quad . \quad . \quad (6)$$

Thus the angular magnification is equal to the ratio of the focal length of the objective (f_1) to that of the eye-piece (f_2). For high angular magnification, it follows from (6) that the objective should have a long focal length and the eye-piece a short focal length.

It will be noted that the distance between the lenses is equal to the sum ($f_1 + f_2$) of their focal lengths. This provides a simple method of setting up two converging lenses to form an astronomical telescope when their focal lengths are known.

The Eye-ring, and Relation to Angular Magnification

As we explained in the case of the microscope, the rays which pass through the telescope from the distant object are those bounded by the objective lens. Fig. 23.10 illustrates three rays from a point on the distant object which pass through the objective, forming an image at Y. The eye-ring, M, the best position for the eye, is the circular image of the objective in the eye-piece L_2, and we can calculate its position as

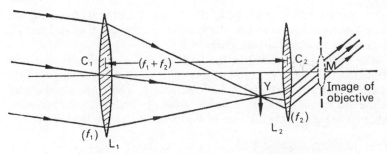

FIG. 23.10. Eye-ring relation to magnification.

$C_1C_2 = f_1 + f_2$, from previous. As the focal length of $L_2 = f_2$, the distance C_2M, or v, is given by $\dfrac{1}{v} + \dfrac{1}{u} = \dfrac{1}{f}$,

i.e., $\quad \dfrac{1}{v} + \dfrac{1}{+(f_1+f_2)} = \dfrac{1}{+(f_2)}$

from which $\quad\quad\quad\quad\quad v = \dfrac{f_2}{f_1}(f_1 + f_2).$

Now the objective diameter: eye-ring diameter $= C_1 C_2 : C_2M$

$$= u:v = (f_1 + f_2) : \frac{f_2}{f_1}(f_1 + f_2)$$

$$= f_1/f_2.$$

But the angular magnification of the telescope $= f_1/f_2$ (p. 534). Thus the angular magnification, M, is also given by

$$M = \frac{\text{diameter of objective}}{\text{diameter of eye-ring}}, \qquad . \qquad . \qquad (7)$$

the telescope being in normal adjustment.

Telescope with Final Image Near Point

When a telescope is used, the final image can be formed at the near point of the eye instead of at infinity. The eye is then "accommodated", and although the image is still clearly seen, the telescope is *not* in normal adjustment (p. 533). Fig. 23.11 illustrates the formation of the final image.

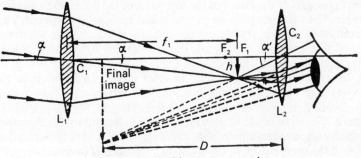

FIG. 23.11. Final image at near point.

The objective forms an image of the distant object at its focus F_1, and the eye-piece is moved so that the image is nearer to it than its focus F_2, thus acting as a magnifying glass.

The angle a subtended at the unaided eye is practically that subtended at the objective L_1. Thus $a = h/f_1$, where h is the length of the image in the objective and f_1 its focal length. The angle a' subtended at the eye by the final image $= h/u$, if the eye is close to the eye-piece, where $u = F_1C_2 =$ the distance of the image at F_1 from the eye-piece.

Thus angular magnification, $M = \dfrac{a'}{a} = \dfrac{h/u}{h/f_1}$

$$\therefore \quad M = \frac{f_1}{u} \quad . \qquad . \qquad . \qquad . \qquad \text{(i)}$$

As the final image is formed at a numerical distance D from the eye-piece L_2, we have $v = -D$ when $f = +f_2$. Thus, from $\dfrac{1}{v} + \dfrac{1}{u} = \dfrac{1}{f}$,

$$\frac{1}{-D} + \frac{1}{u} = \frac{1}{+f_2},$$

from which $u = \dfrac{f_2 D}{f_2 + D}.$

Substituting in (i) for u,

$$\therefore \quad M = \frac{f_1}{f_2}\left(\frac{f_2 + D}{D}\right)$$

$$\therefore \quad M = \frac{f_1}{f_2}\left(1 + \frac{f_2}{D}\right) \quad . \qquad . \qquad . \qquad \text{(8)}$$

The angular magnification when the telescope is in normal adjustment (i.e., final image at infinity) $= \dfrac{f_1}{f_2}$ (p. 534). Hence, from (8), the angular magnification is increased in the ratio $\left(1 + \dfrac{f_2}{D}\right) : 1$ when the final image is formed at the near point.

Terrestrial Telescope

From Fig. 23.11, it can be seen that the top point of the distant object is above the axis of the lens, but the top point of the final image is below the axis. Thus the image in an astronomical telescope is *inverted*. This instrument is suitable for astronomy because it makes little difference if a star, for example, is inverted, but it is useless for viewing objects on the earth or sea, in which case an erect image is required.

A *terrestrial telescope* provides an erect image. In addition to the objective and eye-piece of the astronomical telescope, it has a converging lens L of focal length f between them, Fig. 23.12. L is placed at a distance $2f$ in front of the inverted real image I_1 formed by the objective, in which case, as shown on p. 488, the image I in L of I_1 (i) is inverted, real, and the same size as I_1, (ii) is also at a distance $2f$ from L. Thus the image I is now the same way up as the distant object. If I is

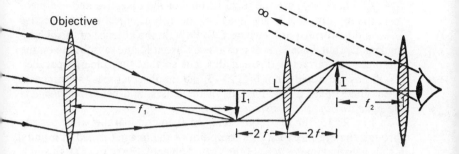

FIG. 23.12. Terrestrial telescope.

at the focus of the eye-piece, the final image is formed at infinity and is also erect.

The lens L is often known as the "erecting" lens of the telescope, as its only function is that of inverting the image I_1. Since the image I produced by L is the same size as I_1, the presence of L does not affect the magnitude of the angular magnification of the telescope, which is thus f_1/f_2 (p. 538). The erecting lens, however, reduces the intensity of the light emerging through the eye-piece, as light is reflected at the lens surfaces. Yet another disadvantage is the increased length of the telescope when L is used; the distance from the objective to the eye-piece is now $(f_1 + f_2 + 4f)$, Fig. 23.12, compared with $(f_1 + f_2)$ in the astronomical telescope.

Galileo's Telescope

About 1610, with characteristic genius, Galileo designed a telescope which provides an erect image of an object with the aid of only two lenses. The *Galilean telescope* consists of an objective which is a converging lens of long focal length, and an eye-piece which is a *diverging* lens of short focal length. The distance between the lenses is equal to the *difference* in the magnitudes of their focal lengths, i.e., $C_1C_2 =$

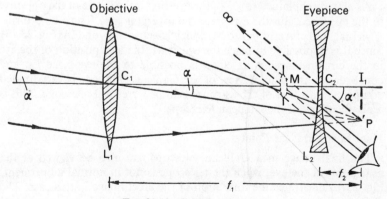

FIG. 23.13. Galilean telescope.

$f_1 - f_2$, where f_1, f_2 are the focal lengths of the objective and eye-piece respectively, Fig. 23.13. The image of the distant object in the objective L_1 would be formed at I_1, where $C_1I_1 = f_1$, in the absence of the diverging lens L_2; but since L_2 is at a distance f_2 from I_1, the rays falling on the eye-piece are refracted through this lens so that they emerge parallel. It will now be noted from Fig. 23.13 that an observer sees the top point of the final image above the axis of the lenses, and hence *the image is the same way up as the distant object.*

In Fig. 23.13, the rays converging to P emerge parallel after passing through the eye-piece L_2. The top point of the image formed at infinity is thus a virtual image in L_2 of the virtual object P. But a ray C_2P through the middle of L_2 passes straight through the lens, and this will also be a ray which passes through the top point of the image at infinity. Thus the three parallel rays shown emerging from the eye-piece in Fig. 23.13 are parallel to the line PC_2. Hence if the eye is placed close to the diverging lens, the angle a' subtended at it by the image at infinity is angle I_1C_2P.

The angle a subtended at the eye by the distant object is practically equal to the angle subtended at the objective, Fig. 23.13. Now $a = h/C_1I_1$ $= h/f_1$, where f_1 is the objective focal length and h is the length I_1P; and $a' = h/C_2I_1 = h/f_2$.

$$\therefore \quad \text{angular magnification, } M = \frac{a'}{a} = \frac{h/f_2}{h/f_1}$$

$$\therefore \qquad M = \frac{f_1}{f_2} \qquad . \qquad . \qquad . \qquad . \qquad (9)$$

Thus for high angular magnification, an objective of long focal length (f_1) and an eye-piece of short focal length (f_2) are required, as in the case of the astronomical telescope (see p. 534).

Advantage and Disadvantage of Galilean Telescope. Opera Glasses

The distance C_1C_2 between the objective and the eye-piece in the Galilean telescope is $(f_1 - f_2)$; the distance between the same lenses in the terrestrial telescope is $(f_1 + f_2 + 4f)$, p. 537. Thus the Galilean telescope is a much shorter instrument than the terrestrial telescope, and is therefore used for *opera glasses.*

As already explained (p. 532), the eye-ring is the image of the objective in the eye-piece. But the eye-piece is a diverging lens. Thus the eye-ring is virtual, and corresponds to M, which is between L_1 and L_2 (Fig. 23.13). Since it is impossible to place the eye at M, the best position of the eye in the circumstances is as close as possible to the eye-piece L_2, and consequently the field of view of the Galilean telescope is very limited compared with that of the astronomical or terrestrial telescope. This is a disadvantage of the Galilean telescope.

Final Image at Near Point

The final image in a Galilean telescope can also be viewed at the near point of the eye, when the telescope is not in normal adjustment. Fig. 23.14 illustrates the formation of the erect image in this case. The distance C_2I_1 is now more than the focal length f_2 of the eye-piece; and since $C_2I_2 = D$, the least distance of distinct vision, we have $v = -D$ (the image in L_2 is virtual) and f_2 is negative. Since $\dfrac{1}{v} + \dfrac{1}{u} = \dfrac{1}{f}$, we obtain

$$\frac{1}{-D} + \frac{1}{u} = \frac{1}{-f_2},$$

assuming f_2 is the *numerical* value of the diverging lens focal length, from which $u = \dfrac{-f_2 D}{D - f_2}$.

With the usual notation, the angular magnification, $M = \dfrac{a'}{a}$.

But $a' = h/u$, $a = h/f_1$. Thus $M = f_1/u$.

But
$$u = \frac{-f_2 D}{D - f_2}$$

$$\therefore \quad M = \frac{f_1}{f_2}\left(\frac{f_2}{D} - 1\right)$$

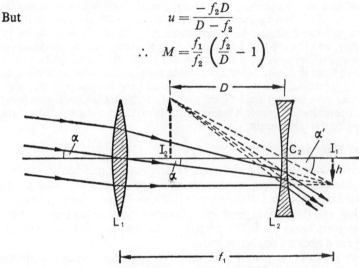

FIG. 23.14. Final image at near point.

Measurement of magnifying power of telescope and microscope

Method 1. *The magnifying power of a telescope* can be measured by placing a well-illuminated large scale S at one end of the laboratory, and viewing it through the telescope at the other end. If the telescope consists of converging lenses O, E acting as objective and eye-piece respectively, the distance

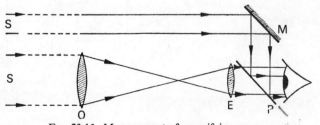

FIG. 23.15. Measurement of magnifying power.

between O, E is the sum of their focal lengths when the instrument is in normal use (p. 534), Fig. 23.15. By means of a plane mirror M and a plane piece of glass P, the divisions on the scale S can be superimposed on its image seen through the telescope, and the ratio of the divisions in equal lengths of the field of view can thus be determined. Since the ratio of the angles subtended at the eye by the image and by the scale is equal to the ratio, a, of the divisions, the magnifying power is equal to a. There must be no parallax between the scale and the final image.

Method 2. The magnifying power of a telescope is the ratio of the diameters of the objective and eye-ring (p. 535). The eye-ring is the image of the objective in the eye-piece, and is obtained by pointing the telescope to the sky and holding a ground glass screen near the eye-piece. A circle, which is the image of the objective in the eye-piece, is observed on the screen, and its diameter, d, is measured. The magnifying power is then given by the ratio d_0/d, where d_0 is the diameter of the objective.

This method is particularly useful when the telescope is fixed in a tube, as it is in practice. When a telescope is set up as in Fig. 23.10, a piece of cardboard with a circular hole can be placed round the objective lens to define its diameter, and the eye-ring found by placing a screen near the eye-piece.

The magnifying power of a microscope can be found by placing two similar scales in front of it, one being 25 cm from the eye. The other scale is placed near the objective, and the eye-piece is moved until the image of this scale coincides with the first scale by the method of no parallax, both eyes being used. The magnifying power is then given by the ratio of the number of divisions occupying the same length.

Prism Binoculars

Prism binoculars are widely used as field glasses, and consist of short astronomical telescopes containing two right-angled isosceles prisms between the objective and eye-piece, Fig. 23.16. These lenses are both converging, and they would produce an inverted image of the distant object if they were alone used. The purpose of the two prisms is to invert the image and obtain a final *erect* image.

One prism A, is placed with its refracting edge vertical, while the other, B, is placed with its refracting edge horizontal. As shown in Fig. 23.11, the image formed by the objective alone is inverted. Prism A,

FIG. 23.16. Prism binoculars.

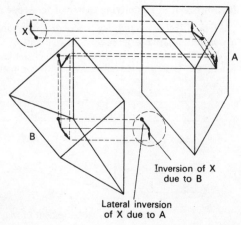

Inversion of X
due to B

Lateral inversion
of X due to A

however, turns it round in a horizontal direction, and prism B inverts it in a vertical direction, both prisms acting as reflectors of light (see p. 449). The image produced after reflection at B is now the same way up, and the same way round, as the original object. Since the eye-piece is a convex lens acting as a magnifying glass, it produces a final image the same way up as the image in front of it, and hence the final image is the same way up as the distant object.

Fig. 23.16 illustrates the path of some rays through the optical system. Since the optical path of a ray is about 3 times the distance d between the objective and the eye-piece, the system is equivalent optically to an astronomical telescope of length $3d$. The focal lengths of the objective and eye-piece in the prism binoculars can thus provide the same angular magnification as an astronomical telescope 3 times as long. The compactness of the prism binocular is one of its advantages; another advantage is the wide field of view obtained, as it is an astronomical telescope (p. 541).

Projection Lantern

The projection lantern is used for showing slides on a screen, and the *essential features* of the apparatus are illustrated in Fig. 23.17. S is a slide whose image is formed on the screen A by adjusting the position of an achromatic objective lens L.

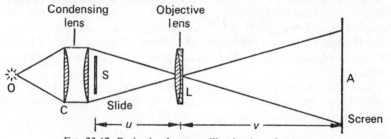

FIG. 23.17. Projection lantern—illumination of screen.

The illumination of the slide must be as high as possible, otherwise the image of it on the screen is difficult to see clearly. For this purpose a very bright point source of light, O, is placed near a *condensing lens* C, and the slide S is placed immediately in front of C. The condensing lens consists of a plano-convex lens arrangement, which concentrates the light energy from O in the direction of S, and it has a short focal length. The lens L and the source O are arranged to be conjugate foci for the lens C (i.e., the image of O is formed at L), in which case (i) all the light passes through L, and (ii) an image of O is not formed on the screen. Fig. 23.17 illustrates the path of the beam of light from O which forms the image of S on the screen.

The linear magnification, m, of the slide is given by $m = \dfrac{v}{u}$, where

v, u are the respective screen and slide distances from L. Now $\dfrac{v}{u} = \dfrac{v}{f} - 1$

(see p. 528). Thus the required high magnification is obtained by using an objective whose focal length is small compared with v.

Pinhole Camera

The *pinhole camera* consists essentially of a closed box with a pinhole in front and a photographic plate at the back on which the image is formed. The principle was first discovered by PORTA about 1600, who found that clear images were formed on a screen at the back of the box when objects were placed in front of the pinhole.

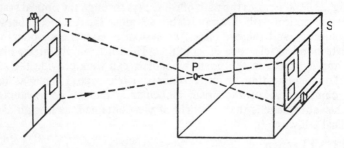

FIG. 23.18. Action of pinhole camera (*not to scale*).

The simple camera utilises the principle that rays of light normally travel in straight lines. As the pinhole P is small, a very narrow cone of rays pass through P from a point T on a house, for example, in front of the box; thus a well-defined image of T is obtained on the photographic plate S, Fig. 23.18. Similarly, other points on the building give rise to clear images on the screen. If the pinhole is enlarged a blurred image is obtained, as the rays from different points on the building then tend to overlap.

The pinhole camera is used by surveyors to photograph the outline of buildings, as the image obtained is free from the distortion produced by the lens in a normal camera.

Photographic Camera; f-number

The photographic camera consists essentially of a *lens system* L, a *light-sensitive film* F at the back, and a *focusing arrangement*, Fig. 23.19.

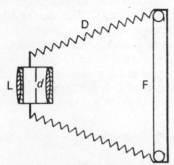

FIG. 23.19. Photographic camera.

The latter is usually a concertina-shaped canvas bag D, which adjusts the distance of the lens L from F. The lens in the camera is an achromatic doublet (p. 515), and the use of two lenses diminishes spherical aberration (p. 518). An *aperture* or *stop* of diameter d is provided, so that the light is incident centrally on the lens, thus diminishing distortion.

The amount of luminous flux falling on the image in a camera is proportional to the area of the lens aperture, or to d^2, where d is the diameter of the aperture. The area of the image formed is proportional to f^2, where f is the focal length of the lens, since the length of the image formed is proportional to the focal length, as illustrated by Fig. 23.20. It therefore follows that the

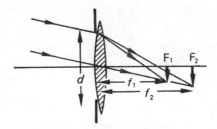

FIG. 23.20. Brightness of image.

luminous flux per unit area of the image, or *brightness B*, of the image, is proportional to d^2/f^2. The time of exposure, t, for activating the chemicals on the given negative is inversely proportional to B. Hence

$$t \propto \frac{f^2}{d^2} \qquad . \qquad . \qquad . \qquad . \qquad (i)$$

The *relative aperture* of a lens is defined as the ratio d/f, where d is the diameter of the aperture and f is the focal length of the lens. The aperture is usually expressed by its *f-number*. If the aperture is f-4, this means that the diameter d of the aperture is $f/4$, where f is the focal length of the lens. An aperture of f-8 means a diameter d equal to $f/8$, which is a smaller aperture than $f/4$.

Since the time t of exposure is proportional to f^2/d^2, from (i), it follows that the exposure required with an aperture f-8 ($d = f/8$) is 16 times that required with an aperture f-2 ($d = f/2$). The f-numbers on a camera are 2, 2·8, 3·5, 4, 4·8, for example. On squaring the values of f/d for each number, we obtain the relative exposure times, which are 4, 8, 12, 16, 20, or 1, 2, 3, 4, 5.

Depth of Field

An object will not be seen by the eye until its image on the retina covers at least the area of a single cone, which transmits along the optic nerve light energy just sufficient to produce the sensation of vision. As a basis of calculation in photography, a circle of finite diameter about 0·25 mm viewed 250 mm away will just be seen by the eye, as a fairly sharp point, and this is known as the *circle of least confusion*. It corresponds to an angle of about 1/1000th radian subtended by an object at the eye.

On account of the lack of resolution of the eye, a camera can take clear pictures of objects at different distances. Consider a point object O in front of a camera lens A which produces a point image I on a film, Fig. 23.21. If XY represents the diameter of the circle of least confusion round I, the eye will see all points in the circle as reasonably sharp points. Now rays from the lens aperture to the edge of XY meet at I_1 beyond I, and also at I_2 in front of I. The point images I_1, I_2 correspond

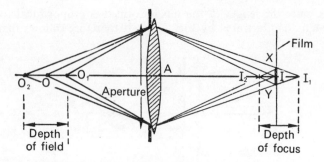

FIG. 23.21. Depth of field.

to point objects O_1, O_2 on either side of O, as shown. Consequently the images of all objects between O_1, O_2 are seen clearly on the film.

The distance O_1O_2 is therefore known as the *depth of field*. The distance I_1I_2 is known as the *depth of focus*. The depth of field depends on the lens aperture. If the aperture is made smaller, and the diameter XY of the circle of least confusion is unaltered, it can be seen from Fig. 23.21 that the depth of field increases. If the aperture is made larger, the depth of field decreases.

The Mount Palomar Telescope

The construction of the largest telescope in the world is one of the most fascinating stories of scientific skill and invention. The major feature of the telescope is a parabolic *mirror*, 5 metres across, which is made of pyrex, a low expansion glass. The glass itself took more than six years to grind and polish, having been begun in 1936, and the front of the mirror is coated with aluminium, instead of being covered with silver, as it lasts much longer. The huge size of the mirror enables enough light from very distant stars and planets to be collected and brought to a focus for them to be photographed. Special cameras are incorporated in the instrument to photograph the universe. This method has the advantage that plates can be exposed for hours, if necessary, to the object to be studied, enabling records to be made. It is used to obtain useful information about the building-up and breaking-down of the elements in space (thus assisting in atomic energy research), to investigate astronomical theories of the universe, and to photograph Mars.

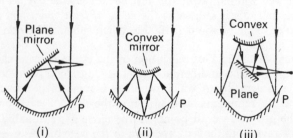

FIG. 23.22. (i). Newton reflector.   (ii). Cassegrain reflector.   (iii). Coudé reflector.

Besides the main parabolic mirror P, seven other mirrors are used in the 5 metre telescope. Some are plane, Fig. 23.22 (i), while others are convex, Fig. 23.22 (ii), and they are used to bring the light to a more convenient focus, where the image can be photographed, or magnified several hundred times by an eye-piece for observation. The various methods of focusing the image were suggested respectively by *Newton*, *Cassegrain*, and *Coudé*, the last being a combination of the former two methods.

EXAMPLES

1. What do you understand by the magnifying power of an astronomical telescope? Illustrate your answer with a ray diagram depicting the use of the instrument to view stars in the heavens. If such a telescope has an object glass of focal length 50 cm and an eye lens of focal length 5 cm, what is its magnifying power? If it is assumed that the eye is placed very close to the eye lens and that the pupil of the eye has a diameter of 3 mm, what will be the diameter of the object glass if all the light passing through the object glass is to emerge as a beam which fills the pupil of the eye? Assume that the telescope is pointed directly at a particular star. (*W.*)

First part. See text.

Second part. Assuming the telescope is in *normal* adjustment, the final image is formed at infinity. The magnifying power of the telescope is then $\frac{50 \text{ cm}}{5 \text{ cm}}$, or 10. See p. 529.

If all the light emerging from the eye-piece fills the pupil of the eye, the pupil is at the eye-ring. See p. 534. The eye-ring is the image of the objective in the eye-piece. Since the distance, u, from the objective to the eye-piece = 50 + 5 = 55 cm, the eye-ring distance, v, is given by

$$\frac{1}{v} + \frac{1}{(+55)} = \frac{1}{(+5)}$$

from which $v = 5 \cdot 5$ cm.

This is the position of the pupil of the eye. The magnification of the objective is given by

$$\frac{\text{eye-ring diameter}}{\text{objective diameter}} = \frac{v}{u} = \frac{5 \cdot 5}{55} = \frac{1}{10}$$

$$\therefore \quad \frac{3 \text{ mm}}{\text{objective diameter}} = \frac{1}{10}$$

$$\therefore \quad \text{objective diameter} = 3 \text{ cm.}$$

2. What do you understand by (*a*) the apparent size of an object, and (*b*) the magnifying power of a microscope? A model of a compound microscope is made up of two converging lenses of 3 and 9 cm focal length at a fixed separation of 24 cm. Where must the object be placed so that the final image may be at infinity? What will be the magnifying power if the microscope as thus arranged is used by a person whose nearest distance of distinct vision is 25 cm? State what is the best position for the observer's eye and explain why. (*L.*)

First part. (*a*) The apparent size of an object is proportional to the visual angle. See p. 525. (*b*) The magnifying power of a microscope is defined on p. 527.

Second part. (i) Suppose the objective A is 3 cm focal length, and the eye-piece B is 9 cm focal length, Fig. 23.23. If the final image is at infinity, the image I_1 in the objective must be 9 cm from B, the focal length of the eye-piece. See p. 531. Thus the image distance LI_1, from the objective A $=$ $24 - 9 = 15$ cm. The object distance OL is thus given by

$$\frac{1}{(+15)} + \frac{1}{u} = \frac{1}{(+3)},$$

from which $\qquad u = OL = 3\frac{3}{4}$ cm.

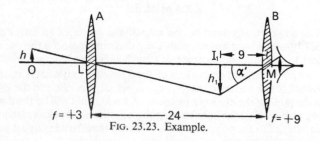

FIG. 23.23. Example.

(ii) The angle α' subtended at the observer's eye is given by $\alpha' = h_1/9$, where h_1 is the height of the image at I_1, Fig. 23.23. Without the lenses, the object subtends an angle α at the eye given by $\alpha = h/25$, where h is the height of the object, since the least distance of distinct vision is 25 cm.

$$\therefore \quad \text{magnifying power } M = \frac{\alpha'}{\alpha} = \frac{h_1/9}{h/25} = \frac{25}{9} \times \frac{h_1}{h}$$

But $\qquad \dfrac{h_1}{h} = \dfrac{LI_1}{LO} = \dfrac{15}{3\frac{3}{4}} = 4$

$$\therefore \quad M = \frac{25}{9} \times 4 = 11\tfrac{1}{9}$$

The best position of the eye is at the eye-ring, which is the image of the objective A in the eye-piece B (p. 534).

3. A Galilean telescope has an object-glass of 12 cm focal length and an eye lens of 5 cm focal length. It is focused on a distant object so that the final image seen by the eye appears to be situated at a distance of 30 cm from the eye lens. Determine the angular magnification obtained and draw a ray diagram. What are the advantages of prism binoculars as compared with field glasses of the Galilean type? (N.)

Suppose I_2 is the final image, distant 30 cm from the eye lens L_2. Fig. 23.24. The corresponding object is I_1. Since I_2 is a virtual image in L_2, $v = I_2L_2$ $= - 30, f = - 5, u = L_2I_2$. From the lens equation for L_2, we have

$$\frac{1}{(-30)} + \frac{1}{u} = \frac{1}{(-5)}$$

from which $\qquad u = - 6$ cm.

Thus I_1 is a virtual object for L_2.

The angular magnification, M, is given by $M = \alpha'/\alpha$. Now $\alpha' = h_1/L_2I_1$, and $\alpha = h_1/L_1I_1$.

$$\therefore \quad M = \frac{h_1/L_2I_1}{h_1/L_1I_1} = \frac{L_1I_1}{L_2I_1}$$

But $L_2I_1 = 6$ cm, from above, and $L_1I_1 =$ focal length of $L_1 = 12$ cm, since the object is distant.

$$\therefore \quad M = \frac{12}{6} = 2$$

The advantages of the prism binoculars are given on p. 541.

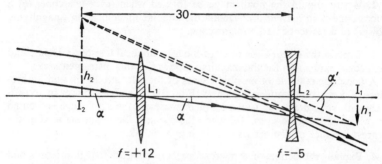

FIG. 23.24. Example.

4. Describe the optical system of a projection lantern. A lantern is required for the projection of slides 7·5 cm square on to a screen 2·1 m square. The distance between the front of the lantern and the screen is to be 20 m. What focal length of projection lens would you consider most suitable?

First part. See text.

Second part. Suppose O is the slide, L is the projection lens, and S is the screen, Fig. 23.25. The linear magnification, m, due to the lens is given by

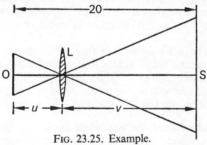

FIG. 23.25. Example.

$$m = \frac{210 \text{ cm}}{7 \cdot 5 \text{ cm}} = 28$$

$$\therefore \quad \text{LS} : \text{LO} = 28 : 1$$

$$\therefore \quad \text{LS} = v = \frac{28}{29} \times 20 \text{ m}$$

and $\quad \text{LO} = u = \frac{1}{29} \times 20 \text{ m}.$

Applying the lens equation,

$$\therefore \quad \frac{1}{560/29} + \frac{1}{20/29} = \frac{1}{f}$$

from which $\qquad f = \frac{560}{841} \text{ m} = 67 \text{ cm (to nearest cm)}$

EXERCISES 23

1. An object is viewed with a normal eye at (i) the least distance of distinct vision, (ii) 40 cm, (iii) 100 cm from the eye. Find the ratio of the visual angles in the three cases, and raw diagrams in illustration.

2. Where should the final image be formed when (*a*) a telescope, (*b*) a microscope is in *normal* use? Define the angular magnification (magnifying power) of a telescope and a microscope.

3. Explain the essential features of the astronomical telescope. Define and deduce an expression for the magnifying power of this instrument.

A telescope is made of an object glass of focal length 20 cm and an eye-piece of 5 cm, both converging lenses. Find the magnifying power in accordance with your definition in the following cases: (*a*) when the eye is focused to receive parallel rays, and (*b*) when the eye sees the image situated at the nearest distance of distinct vision which may be taken as 25 cm. (*L*.)

4. Explain the action of a microscope consisting of two thin lenses and show on a diagram the paths of three rays from a non-axial object point to the eye. Distinguish clearly between rays and construction lines.

A microscope having as eyepiece a thin lens of focal length 5·0 cm is set up by an observer whose least distance of distinct vision is 25·0 cm. An observer with defective eyesight has to withdraw the eyepiece by 0·50 cm in order that the image may be at his least distance of distinct vision. Find the nature of the defect and specify the nature and the focal length of the spectacle lens he needs to make his least distance of distinct vision 25·0 cm. (*L*.)

5. Describe, with the help of diagrams, how (*a*) a single biconvex lens can be used as a magnifying glass, (*b*) two biconvex lenses can be arranged to form a microscope. State (i) one advantage, (ii) one disadvantage, of setting the microscope so that the final image is at infinity rather than at the near point of the eye.

A centimetre scale is set up 5 cm in front of a biconvex lens whose focal length is 4 cm. A second biconvex lens is placed behind the first, on the same axis, at such a distance that the final image formed by the system coincides with the scale itself and that 1 mm in the image covers 2·4 cm in the scale. Calculate the position and focal length of the second lens. (*O. & C.*)

6. Compare and contrast the optical properties of (*a*) an astronomical telescope, (*b*) a Galilean telescope, (*c*) a reflecting telescope. Draw ray diagrams for (*a*) and (*b*) showing the path through each instrument of a non-axial pencil of rays from a distant object.

Describe one method by which the image in an astronomical telescope may be made erect. (*L*.)

7. What is the *eye-ring* of a telescope?

For an astronomical telescope in normal adjustment deduce expressions for the size and position of the eye-ring in terms of the diameter of the object glass and the focal lengths of the object glass and eye-lens.

Discuss the importance of (i) the magnitude of the diameter of the object glass, (ii) the structure of the object glass, (iii) the position of the eye. (*L*.)

8. Show, by means of a ray diagram, how an image of a distant extended object is formed by an astronomical refracting telescope in normal adjustment (i.e. with the final image at infinity).

A telescope objective has focal length 96 cm and diameter 12 cm. Calculate the focal length and minimum diameter of a simple eyepiece lens for use with the telescope, if the magnifying power required is × 24, and all the light transmitted by the objective from a distant point on the telescope axis is to fall on the eyepiece. Derive any formulae you use.

If the eyepiece is an equiconvex lens made from glass of refractive index 1·6, calculate the radius of curvature of its faces and the minimum thickness of the lens at its centre. (*O. & C.*)

9. Draw the path of two rays, from a point on an object, passing through the optical system of a compound microscope to the final image as seen by the eye.

If the final image formed coincides with the object, and is at the least distance of distinct vision (25 cm) when the object is 4 cm from the objective, calculate the focal lengths of the objective and eye lenses, assuming that the magnifying power of the microscope is 14. (*L.*)

10. Define *magnifying power* (*angular magnification*) of an optical instrument.

An astronomical telescope consists of two thin converging lenses which are 25·00 cm apart when the telescope is in normal adjustment. The distance between the lenses is reduced to 24·50 cm and a virtual image of an infinitely distant object is then formed 28·00 cm from the eye lens. Calculate the values of the focal lengths of the two lenses and the magnifying power of the instrument with this adjustment, supposing the eye to be placed at the eye lens.

For this same adjustment show on a labelled diagram (not to scale) the relative positions of the principal foci of the two lenses and the construction lines showing the relation of the final image to the intermediate image. On the separate diagram show the paths through the instrument of two rays from a non-axial point. One of the rays should pass through the centre of the objective and the other through its periphery. (*L.*)

11. Distinguish between the *magnification* produced by, and the *magnifying power* of, an optical system.

Draw a ray diagram showing the action of a simple astronomical telescope (assume two lenses only) in forming separate images of two stars which are close together and near the axis, the final images being at infinity.

If the objective has a diameter of 30 cm and a focal length of 3 metres, and the focal length of the eyepiece is 1·2 cm, calculate (*a*) the magnifying power of the telescope, (*b*) the diameter of the image of the objective formed by the eyepiece. (*O. & C.*)

12. Briefly describe an optical instrument which includes a reflecting prism. What is the function of the prism and what is the principle governing its action?

What are the advantages in using a prism rather than a silvered mirror in the apparatus you describe?

Parallel rays of light fall normally on the face *AC* of a total reflection prism *ABC* of refractive index 1·5 which has angle *A* exactly 45°, angle *B* approximately 90° and angle *C* approximately 45°. After total internal reflections in the prism two beams of parallel light emerge from the hypotenuse face, the angle between them being 6°. Calculate the value of angle *B*. You may assume that for small angles sin *i*/sin *r* equals *i*/*r*.

How would you discover whether the angle is more or less than 90°? (*O. & C.*)

13. Give a detailed description of the optical system of the compound microscope, explaining the problems which arise in the design of an object lens for a microscope.

A compound microscope has lenses of focal length 1 cm and 3 cm. An object is placed 1·2 cm from the object lens; if a virtual image is formed 25 cm from the eye, calculate the separation of the lenses and the magnification of the instrument. (*O. & C.*)

14. A projection lantern contains a condensing lens and a projection lens. Show clearly in a ray diagram the function of these lenses.

A lantern has a projection lens of focal length 25 cm and is required to be able to function when the distance from lantern to screen may vary from 6 m to 12 m. What range of movement for the lens must be provided in the focusing arrangement? What is the approximate value of the ratio of the magnifications at the two extreme distances? (*W.*)

15. A converging lens of focal length 20 cm and a diverging lens of focal length 10 cm are arranged for use as an opera glass. Draw a ray diagram to scale showing how the final image at infinity is produced, describing briefly how you do this, and derive the magnifying power.

When an object is placed 60 cm in front of the converging lens and the lenses are separated by a distance x, a real image is formed 30 cm beyond the diverging lens. Calculate x. (*C.*)

16. An astronomical telescope consisting of an objective focal length 60 cm and an eyepiece of focal length 3 cm is focused on the moon so that the final image is formed at the minimum distance of distinct vision (25 cm) from the eyepiece. Assuming that the diameter of the moon subtends an angle of $\frac{1}{2}°$ at the objective, calculate (*a*) the angular magnification, (*b*) the actual size of the image seen.

How, with the same lenses, could an image of the moon, 10 cm in diameter, be formed on a photographic plate? (*C.*)

17. Explain, with the aid of a ray diagram, how a simple astronomical telescope employing two converging lenses may form an apparently enlarged image of a distant extended object. State with reasons where the eye should be placed to observe the image.

A telescope constructed from two converging lenses, one of focal length 250 cm, the other of focal length 2 cm, is used to observe a planet which subtends an angle of 5×10^{-5} radian. Explain how these lenses would be placed for normal adjustment and calculate the angle subtended at the eye of the observer by the final image.

How would you expect the performance of this telescope for observing a star to compare with one using a concave mirror as objective instead of a lens, assuming that the mirror had the same diameter and focal length as the lens. (*O. & C.*)

chapter twenty-four

Velocity of light.
Photometry

FOR many centuries the velocity of light was thought to be infinitely large; from about the end of the seventeenth century, however, evidence began to be obtained which showed that the speed of light, though enormous, was a finite quantity. Galileo, in 1600, attempted to measure the velocity of light by covering and uncovering a lantern at night, and timing how long the light took to reach an observer a few miles away. Owing to the enormous speed of light, however, the time was too small to measure, and the experiment was a failure. The first successful attempt to measure the velocity of light was made by RÖMER, a Danish astronomer, in 1676.

Römer's Astronomical Method

Römer was engaged in recording the eclipses of one of Jupiter's satellites or moons, which has a period of 1·77 days round Jupiter. The period of the satellite is thus very small compared with the period of the earth round the sun (one year), and the eclipses of the satellite occur very frequently while the earth moves only a very small distance in its orbit. Thus the eclipses may be regarded as *signals sent out from Jupiter* at comparatively short intervals, and observed on the earth; almost like a bright lamp covered at regular intervals at night and viewed by a distant observer.

The earth makes a complete revolution round the sun, S, in one year. Jupiter makes a complete revolution round the sun in about $11\frac{3}{4}$ years, and we shall assume for simplicity that the orbits of the earth and Jupiter are both circular, Fig. 24.1. At some time, Jupiter (J_1) and the earth (E_1) are on the same side of the sun, S, and in line with each other, and the earth and Jupiter are then said to be in *conjunction**. Suppose that an eclipse, or "signal", is now observed on the earth E_1. If $E_1J_1 = x$, and c is the velocity of light, the time taken for the "signal" to reach E_1 is x/c; and if the actual time when the eclipse occurred was a (which is not known), the time T_1 of the eclipse *recorded on the earth* is given by

$$T_1 = a + \frac{x}{c} \qquad . \qquad . \qquad . \qquad . \qquad \text{(i)}$$

The earth and Jupiter now move round their respective orbits, and

* Astronomers use the terms 'conjunction' and 'opposition' in relation to the positions of the *sun* and Jupiter. With this usage, the latter are in 'opposition' here.

at some time, about $6\frac{1}{2}$ months later, the earth (E_2) and Jupiter (J_2) are on opposite sides of the sun S and in line with each other, Fig. 24.1. The earth and Jupiter are then said to be in *opposition*. During the $6\frac{1}{2}$ months

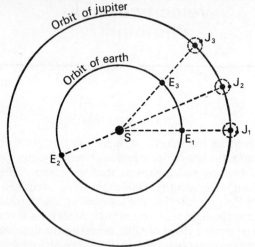

FIG. 24.1. Römer's method.

suppose that m eclipses have occurred at regular intervals T, i.e., T is the actual time between successive eclipses: the time for the interval between the 1st and mth eclipses is then $(m - 1) T$. In the position J_2, E_2, however, the light travels a distance J_2E_2, or $(x + d)$, from Jupiter to the earth, where d is the diameter of the earth's orbit round the sun. The time taken to travel this distance $= (x + d)/c$. Thus the time T' recorded on the earth at E_2 when the mth eclipse occurs is given by

$$T' = a + (m - 1) T + \frac{x + d}{c} \qquad . \qquad . \qquad . \qquad \text{(ii)}$$

But, from (i), $T_1 = a + \dfrac{x}{c}$

Subtracting, $T' - T_1 = I = (m - 1) T + \dfrac{d}{c}$,   .   .   .   (iii)

where I is the interval recorded on the earth for the time of m eclipses, from the position of conjunction of Jupiter and the earth to their position of opposition. By similar reasoning to the above, the interval I_1 recorded on the earth for m eclipses from the position of opposition (J_2, E_2) to the next position of conjunction of (J_3, E_3) is given by

$$I_1 = (m - 1) T - \frac{d}{c} \qquad . \qquad . \qquad . \qquad . \qquad \text{(iv)}$$

The reason why I_1 is less than I is that the earth is moving towards Jupiter from E_2 to E_3, and *away* from Jupiter from E_1 to E_2.

Römer observed that the mth eclipse between the position J_1, E_1 to the position J_2, E_2 occurred later than he expected by about $16\frac{1}{2}$ minutes; and he correctly deduced that the additional time was due to the time taken by light to travel across the earth's orbit. In (iii), $(m - 1) T$ was the time expected for $(m - 1)$ eclipses, and d/c was the extra time ($16\frac{1}{2}$ minutes) recorded on the earth. Since $d = 300\,000\,000$ km approximately, the velocity of light, c, $= 300\,000\,000/(16\frac{1}{2} \times 60) = 300\,000$ km per second approximately.

Römer also recorded that the time I for m eclipses from the position E_1, J_1 to the position E_2, J_2 was about 33 minutes more than the time I_1, for m eclipses between position E_2, J_2 to the position E_3, J_3. But, subtracting (iv) from (iii),

$$I - I_1 = \frac{2d}{c} \qquad \qquad \qquad (1)$$

$$\therefore \ \frac{2d}{c} = 33 \text{ mins} = 33 \times 60 \text{ secs}$$

$$\therefore \quad c = 2d/(33 \times 60) = 2 \times 300\,000\,000/(33 \times 60)$$

$$\therefore \quad c = 300\,000 \text{ km per second (approx.)}$$

Maximum and minimum observed periods. When the earth E_1 is moving directly away from Jupiter at J_a, the apparent period T' of the satellite is a maximum. Suppose the earth moves from E_1 to E_2 in the time T', Fig. 24.2.

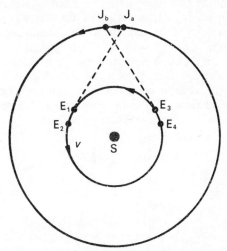

Fig. 24.2. Maximum and minimum periods.

Then if T is the actual period, v is the velocity of the earth, and c is the velocity of light, it follows that

$$T' = T + \frac{E_1 E_2}{c} = T + \frac{vT'}{c}$$

$$\therefore \ T'\left(1 - \frac{v}{c}\right) = T \qquad \qquad (i)$$

When the earth is moving directly towards Jupiter at J_b, the apparent period T'' is a minimum. Suppose the earth moves from E_3 to E_4 in this time. Then

$$T'' = T - \frac{E_3 E_4}{c} = T - \frac{vT''}{c}$$

$$\therefore \quad T'' \left(1 + \frac{v}{c}\right) = T \quad . \quad . \quad . \quad \text{(ii)}$$

From (i) and (ii), it follows that

$$T' \left(1 - \frac{v}{c}\right) = T'' \left(1 + \frac{v}{c}\right)$$

$$\therefore \quad \frac{T'}{T''} = \frac{1 + \dfrac{v}{c}}{1 - \dfrac{v}{c}} = \begin{array}{l}\text{ratio of maximum and minimum}\\ \text{observed periods}\end{array}$$

Bradley's Aberration Method

Römer's conclusions about the velocity of light were ignored by the scientists of his time. In 1729, however, the astronomer BRADLEY observed that the angular elevation of a "fixed" star varied slightly according to the position of the earth in its orbit round the sun. For some time he was puzzled by the observation. But while he was being rowed across a stream one day, he noticed that the boat drifted slightly downstream; and he saw immediately that the difference between the actual and observed angular elevation of the star was due to a combination of the velocity of the earth in its orbit (analogous to the velocity of the stream) with that of the velocity of light (analogous to the velocity of the boat). Thus if the earth were stationary, a telescope T would have to point in the true direction AS of a star S to observe it; but since the earth is moving in its orbit round the sun with a velocity v, T would have to be directed along MN to observe the star, where MN makes a small angle α with the direction SA, Fig. 24.3 (i).

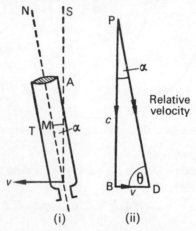

FIG. 24.3. Bradley's aberration method.

The direction of MN is that of the *relative velocity* between the earth and the light from S, which is found by subtracting the velocity v from the velocity c of the light. This is easily done by drawing the triangle of velocities PBD, in which PB, parallel to SA, represents c in magnitude and direction, while BD represents a velocity equal and opposite to v, Fig. 24.3 (ii). The resultant of PB and BD is then PD, which is parallel to NM, and PD represents the relative velocity.

The angle α between the true and apparent directions of the star is known as the *aberration*. From Fig. 24.3 (ii), it follows that

$$\frac{v}{c} = \frac{\sin \alpha}{\sin \theta},$$

where θ is the apparent altitude of the star.

$$\therefore \quad c = \frac{v \sin \theta}{\sin \alpha} \qquad . \qquad . \qquad . \qquad . \qquad (2)$$

Since α is very small, $\sin \alpha$ is equal to α in radians. By using known values of v, θ, and α, Bradley calculated c, the velocity of light, and obtained a value close to Römer's value. The aberration α is given by half the difference between the maximum and minimum values of the apparent altitude, θ, of the star.

Fizeau's Rotating Wheel Method. A Terrestrial Method

In 1849 FIZEAU succeeded in measuring the velocity of light with apparatus on the earth, for the first time. His method, unlike Römer's and Bradley's method, is thus known as a *terrestrial method*.

Fizeau's apparatus is illustrated in Fig. 24.4. A bright source at O emits light which is converged to a point H by means of the lens and the plane sheet of glass F, and is then incident on a lens B. H is at the focus of B, and the light thus emerges parallel after refraction through the latter and travels several miles to another lens C. This lens brings the light to a focus at M, where a silvered plane mirror is positioned, and the light is now returned back along its original path to the glass plate F. An image of O can thus be observed through F by a lens E.

The rim of a toothed wheel, W, which can rotate about a horizontal axis Q, is placed at H, and is the important feature of Fizeau's method.

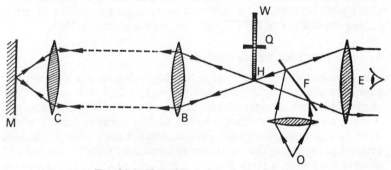

FIG. 24.4. Fizeau's rotating wheel method.

The teeth and the gaps of W have the same width, Fig. 24.5. As W is rotated, an image is observed through E as long as the light passes through the wheel towards E. When the speed of rotation exceeds about 10 cycles per second, the succession of images on the retina causes an image of O to be seen continuously. As the speed of W is further increased, however, a condition is reached when the returning light passing through a gap of W and reflected from M arrives back at the wheel to find that the tooth *next* to the gap has taken the place of the

gap. Assuming the wheel is now driven at a constant speed, it can be seen that light continues to pass through a gap in the wheel towards M but always arrives back at W to find its path barred by the neighbouring tooth. The field of view through E is thus dark. If the speed of W is now doubled a bright field of view is again observed, as the light passing through a gap arrives back from M to find the next *gap* in its place, instead of a tooth as before.

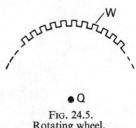

FIG. 24.5.
Rotating wheel.

Fizeau used a wheel with 720 teeth, and first obtained a dark field of view through E when the rate of revolution was 12·6 revs per second. The distance from H to M was 8633 metres. Thus the time taken for the light to travel from H to M and back $= 2 \times 8633/c$ seconds, where c is the velocity of light in metres per second. But this is the time taken by a tooth to move to a position corresponding to a neighbouring gap. Since there are 2×720 teeth and gaps together, the time $= 1/(2 \times 720)$ of the time taken to make one revolution, or $1/(2 \times 720 \times 12\cdot6)$ seconds, as 12·6 revs are made in one second.

$$\therefore \quad \frac{2 \times 8633}{c} = \frac{1}{2 \times 720 \times 12\cdot6}$$
$$\therefore \quad c = 2 \times 8633 \times 2 \times 720 \times 12\cdot6$$
$$= 3\cdot1 \times 10^8 \text{ metres per second.}$$

The disadvantage of Fizeau's method is mainly that the field of view can never be made perfectly dark, owing to the light diffusedly reflected at the teeth towards E. To overcome this disadvantage the teeth of the wheel were bevelled, but a new and more accurate method of determining the velocity of light was devised by FOUCAULT in 1862.

Foucault's Rotating Mirror Method

In Foucault's method a plane mirror M_1 is rotated at a high constant angular velocity about a vertical axis at A, Fig. 24.6. A lens L is placed so that light from a bright source at O_1 is reflected at M_1 and comes to a focus at a point P on a concave mirror C. The centre of curvature of C is at A, and consequently the light is reflected back from C along its original path, giving rise to an image coincident with O_1. In order to observe the image, a plate of glass G is placed at 45° to the axis of the lens, from which the light is reflected to form an image at B_1.

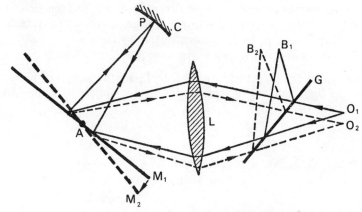

FIG. 24.6. Foucault's rotating mirror method.

Suppose the plane mirror M_1 begins to rotate. The light reflected by it is then incident on C for a fraction of a revolution, and if the speed of rotation is 2 revs per sec, an intermittent image is seen. As the speed of M_1 is increased to about 10 revs per sec the image is seen continuously as a result of the rapid impressions on the retina. As the speed is increased further, the light reflected from the mirror flashes across from M_1 to C, and returns to M_1 to find it displaced by a very small angle θ to a new position M_2. An image is now observed at B_2, and by measuring the displacement, B_1B_2, of the image Foucault was able to calculate the velocity of light.

Theory of Foucault's Method

Consider the point P on the curved mirror from which the light is always reflected back to the plane mirror, Fig. 24.7. When the plane mirror is at M_1, the image of P in it is at I_1, where $AI_1 = AP = a$, the radius of curvature of C (see p. 396). The rays incident on the lens L from the plane mirror appear to come from I_1. When the mirror is at M_2 the image of P in it is at I_2, where $AI_2 = AP = a$, and the rays incident on L from the mirror now appear to come from I_2. Now the

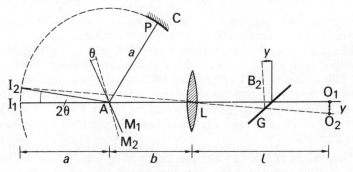

FIG. 24.7. Explanation of Foucault's method.

mirror has rotated through an angle θ from M_1 to M_2, and the direction PA of the light incident on it is constant. The angle between the reflected rays is thus 2θ (see p. 393), and hence $I_2AI_1 = 2\theta$.

$$\therefore \quad I_2I_1 = a \times 2\theta = 2a\theta \quad . \qquad . \qquad . \qquad . \quad \text{(i)}$$

The images O_1, O_2 formed by the *lens*, L, are the images of I_1, I_2 in it, as the light incident on L from the mirror appear to come from I_1, I_2. Hence $I_2I_1 : O_1O_2 = I_1L : LO_1$.

$$\therefore \quad \frac{I_2I_1}{y} = \frac{(a+b)}{l}$$

where $y = O_1O_2$, AL $= b$, and $LO_1 = l$.

$$\therefore \quad I_2I_1 = \frac{(a+b)\,y}{l} \qquad . \qquad . \qquad . \qquad . \quad \text{(ii)}$$

From (i) and (ii), it follows that

$$2a\theta = \frac{(a+b)\,y}{l}$$

$$\therefore \qquad \theta = \frac{(a+b)\,y}{2al} \qquad . \qquad . \qquad . \qquad . \quad \text{(iii)}$$

The angle θ can also be expressed in terms of the velocity of light, c, and the number of revolutions per second, m, of the plane mirror. The angular velocity of the mirror is $2\pi m$ radians per second, and hence the time taken to rotate through an angle θ radians is $\theta/2\pi m$ secs. But this is the time taken by the light to travel from the mirror to C and back, which is $2a/c$ secs.

$$\therefore \quad \frac{\theta}{2\pi m} = \frac{2a}{c}$$

$$\therefore \qquad \theta = \frac{4\pi ma}{c} \qquad . \qquad . \qquad . \qquad . \quad \text{(iv)}$$

From (iii) and (iv), we have

$$\frac{(a+b)\,y}{2al} = \frac{4\pi ma}{c}$$

$$\therefore \qquad c = \frac{8\pi ma^2 l}{(a+b)\,y} \qquad . \qquad . \qquad . \quad \text{(3)}$$

As m, a, l, b are known, and the displacement $y = O_1O_2 = B_1B_2$ and can be measured, the velocity of light c can be measured.

The disadvantage of Foucault's method is mainly that the image obtained is not very bright, making observation difficult. Michelson (p. 559) increased the brightness of the image by placing a large lens between the plane mirror and C, so that light was incident on C for a greater fraction of the mirror's revolution. Since the distance a was increased at the same time, Fig. 24.7, the displacement of the image was also increased.

The velocity of light in water was observed by Foucault, who placed a pipe of water between the plane mirror and C. He found that, with the number of revolutions per second of the mirror the same as when air was used, the displacement y of the image B was *greater*. Since the velocity of light $= 8\pi ma^2 l/(a + b)\, y$, from (3), it follows that the velocity of light in water is *less* than in air. Newton's "corpuscular theory" of light predicted that light should travel faster in water than in air (p. 680), whereas the "wave theory" of light predicted that light should travel slower in water than in air. The direct observation of the velocity of light in water by Foucault's method showed that the corpuscular theory of Newton could not be true.

Michelson's Method for the Velocity of Light

The velocity of light, c, is a quantity which appears in many fundamental formulæ in advanced Physics, especially in connection with the theories concerning particles in atoms and calculations on atomic (nuclear) energy. EINSTEIN has shown, for example, that the energy W released from an atom is given by $W = mc^2$ joules, where m is the decrease in mass of the atom in kilogrammes and c the velocity of light in metres per second. A knowledge of the magnitude of c is thus important. A. A. MICHELSON, an American physicist, spent many years of his life in measuring the velocity of light, and the method he devised is regarded as one of the most accurate.

The essential features of Michelson's apparatus are shown in Fig. 24.8. X is an equiangular octagonal steel prism which can be rotated at constant speed about a vertical axis through its centre. The faces of the prism are highly polished, and the light passing through a slit from a very bright source O is reflected at the surface A towards a plane mirror B. From B the light is reflected to a plane mirror L, which is placed so that the image of O formed by this plane mirror is at the focus of a large concave mirror HD. The light then travels as a parallel

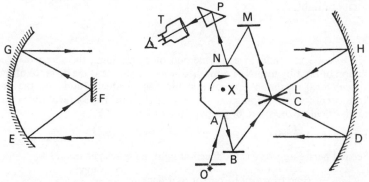

FIG. 24.8. Michelson's rotating prism method.

beam to another concave mirror GE a long distance away, and it is reflected to a plane mirror F at the focus of GE. The light is then reflected by the mirror, travels back to H, and is there reflected to a plane mirror C placed just below L and inclined to it as shown. From C the light is reflected to a plane mirror M, and is then incident on the face N of the octagonal prism opposite to A.

OPTICS, SOUND AND WAVES

The final image thus obtained is viewed through T with the aid of a totally reflecting prism P.

The image is seen by light reflected from the top surface of the octagonal prism X. When the latter is rotated the image disappears at first, as the light reflected from A when the prism is just in the position shown in Fig. 24.8 arrives at the opposite face to find this surface in some position inclined to that shown. When the speed of rotation is increased and suitably adjusted however, the image reappears and is seen in the same position as when the prism X is at rest. *The light reflected from A now arrives at the opposite surface in the time taken for the prism to rotate through 45°, or ⅛th of a revolution,* as in this case the surface on the left of N, for example, will occupy the latter's position when the light arrives at the upper surface of X.

Suppose *d* is the total distance in metres travelled by the light in its journey from A to the opposite face; the time taken is then *d/c*, where *c* is the velocity of light. But this is the time taken by X to make ⅛th of a revolution, which is 1/8*m* secs if the number of revolutions per second is *m*.

$$\therefore \quad \frac{1}{8m} = \frac{d}{c}$$

$$\therefore \quad c = 8md . \qquad . \qquad . \qquad . \qquad (4)$$

Thus *c* can be calculated from a knowledge of *m* and *d*.

Michelson performed the experiment in 1926, and again in 1931, when the light path was enclosed in an evacuated tube 1·6 km long. Multiple reflections were obtained to increase the effective path of the light. A prism with 32 faces was also used, and Michelson's result for the velocity of light *in vacuo* was 299 774 kilometres per second. Michelson died in 1931 while he was engaged in another measurement of the velocity of light.

EXAMPLES

1. Describe carefully Fizeau's method of determining the speed of propagation of light by means of a toothed wheel. Given that the distance of the mirror is 8000 m, that the revolving disc has 720 teeth and that the first eclipse occurs when the angular velocity of the disc is 13¾ revolutions per second, calculate the speed of propagation of light. (*W.*)

First part. See text.

Second part. Suppose *c* is the speed of light in metres per second.

$$\therefore \quad \text{time to travel to mirror and back} = \frac{2 \times 8000}{c} \text{ s.}$$

But time for one tooth to occupy the next gap's position $= \frac{1}{13\frac{3}{4}} \times \frac{1}{2 \times 720}$ s.

$$\therefore \quad \frac{2 \times 8000}{c} = \frac{1}{13\frac{3}{4} \times 2 \times 720}$$

$$\therefore \qquad c = 2 \times 8000 \times 13\tfrac{3}{4} \times 2 \times 720$$
$$= 3\cdot 2 \times 10^8 \text{ m s}^{-1}.$$

2. A beam of light is reflected by a rotating mirror on to a fixed mirror, which sends it back to the rotating mirror from which it is again reflected, and then makes an angle of 18° with its original direction. The distance between the two mirrors is 10^4 m, and the rotating mirror is making 375 revolutions per sec. Calculate the velocity of light. (*L.*)

Suppose OA is the original direction of the light, incident at A on the mirror in the position M_1, B is the fixed mirror, and AC is the direction of the light reflected from the rotating mirror when it reaches the position M_2, Fig. 24.9.

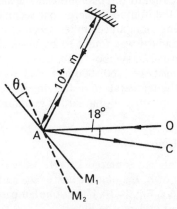

FIG. 24.9. Example.

The angle θ between M_1, M_2 is $\frac{1}{2} \times 18°$, since the angle of rotation of a mirror is half the angle of deviation of the reflected ray when the incident ray (BA in this case) is kept constant. Thus $\theta = 9°$.

$$\text{Time taken by mirror to rotate } 360° = \frac{1}{375} \text{ s.}$$

$$\therefore \quad \text{time taken to rotate } 9° = \frac{9}{360} \times \frac{1}{375} \text{ s.}$$

But this is also the time taken by the light to travel from A to B and back, which is given by $2 \times 10^4/c$, where c is the velocity of light in m s^{-1}.

$$\therefore \quad \frac{2 \times 10^4}{c} = \frac{9}{360} \times \frac{1}{375}$$

$$\therefore \quad c = \frac{2 \times 10^4 \times 360 \times 375}{1} = 3 \times 10^8 \text{ m s}^{-1}.$$

Photometry

Standard Candle. The Candela

Light is a form of energy which stimulates the sensation of vision. The sun emits a continuous stream of energy, consisting of ultra-violet, visible, and infra-red radiations (p. 456), all of which enter the eye; but only the energy in the visible radiations, which is called *luminous energy*, stimulates the sensation of vision. In Photometry we are concerned only with the luminous energy emitted by a source of light.

Years ago the luminous energy per second from a candle of specified wax material and wick was used as a standard of luminous energy. This was called the *British Standard Candle*. The luminous energy per second from any other source of light was reckoned in terms of the standard candle, and its value was given at 10 candle-power (10 c.p.) for example. As the standard candle was difficult to reproduce exactly, the standard was altered. It was defined as one-tenth of the intensity of the flame of the *Vernon Harcourt pentane lamp*, which burns a mixture of air and pentane vapour under specified conditions. Later it was agreed to use as a standard the *international standard candle*, which is defined in terms of the luminous energy per second from a particular electric lamp filament maintained under specified conditions, but the precision of this standard was found to be unsatisfactory. In 1948, a unit known as the *candela*, symbol "cd", was adopted. This is defined as the luminous intensity of $1/600\ 000$ metre^2 ($1/60$ cm^2) of the surface of a black body at the temperature of freezing platinum under 101 325 newtons per metre^2 pressure. A standard is maintained at the National Physical Laboratory.

Illumination and its Units

If a lamp S of 1 candela is placed 1 metre away from a small area A and directly in front of it, the *illumination* of the surface of A is said to be 1 *metre-candle* or *lux*, Fig. 24.10. If the same lamp is placed 1 centimetre away from A, instead of 1 m, the illumination of the surface is said to be 1 *cm-candle* (or 1 *phot*). The SI unit of illumination is the lux (see also p. 564). The "foot-candle" has been used as a unit; the distance of 1 metre in Fig. 24.10 is replaced by 1 foot. 1 lux $= 10^{-4}$ phot $= 9\cdot 3 \times 10^{-2}$ foot-candle. It is

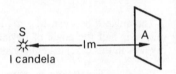

FIG. 24.10. Metre-candle (lux).

recommended that offices should have an intensity of illumination of about 90 lux, and that the intensity of illumination for sewing dark materials in workrooms should be about 200 lux.

Luminous Flux, F

In practice a source of light emits a continuous stream of energy, and the name *luminous flux* has been given to the *luminous energy emitted per second*. The unit of luminous flux is the **lumen**, lm. Since a lumen is a certain amount of "energy per second", or "power", there must be a relation between the lumen and the watt, the mechanical unit of power; and experiment shows that 621 lumens of a green light of wavelength $5\cdot 540 \times 10^{-10}$ m is equivalent to 1 watt.

Solid Angle

A lamp radiates luminous flux in all directions round it. If we think of a particular small lamp and a certain direction from it, for example that of the corner of a table, we can see that the flux is radiated towards the corner in a cone whose apex is the lamp. A thorough study of photometry must therefore include a discussion of the measurement of an angle in three dimensions, such as that of a cone, which is known as a *solid angle*.

An angle in two dimensions, i.e., in a plane, is given in radians by the ratio s/r, where s is the length of the arc cut off by the bounding lines of the angle on a circle of radius r. In an analogous manner, the solid angle, ω, of a cone is defined by the relation

$$\omega = \frac{S}{r^2} \qquad . \qquad . \qquad . \qquad . \qquad (5)$$

where S is the area of the surface of a sphere of radius r cut off by the bounding (generating) lines of the cone, Fig. 24.11 (i). Since S and r^2 both have the dimensions of (length)^2, the solid angle ω is a ratio.

When $S = 1$ m^2, and $r = 1$ m, then $\omega = 1$ from equation (5). Thus *unit solid angle* is subtended at the centre of a sphere of radius

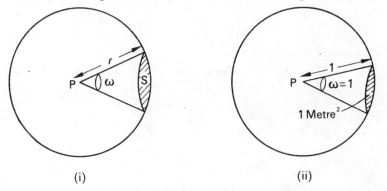

(i) (ii)

FIG. 24.11. (i). Solid angle at P. (ii). Unit solid angle at P.

1 m by a cap of surface area 1 m^2, Fig. 24.11 (ii). It is called "1 steradian", sr. The solid angle all round a point is given from (5) by

$$\frac{\text{total surface area of sphere}}{r^2}$$

i.e., by $\qquad \dfrac{4\pi r^2}{r^2}$, or 4π.

Thus the solid angle all round a point is 4π sr. The solid angle all round a point on one side of a plane is thus 2π sr.

Luminous Intensity of Source, I

Experiment shows that the luminous flux from a source of light varies in different directions; to be accurate, we must therefore consider the luminous flux emitted in a particular direction. Suppose that we consider a small lamp P, and describe a cone PCB of small solid angle ω about a particular direction PD as axis, Fig. 24.12. The *luminous intensity, I, of the source in this direction* is then defined by the relation

$$I = \frac{F}{\omega} \qquad . \qquad . \qquad . \qquad . \qquad . \qquad (6)$$

where F is the luminous flux contained in the small cone. Thus the luminous intensity of the source is the *luminous flux per unit solid angle* in the particular direction. It can now be seen that "luminous intensity" is a measure of the "luminous flux density" in the direction concerned.

The unit of luminous intensity of a source is the *candela*, defined on p. 562, and the luminous intensity was formerly known as the *candle-power* of the source. When the luminous flux, F, in the cone in Fig. 24.12 is 1 lumen (the unit of luminous flux), and the solid angle, ω, of the cone is 1 unit, it follows from equation (6) that $I = 1$ candela. Thus *the lumen can be defined as the luminous flux radiated within unit solid angle by a uniform source of one candela.* A small source of I candela radiates $4\pi I$ lumens all round it.

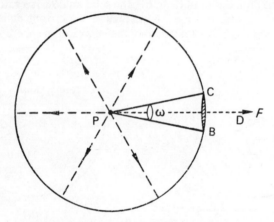

Fig. 24.12. Luminous intensity of source.

Illumination of Surface

On p. 562 we encountered various units of *illumination of a surface*; these were the metre-candle or the lux (lx), which is the SI unit, the cm-candle and the foot-candle. The illumination is defined generally as the *luminous flux per unit area* falling on the part of the surface under consideration. Thus if F is the luminous flux incident on a small area A, the intensity of illumination, E, is given by

$$E = \frac{F}{A} \qquad . \qquad . \qquad . \qquad . \qquad . \qquad (7)$$

When $F = 1$ lumen and $A = 1$ m², then $E = 1$ lm m^{-2}. Thus 1 m-candle of illumination is equivalent to an illumination of 1 lumen per m² of the surface. Similarly, 1 cm-candle (p. 562) is equivalent to an illumination of 1 lumen per cm² of the surface. 1 lux $= 1$ lm m^{-2}.

Suppose a lamp of 50 candela illuminates an area of 2 m² at a distance 20 m away. The flux F emitted all round the lamp $= I\omega = I \times 4\pi = 50 \times 4\pi = 200\pi$ lumens. The flux per unit area at a distance of 20 m away is given by

$$\frac{200\pi}{4\pi r^2} = \frac{200\pi}{4\pi \times 20^2} \text{ lumens per m}^2$$

∴   flux falling on an area of 2 m²

$$= 2 \times \frac{200\pi}{4\pi \times 20^2} = 0.25 \text{ lumens.}$$

The reader should take pains to distinguish carefully between the meaning of "luminous intensity" and "illumination" and their units. The former refers to the *source* of light and is measured in candelas (formerly candle-power); the latter refers to the *surface* illuminated and is measured in lux or metre-candles. Further, "luminous intensity" is defined in terms of unit solid angle, which concerns three dimensions whereas "illumination" is defined in terms of unit area, which concerns two dimensions.

Relation between Luminous Intensity (I) and Illumination (E)

Consider a *point* source of light of uniform intensity and a small part X of a surface which it illuminates. If the source of light is doubled, the illumination of X is doubled because the luminous flux incident on it is twice as much. Thus

$$E \propto I \quad . \qquad . \qquad . \qquad . \qquad . \qquad \text{(i)}$$

where E is the illumination due to a point source of luminous intensity I at a given place.

The illumination of the surface also depends on the distance of X from the source. Suppose two spheres of radii r_1, r_2 are drawn round a point source of intensity I, such as S in Fig. 24.13 (i). The same amount

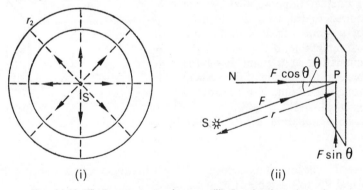

FIG. 24.13. (i). Inverse square law.   (ii). Lambert's cosine rule.

of luminous flux F spreads over the surface area ($4\pi r^2$) of both spheres, and hence $E_1 : E_2 = F/4\pi r_1^2$, where E_1, E_2 are the values of the illumination at the surface of the smaller and larger spheres respectively. Thus $E_1 : E_2 = r_2^2 : r_1^2$. It can hence be seen that the illumination due to a given point source varies inversely as the square of the distance from the source, i.e.,

$$E \propto \frac{1}{r^2} \quad . \qquad . \qquad . \qquad . \qquad . \qquad \text{(ii)}$$

In the eighteenth century LAMBERT showed that the illumination round a particular point P on a surface is proportional to cos θ, where θ is the angle between the normal PN at P to the surface and the line SP joining the source S to P, Fig. 24.13 (ii). A rigid proof of Lambert's law is given shortly, but we can easily see qualitatively why the cosine rule is true. The luminous flux F illuminating P is in the direction SP, and thus has a component $F \cos \theta$ along NP and a component $F \sin \theta$ parallel to the surface, Fig. 24.13. The latter does not illuminate the surface. Hence the effective part of the flux F is $F \cos \theta$.

From equations (i) and (ii) and the cosine law, it follows that the illumination E round P due to the source S of luminous intensity I is given by

$$E = \frac{I \cos \theta}{r^2} \qquad . \qquad . \qquad . \qquad . \qquad (8)$$

where $SP = r$. This is a fundamental equation in Photometry, and it is proved rigidly on p. 567. In applying it in practice one has to take into account that (i) a "point source" is difficult to realise, (ii) the area round the point considered on the surface should be very small so that the flux incident all over it can be considered the same, (iii) the intensity I of a source varies in different directions, (iv) the actual value of illumination round a point on a table, for example, is not only due to the electric lamp above it but also to the luminous flux diffusely reflected towards the point from neighbouring objects such as walls.

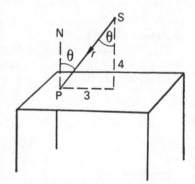

Example: Suppose that we are required to calculate the intensity I of a lamp S fixed 4 m above a horizontal table, if the value of the illumination at a point P on the table 3 m to one side of the vertical through the lamp is 6 m-candles, Fig. 24.14.

FIG. 24.14. Example.

The illumination, E, at P is given by

$$E = \frac{I \cos \theta}{r^2} \qquad . \qquad . \qquad . \qquad . \qquad . \qquad (i)$$

where I is the luminous intensity of S, θ is the angle between SP and the normal PN to the table at P, and $r = SP$.

But $r^2 = 4^2 + 3^2 = 25$, i.e., $r = 5$,

$$\cos \theta = \frac{4}{r} = \frac{4}{5},$$

and $E = 6$ m-candles.

Substituting in (i),

$$\therefore \quad 6 = \frac{I \times \frac{4}{5}}{25}$$

$$\therefore \quad I = 187 \cdot 5 \text{ candelas}$$

Proof of $E = I \cos \theta / r^2$. Consider a source S of intensity I illuminating a very small area A round a point P on a surface, Fig. 24.15. If ω is the solid angle at S in the cone obtained by joining S to the boundary of the area, the illumination at P is given by

$$E = \frac{F}{A} = \frac{I\omega}{A}$$

as $I = \dfrac{F}{\omega}$ (p. 563). Now, by definition,

$$\omega = \frac{A_1}{r_2},$$

where A_1 is the area cut off on a sphere of centre S and radius r ($=$ SP) by the generating lines of the cone.

$$\therefore \quad E = \frac{I\omega}{A} = \frac{I A_1}{r^2 A},$$

But $\qquad A_1 = A \cos \theta,$

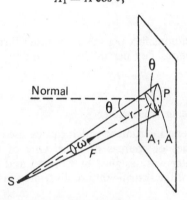

FIG. 24.15. Proof of $E = I \cos \theta / r^2$.

since A_1 is the projection of A on the sphere of centre S, and θ is the angle between the two areas as well as the angle made by SP with the normal to the surface, Fig. 24.15.

$$\therefore \quad E = \frac{I A \cos \theta}{r^2 A} = \frac{I \cos \theta}{r^2}$$

Luminance of a Surface: Reflection and Transmission Factors

The *luminance* of a surface in a given direction is defined as the luminous flux per unit area *coming from* the surface in the particular direction. The luminance of white paint on the wall of a room is considerably higher than the luminance of a brown-painted panel in the middle of the wall; the luminance of a steel nib is much greater than that of a dark ebonite penholder.

The "luminance" of a particular surface should be carefully distinguished from the "illumination" of the surface, which is the luminous flux per sq m *incident on* the surface. Thus the illumination of white chalk on a particular blackboard is practically the same as that of the neighbouring points on the board itself, whereas the luminance of the chalk is much greater than that of the board. The difference in luminance is due to the difference in the *reflection factor*, r, of the chalk and board, which is defined by the relation

$$r = \frac{B}{E} \qquad . \qquad . \qquad . \qquad . \qquad . \qquad (9)$$

where B is the luminance of the surface and E is the illumination of the surface. Thus the luminance, B, is given by

$$B = rE \qquad . \qquad . \qquad . \qquad . \qquad . \qquad (10)$$

Besides reflection, the luminance of a surface may be due to the transmission of luminous flux through it. The luminance of a pearl lamp, for example, is due to the transmission of luminous flux through its surface. The *transmission factor*, t, of a substance is a ratio which is defined by

$$t = \frac{F}{E} \qquad . \qquad . \qquad . \qquad . \qquad . \qquad (11)$$

where F is the luminous flux per m² transmitted through the substance and E is the luminous flux per m² incident on the substance. Thus

$$F = tE \qquad . \qquad . \qquad . \qquad . \qquad . \qquad (12)$$

The Lummer–Brodhun Photometer

A *photometer* is an instrument which can be used for comparing the luminous intensities of sources of light. One of the most accurate forms of photometer was designed by LUMMER and BRODHUN, and the essential features of the instrument are illustrated in Fig. 24.16 (i).

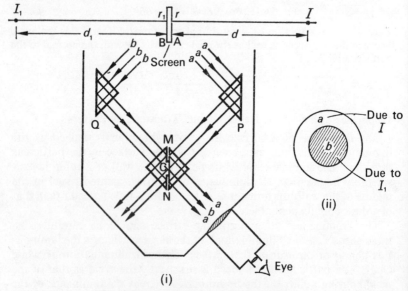

FIG. 24.16. Lummer–Brodhun photometer.

Lamps of luminous intensities I_1 and I_2 respectively are placed on opposite sides of a white opaque screen, and some of the diffusely-reflected light from the opposite surfaces A, B is incident on two identical totally reflecting prisms P, Q, Fig. 24.16 (i). The light reflected from the prisms then passes towards the "Lummer-Brodhun cube", which is the main feature of the photometer. This consists of two right-angled iso-sceles prisms in optical contact at their central portion C, but with the edges of one cut away so that an air-film exists at M, N all round C between the prisms. The rays leaving the prism P are thus transmitted through the central portion C of the "cube", but totally reflected at the edges. Similarly, the rays reflected from the prism Q towards the "cube" are transmitted through the central portion but totally reflected at the edges. An observer of the "cube" thus sees a central circular patch b of light due initially to the light from the source of intensity I_1, and an outer portion α due initially to the light from the source of intensity I_2 Fig. 24.16 (ii).

Comparison of Luminous Intensities

In general, the brightness of the central and outer portions of the field of view in a Lummer-Brodhun photometer is different, so that one appears darker than the other. By moving one of the sources, however, a position is obtained when both portions appear equally bright, in which case they cannot be distinguished from each other and the field of view is uniformly bright. A "photometric balance" is then said to exist.

Suppose that the distances of the sources I_1 and I from the screen are d_1, d respectively, Fig. 24.16. The intensity of illumination, E, due to the source I is generally given by $E = I \cos \theta / d^2$ (p. 566). But $\theta = 0$ in this case, as the line joining the source to the screen is normal to the screen. Hence, since $\cos 0° = 1$, $E = I/d^2$. Similarly, the intensity of illumination, E_1, at the screen due to the source I_1 is given by $E_1 = I_1/d_1{}^2$. Now the luminance of the surface $B = r_1 E_1$, where r_1 is the reflection factor of the surface (p. 568); and the luminance of the surface $A = rE$, where r is the reflection factor of this surface. Hence, for a photometric balance,

$$r_1 E_1 = rE$$

$$\therefore \quad \frac{r_1 I_1}{d_1{}^2} = \frac{rI}{d^2} \quad \cdot \qquad \cdot \qquad \cdot \qquad \cdot \qquad \cdot \qquad \text{(i)}$$

If the reflection factors r_1, r of A, B are equal, equation (i) becomes

$$\frac{I_1}{d_1{}^2} = \frac{I}{d^2}$$

$$\therefore \quad \frac{I_1}{I} = \frac{d_1{}^2}{d^2}$$

The ratio of the intensities are hence proportional to the squares of the corresponding distances of the sources from the screen.

The reflection factors r_1, r are not likely to be exactly equal, however, in which case another or *auxiliary lamp* is required to compare the candle-powers I_1, I. The auxiliary lamp, I_2, is placed on the right side, say, of the screen at a distance d_2, and *one* of the other lamps is placed on the other side. A photometric balance is then obtained,

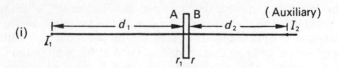

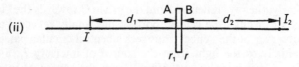

FIG. 24.17. Comparison of luminous intensities.

Fig. 24.17 (i). In this case, if I_1 is the intensity of the lamp and d_1 is its distance from the screen,

$$r_1 \frac{I_1}{d_1^2} = r \frac{I_2}{d_2^2} \quad \cdots \cdots \cdots \quad \text{(ii)}$$

The remaining lamp I is then used instead of the lamp I_1, and a photometric balance is again obtained by moving this lamp, *keeping the position of the lamp I_2 unaltered*, Fig. 24.17 (ii). Suppose the distance of the lamp I from the screen is d. Then

$$r_1 \frac{I}{d^2} = r \frac{I_2}{d_2^2} \quad \cdots \cdots \cdots \quad \text{(iii)}$$

From (ii) and (iii), it follows that

$$r_1 \frac{I_1}{d_1^2} = r_1 \frac{I}{d^2}$$

$$\therefore \quad \frac{I_1}{d_1^2} = \frac{I}{d^2}$$

$$\therefore \quad \frac{I_1}{I} = \frac{d_1^2}{d^2} \quad \cdots \cdots \cdots \quad \text{(13)}$$

The intensities I_1, I are hence proportional to the squares of the corresponding lamp distances from the screen. It should be noted from (13) that the auxiliary lamp's intensity is not required in the comparison of I_1 and I, nor is its constant distance d_2 from the screen required.

Measurement of Illumination

It was pointed out at the beginning of the chapter that the maintenance of standards of illumination plays an important part in safeguarding our health. It is recommended that desks in class-rooms and offices should have an illumination of 60–110 lux, and workshops an illumination of 120–180 lux; for sewing dark materials an intensity of 180–300 lux is recommended, while 1200 lux is suggested for the operating table in a hospital.

Photovoltaic cell. There are two types of meters for measuring the illumination of a particular surface. A modern type is the photovoltaic cell, which may consist of a cuprous oxide and copper plate, made by

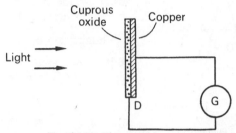

FIG. 24.18. Photovoltaic cell.

oxidising one side of a copper disc D, Fig. 24.18. When the oxide surface is illuminated, electrons are emitted from the surface whose number is proportional to the incident luminous energy, and a current flows in the microammeter or sensitive moving-coil galvanometer G which is proportional to the illumination. The galvanometer is previously calibrated by placing a standard lamp at known distances from the disc D, and its scale reads lux directly. Fig. 24.19 illustrates an "AVO Lightmeter", which operates on this principle; it is simply laid on the surface whose illumination is required and the reading is then taken.

FIG. 24.19. A V O light-meter.

Illumination due to plane mirror.
Consider a small lamp of I candela at O in front of a plane mirror M, Fig. 24.20. The flux F from O in a small cone OBC of solid angle ω is $I\omega$, and falls on an area A of the mirror. This flux is reflected to illuminate an area A_1, or HK, on a screen S in front of the mirror. Assuming the reflection factor of the mirror is unity,

illumination of S, $E = \dfrac{F}{A_1} = \dfrac{I\omega}{A_1}.$

But $\omega = $ area A/d^2, where d is distance of O from the mirror.

$$\therefore \quad E = \frac{IA}{d^2 A_1}.$$

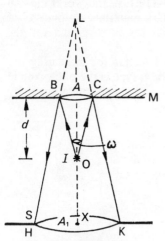

FIG. 24.20. Plane mirror illumination

But $A/A_1 = d^2/LX^2$, from similar triangles LBC, LHK.

$$\therefore \quad E = \frac{I}{LX^2} \quad . \qquad . \qquad . \qquad . \qquad . \qquad . \qquad \text{(i)}$$

The image of O in the plane mirror is at L, which is at a distance LX from the screen S. It follows from (i) that, owing to light reflected from the mirror, the illumination of S is the same as that obtained by a lamp of I candela at the position of the image of O.

EXAMPLE

Define *lumen* and *lux*, and show how they are related. Describe and explain how you would make an accurate comparison of the illuminating powers of two lamps of the same type. A photometric balance is obtained between two lamps A and B when B is 100 cm from the photometer. When a block of glass G is placed between A and the photometer, balance is restored by moving B through 5 cm. Where must B be placed in order to maintain the balance when *two more* blocks, identical with G, are similarly placed between A and the photometer. (*L.*)

First part. See text.

Second part. Suppose I_1, I_2 are the intensities of A, B respectively, and d is the distance of A from the photometer P. Then, originally,

$$\frac{I_1}{d^2} = \frac{I_2}{100^2} \quad . \qquad . \qquad . \qquad . \qquad . \qquad \text{(i)}$$

When G is placed in position, the balance is restored by moving B 105 cm. from P, Fig. 24.21 (i). If t is the transmission factor of G, the effective intensity of A is tI_1, and hence

$$\frac{t I_1}{d^2} = \frac{I_2}{105^2} \quad . \qquad . \qquad . \qquad . \qquad . \qquad \text{(ii)}$$

Dividing (ii) by (i),

$$\therefore \quad . \; t = \frac{100^2}{105^2} \quad . \qquad . \qquad . \qquad . \qquad . \qquad \text{(iii)}$$

When two more blocks are placed beside G, the effective intensity of $A = t \times t \times tI_1 = t^3I_1$, Fig. 24.21 (ii). Thus if the distance of B from P is now x.

$$\frac{t^3I_1}{d^2} = \frac{I_2}{x^2} \qquad . \qquad . \qquad . \qquad . \qquad . \qquad \text{(iv)}$$

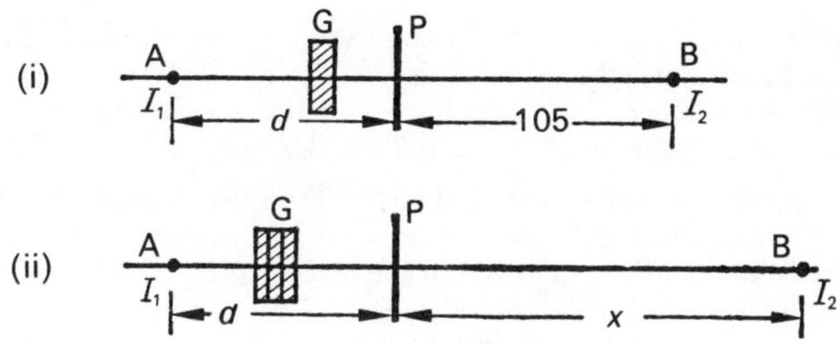

Fig. 24.21. Example

Dividing (iv) by (i)

$$\therefore \quad t^3 = \frac{100^2}{x^2}$$

From (iii)

$$\therefore \quad \left(\frac{100^2}{105^2}\right)^3 = \frac{100^2}{x^2}$$

$$\therefore \quad x^2 = \frac{100^2 \times 105^6}{100^6} = \frac{105^6}{100^4}$$

$$\therefore \quad x = \frac{105^3}{100^2} = 116 \text{ cm.}$$

EXERCISES 24

Velocity of Light

1. In Fizeau's rotating wheel experiment the number of teeth was 720 and the distance between the wheel and reflector was 8633 metres. Calculate the number of revolutions per second of the wheel when extinction first occurs, assuming the velocity of light is $3 \cdot 13 \times 10^8$ metres per second. What are the disadvantages of Fizeau's method?

2. Draw a diagram of Foucault's method of measuring the velocity of light. How has the velocity of light in water been shown to be less than in air? The radius of curvature of the curved mirror is 20 metres and the plane mirror is rotated at 20 revs per second. Calculate the angle in degrees between a ray incident on the plane mirror and then reflected from it after the light has travelled to the curved mirror and back to the plane mirror (velocity of light = 3×10^8 m s^{-1}).

3. Draw a diagram showing the arrangement of the apparatus and the path of the rays of light in Fizeau's toothed wheel method for measuring the velocity of light. What are the chief difficulties met with in carrying out the experiment?

If the wheel has 150 teeth and 150 spaces of equal width and its distance from the mirror be 12 kilometres, at what speed, in revolutions per minute, will the first eclipse occur? (N.)

4. Explain how the velocity of light was first determined, and describe one more recent method of measuring it.

A beam of light after reflection at a plane mirror, rotating 2000 times per minute, passes to a distant reflector. It returns to the rotating mirror from which it is reflected to make an angle of 1° with its original direction. Assuming that the velocity of light is 300 000 km s^{-1}, calculate the distance between the mirrors. (*L.*)

5. Describe and explain in detail one accurate terrestrial method for measuring the speed of light in air.

How does the speed of light depend on (*a*) the wavelength of the light, (*b*) the medium through which it travels? (*O. & C.*)

6. Explain why the velocity of light is difficult to measure by a direct terrestrial method; illustrate your answer with an estimate of the orders of magnitude of the quantities involved in the assessment.

Describe, with the aid of a labelled diagram, Michelson's method for the determination of the velocity of light.

What are the advantages of this method over the earlier one developed by Foucault? (*N.*)

7. Describe any *one* terrestrial method by which the velocity of light has been determined. Why is a knowledge of the value of this velocity so important?

A beam of light can be interrupted $(1 \cdot 0 \pm 0 \cdot 0002) \times 10^7$ times each second on passing through a certain crystal device. Such a beam is reflected from a distant mirror and returned through the same crystal. It is found that for certain positions of the mirror very little light emerges from the crystal a second time. Explain this. One such position of the mirror is known to be between 18 and 26 metres from the crystal. Calculate a more accurate value for this distance, assuming the velocity of light to be $(3 \cdot 0 \pm 0 \cdot 002) \times 10^{10}$ cm s^{-1}. Estimate the error in this calculated value. (*C.*)

8. Describe a method of measuring the speed of light. Explain precisely what observations are made and how the speed is calculated from the experimental data.

A horizontal beam of light is reflected by a vertical plane mirror A, travels a distance of 250 metres, is then reflected back along the same path and is finally reflected again by the mirror A. When A is rotated with constant angular velocity about a vertical axis in its plane, the emergent beam is deviated through an angle of 18 minutes. Calculate the number of revolutions per second made by the mirror.

If an atom may be considered to radiate light of wavelength 5000 Å for a time of 10^{-10} second, how many cycles does the emitted wave train contain? (*O. & C.*)

9. Give one reason why it is important to know accurately the velocity of light.

A glass prism whose cross-section is a regular polygon of n sides has its faces silvered and is mounted so that it can rotate about a vertical axis through the centre of the cross-section, with the silvered faces vertical. An intense narrow horizontal beam of light from a small source is reflected from one face of the prism to a small distant mirror and back to the same face, where it is reflected and used to form an image of the source which can be observed. Describe the behaviour of this image if the prism is set into rotation and slowly accelerated to a high speed. If the distance between the mirrors is D and the velocity of light is c, find an expression for the values of the angular velocity of the prism at which the image will be formed in its original position.

If $c = 3 \times 10^8$ m s^{-1}, $D = 30$ km and the maximum safe speed for the mirror is 700 rev per second, what would be a suitable value for n? (*O. & C.*)

10. Describe a terrestrial method by which the velocity of light has been measured. How could the method be modified to show that the velocity of light in water is less than that in air? Briefly discuss the theoretical importance of this fact. (*C*.)

11. Describe a terrestrial method of determining the velocity of light in air, explaining (without detailed calculation) how the result is obtained.

Why do we conclude that in free space red and blue light travel with the same speed, but that in glass red light travels faster than blue? (*L*.)

Photometry

12. A 30 candelas (cd) lamp X is 40 cm in front of a photometer screen. What is the illumination directly in front of X on the screen? Calculate the cd of a lamp Y which provides the same illumination when placed 60 cm from the screen.

13. A lamp of 800 cd is suspended 16 m above a road. Find the illumination on the road (i) at a point A directly below the lamp, (ii) at a point B 12 m from A.

14. Define the terms *luminous intensity* and *illumination*. Describe an accurate method of comparing the luminous intensities of two sources of light.

A lamp is fixed 4 m above a horizontal table. At a point on the table 3 m to one side of the vertical through the lamp, a light-meter is placed flat on the table. It registers 4 m candles. Calculate the intensity of the lamp. (*C*.)

15. What is meant by *luminous intensity* and *illumination*? How are they related to each other?

A small source of 32 cd giving out light equally in all directions is situated at the centre of a sphere of 8 m diameter, the inner surface of which is painted black. What is the illumination of the surface?

If the inner surface is repainted with a matt white paint which causes it to reflect diffusely 80 per cent of all lighting falling on it, what will the illumination be? (*L*.)

16. Describe an accurate form of photometer for comparing the luminous intensities of lamps.

A lamp is 100 cm from one side of a photometer and produces the same illumination as a second lamp placed at 120 cm on the opposite side. When a lightly smoked glass plate is placed before the weaker lamp, the brighter one has to be moved 50 cm to restore the equality of illumination. Find what fraction of the incident light is transmitted by the plate. (*L*.)

17. How would you compare the luminous intensities of two small lamps?

A small 100 cd lamp is placed 10 m above the centre of a horizontal rectangular table measuring 6 m by 4 m. What are the maximum and minimum values of the illumination on the table due to direct light?

How would your results be changed by the presence of a large horizontal mirror, placed 2 m above the lamp, so as to reflect light down on to the table, assuming that only 80 per cent of the light incident on the mirror is reflected? (*W*.)

18. Describe one form of a photometer, and explain how you would measure the light loss which results from enclosing a light source by a glass globe. Two small 16 cd lamps are placed on the same side of a screen at distances of 2 and 5 m from it. Calculate the distance at which a single 32 cd lamp must be placed in order to give the same intensity of illumination on the screen. (*N.*)

19. Define *lumen*, *metre-candle*. Describe the construction and use of a Lummer-Brodhun photometer.

Twenty per cent of the light emitted by a source of 500 cd is evenly distributed over a circular area 5 m in diameter. What is the illumination at points within this area? (*L.*)

20. Distinguish between *luminous intensity* and *illumination*. How may the two sources be accurately compared?

A surface receives light normally from a source at a distance of 3 m. If the source is moved closer until the distance is only 2 m, through what angle must the surface be turned to reduce the illumination to its original value? (*O. & C.*)

21. Define *illumination of a screen*, *luminous intensity of a source of light*. Indicate units in which each of these quantities may be measured.

Describe a reliable photometer, and explain how you would use it to compare the reflecting powers of plaster of Paris and ground glass. (*C.*)

22. Describe an accurate form of photometer for comparing the luminous intensities of two sources of light.

Two electric lamps, A and B, are found to give equal illuminations on the two sides of a photometer when their distances from the photometer are in the ratio 4 : 5. A sheet of glass is then placed in front of B, and it is found that equality of illumination is obtained when the distances of A and B are in the ratio 16 : 19. Find the percentage of light transmitted by the glass. (*C.*)

WAVES AND SOUND

chapter twenty-five

Oscillations and Waves. Sound Waves

In this chapter we shall study the properties of oscillations and waves in general. Topics in waves which concern particular branches of the subject are discussed elsewhere in this book. We begin with a summary of the results relating to simple harmonic motion, which is the simplest form of oscillatory motion.

S.H.M.

Simple harmonic motion (S.H.M.) occurs when the force acting on an object or system is directly proportional to its displacement x from a fixed point and is always directed towards this point. If the object executes S.H.M., then the variation of the displacement x with time t can be written as

$$x = a \sin \omega t. \qquad . \qquad . \qquad . \qquad . \quad (1)$$

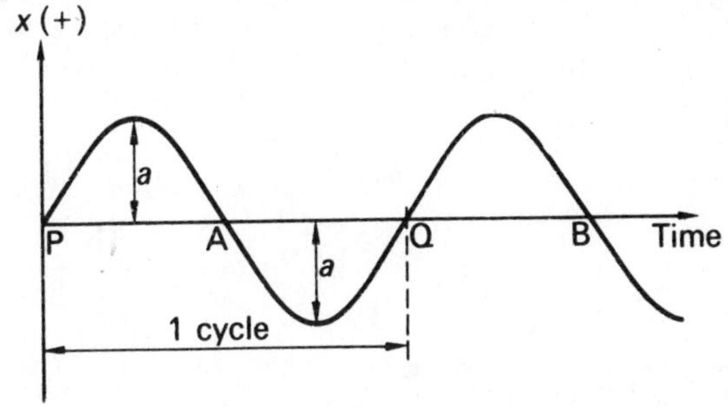

Fig. 25.1 Sine curve.

Here a is the greatest displacement from the mean or equilibrium position; a is the *amplitude*, Fig. 25.1. The constant ω is the 'angular frequency', and $\omega = 2\pi f$ where f is the *frequency* of vibration or number of cycles per second. The period T of the motion, or time to undergo one complete cycle is equal to $1/f$, so that $\omega = 2\pi/T$.

The small oscillation of a pendulum bob or vibrating layer of air is a *mechanical oscillation*, so that x is a displacement from a mean fixed position. Later, *electrical oscillations* are considered. x may then represent the instantaneous charge on the plates of a capacitor when the charge alternates about a mean value of zero. In an *electromagnetic wave*, x may represent the component of the electric or magnetic field vectors at a particular place.

Energy in S.H.M.

A calculation shows that the sum of the potential and kinetic energies of a body moving with S.H.M. is *constant* and equal to the total energy in the vibration. Further, it can be shown that the time averages of the potential energy (P.E.) and kinetic energy (K.E.) are equal; each is half the total energy. In any mechanical oscillation, *there is a continuous interchange or exchange of energy from P.E. to K.E. and back again.*

For vibrations to occur, therefore, an agency is required which can possess and store P.E. and another which can possess and store K.E. This is the case for a mass oscillating on the end of a spring which undergoes S.H.M. The mass stores K.E. and the spring stores P.E.; and interchange occurs continuously from one to the other as the spring is compressed and released alternately. In the oscillations of a simple pendulum, the mass stores K.E. as it swings downwards from the end of an oscillation, and this is changed to P.E. as the height of the bob increases above its mean position.

Note that some agency is needed to accomplish the transfer of energy. In the case of the mass and spring, the force in the spring causes the transfer. In the case of the pendulum, the component of the weight along the arc of the circle causes the change from P.E. to K.E.

Electrical Oscillations

So far we have dealt with mechanical oscillations and energy. The energy in electrical oscillations takes a different form. There are still two types of energy. One is the energy stored in the electric field, and the other that stored in the magnetic field. To obtain electrical oscillations,

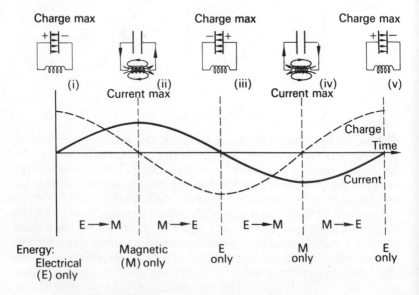

FIG. 25.2   Electrical oscillations – energy exchanges.

an inductor (coil) is used to produce the magnetic field and a capacitor to produce the electric field (see also p. 925).

Suppose the capacitor is charged and there is no current at this moment, Fig. 25.2 (i). A p.d. then exists across the capacitor and an electric field is present between the plates. At this instant all the energy is stored in the electric field, and since the current is zero there is no magnetic energy. Because of the p.d. a current will begin to flow and magnetic energy will begin to be stored in the inductor. Thus there will be a change from electric to magnetic energy. The p.d. is the agency which causes the transfer of energy.

One quarter of a cycle later the capacitor will be fully discharged and the current will be at its greatest, so that the energy is now entirely stored in the magnetic field, Fig. 25.2 (ii). The current continues to flow for a further quarter-cycle until the capacitor is fully charged in the opposite direction, when the energy is again completely stored in the electric field, Fig. 25.2 (iii). The current then reverses and the processes occur in reverse order, Fig. 25.2 (iv), after which the original state is restored and a complete oscillation has taken place, Fig. 25.2 (v). The whole process then repeats, giving continuous oscillations.

Phase of vibrations

Consider an oscillation given by $x_1 = a \sin \omega t$. Suppose a second oscillation has the same amplitude, a, and angular frequency, ω, but reaches the end of its oscillation a fraction, β, of the period T later than the first one. The second oscillation thus *lags behind* the first by a time βT, and so its displacement x_2 is given by

$$x_2 = a \sin \omega(t - \beta T)$$
$$= a \sin (\omega t - \varphi), \qquad . \qquad . \qquad . \qquad (2)$$

where $\varphi = \omega \beta T = 2\pi \beta T / T = 2\pi \beta$. If the second oscillation *leads* the first by a time βT, the displacement is given by

$$x_2 = a \sin (\omega t + \varphi). \qquad . \qquad . \qquad . \qquad (3)$$

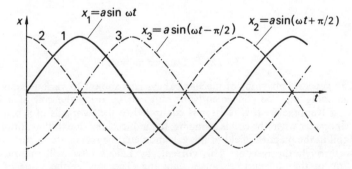

FIG. 25.3   Phase difference.

φ is known as the *phase angle* of the oscillation. It represents the *phase difference* between the oscillations $x_1 = a \sin \omega t$ and $x_2 = a \sin (\omega t - \varphi)$. Graphs of displacement v. time are shown in Fig. 25.3.

Curve 1 represents $x_1 = a \sin \omega t$. Curve 2 represents $x_2 = a \sin (\omega t + \pi/2)$, so that its phase lead is $\pi/2$; this is a lead of one quarter of a period. Curve 3 represents $x_3 = a \sin (\omega t - \pi/2)$ so that its phase lag is $\pi/2$; this is a lag of one quarter of a period on curve 1. If the phase difference is 2π, the oscillations are effectively in phase.

Note that if the phase difference is π, the displacement of one oscillation reaches a positive maximum value at the same instant as the other oscillation reaches a *negative* maximum value. The two oscillations are thus sometimes said to be 'antiphase'.

Damped Vibrations

In practice, the amplitude of vibration in simple harmonic motion does not remain constant but becomes progressively smaller. Such a vibration is said to be *damped*. The diminution of amplitude is due to loss of energy; for example, the amplitude of the bob of a simple pendulum diminishes slowly owing to the viscosity (friction) of the air. This is shown by curve 1 in Fig. 25.4.

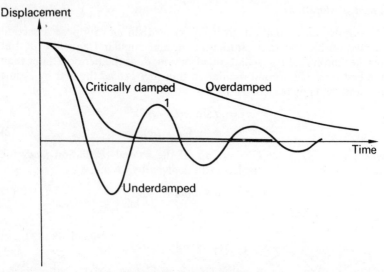

FIG. 25.4   Damped motion.

The general behaviour of mechanical systems subject to various amounts of damping may be conveniently investigated by using the coil of a ballistic galvanometer. If a resistor is connected to the terminals of a ballistic galvanometer when the coil is swinging, the induced emf due to the motion of the coil in the magnetic field of the galvanometer magnet causes a current to flow through the resistor. This current, by Lenz's Law, will oppose the motion of the coil and so causes damping. The smaller the value of the resistor, the greater is the degree of damping. The galvanometer coil is set swinging by discharging a capacitor through it. The time period, and the time taken for the amplitude to be reduced to a certain fraction of its original value, are then measured. The experiment can then be repeated using different values of resistor connected to the terminals.

It is found that as the damping is increased the time period increases and the oscillations die away more quickly. As the damping is increased further there is a value of resistance which is just sufficient to prevent the coil from vibrating past its rest position. This degree of damping, called the *critical damping, reduces the motion to rest in the shortest possible time*. If the resistance is lowered further, to increase the damping, no vibrations occur but the coil takes a longer time to settle down to its rest position. Graphs showing the displacement against time for 'underdamped', 'critically damped', and 'over-damped' motion are shown in Fig. 25.4.

When it is required to use a galvanometer as a current-measuring instrument, rather than ballistically to measure charge, it is generally critically damped. The return to zero is then as rapid as possible.

These results, obtained for the vibrations of a damped galvanometer coil, are quite general. All vibrating systems have a certain critical damping, which brings the motion to rest in the shortest possible time.

Forced Oscillations. Resonance

In order to keep a system, which has a degree of damping, in continuous oscillatory motion, some external periodic force must be used. The frequency of this force is called the *forcing frequency*. In order to see how systems respond to a forcing oscillation, we may use an electrical circuit comprising a coil L, capacitor C and resistor R, shown in Fig. 25.5.

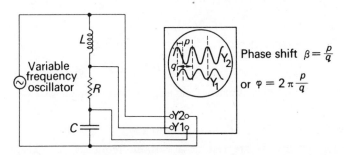

FIG. 25.5   Demonstration of oscillations.

The applied oscillating voltage is displayed on the Y_2 plates of a double-beam oscilloscope (p. 1011). The voltage across the resistor R is displayed on the Y_1 plates. Since the current I through the resistor is given by $I = V/R$, the voltage across R is a measure of the current through the circuit. The frequency of the oscillator is now set to a low value and the amplitude of the Y_1 display is recorded. The frequency is then increased slightly and the amplitude again measured. By taking many such readings, a graph can be drawn of the current through the circuit as the frequency is varied. A typical result is shown in Fig. 25.6 (i).

The phase difference, φ, between the Y_1 and Y_2 displays can be found by measuring the horizontal shift p between the traces, and the length q occupied by one complete waveform. φ is given by $(p/q) \times 2\pi$. A graph of the variation of phase difference between current and applied voltage can then be drawn. Fig. 25.6 (ii) shows a typical curve.

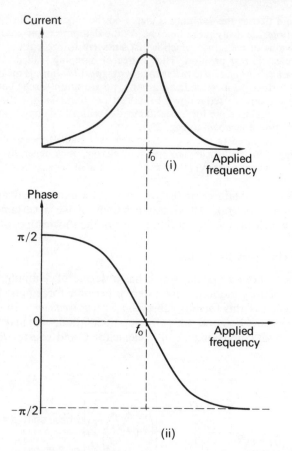

FIG. 25.6   Amplitude and phase in forced vibrations.

The following observations may be made:

1. The current is greatest at a certain frequency f_0. This is the frequency of undamped oscillations of the system, when it is allowed to oscillate on its own. f_0 is called the *natural frequency* of the system. When the forcing frequency is equal to the natural frequency, *resonance* is said to occur. The largest current is then produced.

2. At resonance, the current and voltage are in phase. Well below resonance, the current leads the voltage by $\pi/2$; at very high frequencies the current lags by $\pi/2$. The behaviour of other resonant systems is similar.

3. The forced oscillations always have the same frequency as the forcing oscillations.

Examples of resonance occur in sound and in optics. These are discussed later. It should be noted that considerable energy is absorbed at the resonant frequency from the system supplying the external periodic force.

Waves and Wave-motion

A *wave* allows energy to be transferred from one point to another some distance away without any particles of the medium travelling between the two points. For example, if a small weight is suspended by a string, energy to move the weight may be obtained by repeatedly shaking the other end of the string up and down through a small distance. Waves, which carry energy, then travel along the string from the top to the bottom. Likewise, water waves may spread along the surface from one point A to another point B, where an object floating on the water will be disturbed by the wave. No particles of water at A actually travel to B in the process. The energy in the electromagnetic spectrum, comprising X-rays and light waves, for example, may be considered to be carried by electromagnetic waves from the radiating body to the absorber. Again, sound waves carry energy from the source to the ear by disturbance of the air (p. 585).

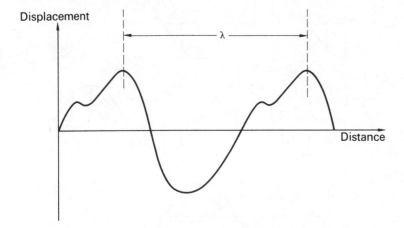

FIG. 25.7   Wave and wavelength.

If the source or origin of the wave oscillates with a frequency f, then each point in the medium concerned oscillates with the same frequency. A snapshot of the wave profile or waveform may appear as in Fig. 25.7 at a particular instant. The source repeats its motion f times per second, so a repeating *waveform* is observed spreading out from it. The distance between corresponding points in successive waveforms, such as two successive crests or two successive troughs, is called the *wavelength*, λ. Each time the source vibrates once, the waveform moves forward a distance λ. Thus in one second, when f vibrations occur, the wave moves forward a distance $f\lambda$. Hence the velocity v of the waves, which is the distance the profile moves in one second, is given by:

$$v = f\lambda.$$

This equation is true for all wave motion, whatever its origin, that is, it applies to sound waves, electromagnetic waves and mechanical waves.

Transverse Waves

A wave which is propagated by vibrations *perpendicular* to the direction of travel of the wave is called a *transverse* wave. Examples of transverse waves are waves on plucked strings and on water. Electromagnetic waves, which include light waves, are transverse waves.

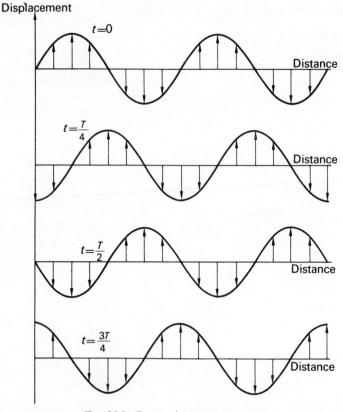

FIG. 25.8   Progressive transverse wave.

The propagation of a transverse wave is illustrated in Fig. 25.8. Each particle vibrates perpendicular to the direction of propagation with the same amplitude and frequency, and the wave is shown successively at $t = 0$, $T/4$, $T/2$, $3T/4$, in Fig. 25.8, where T is the period.

Longitudinal Waves

In contrast to a transverse wave, a *longitudinal* wave is one in which the vibrations occur in the *same* direction as the direction of travel of the wave. Fig. 25.9 illustrates the propagation of a longitudinal wave. The row of dots shows the actual positions of the particles whereas the *graph* shows the *displacement* of the particles from their equilibrium positions. The positions at time $t = 0$, $t = T/4$, $t = T/2$ and $t = 3T/4$

are shown. The diagram for $t = T$ is, of course, the same as $t = 0$. With displacements to R (right) and to L (left), note that:

(i) The displacements of the particles cause regions of high density (*compressions* C) and of low density (*rarefactions* R) to be formed.

(ii) These regions move along with the speed of the wave, as shown by the broken diagonal line.

(iii) Each particle vibrates about its mean position with the same amplitude and frequency.

(iv) The regions of greatest compression are one-quarter wavelength ahead of the greatest displacement in the direction of the wave. This result is important in understanding some processes involving sound waves.

The most common example of a longitudinal wave is a sound wave. This is propagated by alternate compressions and rarefactions of the air.

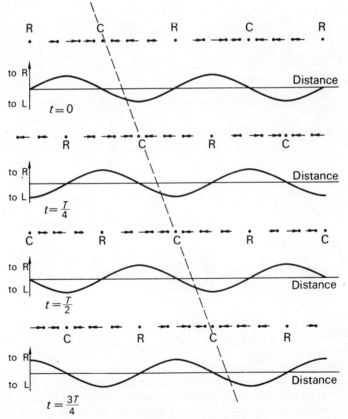

FIG. 25.9   Progressive longitudinal wave.

Progressive Waves

Both the transverse and longitudinal waves described above are *progressive*. This means that the wave profile moves along with the speed of the wave. If a snapshot is taken of a progressive wave, it repeats at equal distances. The repeat distance is the *wavelength* λ. If one point is taken, and the profile is observed as it passes this point, then the profile is seen to repeat at equal intervals of time. The repeat time is the *period, T*.

The vibrations of the particles in a progressive wave are of the same amplitude and frequency. But *the phase of the vibrations changes for different points along the wave*. This can be seen by considering Figs. 25.8 and 25.9. The phase difference may be demonstrated by the following experiment.

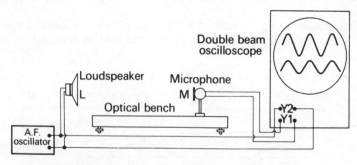

FIG. 25.10 Demonstration of phase in progressive wave.

An audio-frequency (af) oscillator is connected to the loudspeaker L and to the Y_2 plates of a double-beam oscilloscope, Fig. 25.10. A microphone M,

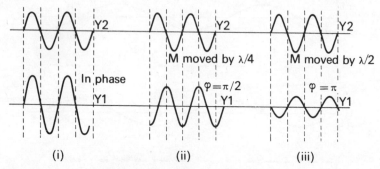

FIG. 25.11   Phase difference and wavelength.

mounted on an optical bench, is connected to the Y_1 plates. When M is moved away from or towards L, the two traces on the screen are as shown in Fig. 25.11 (i) at one position. This occurs when the distance LM is equal to a whole number of wavelengths, so that the signal received by M is in phase with that sent out by L. When M is now moved further away from L through a distance $\lambda/4$, where λ is the wavelength, the appearance on the screen changes to that shown in Fig. 25.11 (ii). The resultant phase change is $\pi/2$, so that the signal now arrives a quarter of a period later. When M is moved a distance $\lambda/2$ from its 'in-phase' position, the signal arrives half a period later, a phase change of π, Fig. 25.11 (iii).

Velocity of Sound in Free Air

The velocity of sound in free air can be found from this experiment. Firstly, a position of the microphone M is obtained when the two signals on the screen are in phase, as in Fig. 25.10. The reading of the position of M on the optical bench is then taken. M is now moved slowly until the phase of the two signals on the screen is seen to change through $\pi/2$ to π and then to be in phase again. The shift of M is then measured. It is equal to λ, the wavelength. From several measurements the average value of λ is found, and the velocity of sound is calculated from $v = f\lambda$, where f is the frequency obtained from the oscillator dial.

Progressive Wave Equation

An equation can be formed to represent generally the displacement y of a vibrating particle in a medium which a *wave* passes. Suppose the wave moves from left to right and that a particle at the origin O then vibrates according to the equation $y = a \sin \omega t$, where t is the time and $\omega = 2\pi f$ (p. 577).

At a particle P at a distance x from O to the right, the phase of the

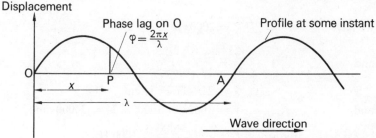

FIG. 25.12   Progressive wave equation.

vibration will be different from that at O, Fig. 25.12. A distance λ from O corresponds to a phase difference of 2π (p. 580). Thus the phase difference φ at P is given by $(x/\lambda) \times 2\pi$ or $2\pi x/\lambda$. Hence the displacement of any particle at a distance x from the origin is given by

$$y = a \sin (\omega t - \varphi)$$

or
$$y = a \sin \left(\omega t - \frac{2\pi x}{\lambda} \right). \qquad . \qquad . \qquad . \qquad (4)$$

Since $\omega = 2\pi f = 2\pi v/\lambda$, where v is the velocity of the wave, this equation may be written:

$$y = a \sin \left(\frac{2\pi vt}{\lambda} - \frac{2\pi x}{\lambda} \right)$$

or
$$y = a \sin \frac{2\pi}{\lambda} (vt - x). \qquad . \qquad . \qquad . \qquad (5)$$

Also, since $\omega = 2\pi/T$, equation (4) may be written:

$$y = a \sin 2\pi \left(\frac{t}{T} - \frac{x}{\lambda} \right). \qquad . \qquad . \qquad . \qquad (6)$$

Equations (5) or (6) represent a *plane-progressive wave*. The negative sign in the bracket indicates that, since the wave moves from left to right, the vibrations at points such as P to the right of O will lag on that at O. A wave travelling in the *opposite direction*, from right to left, arrives at P before O. Thus the vibration at P leads that at O. Consequently a wave travelling in the opposite direction is given by

$$y = a \sin 2\pi\left(\frac{t}{T} + \frac{x}{\lambda}\right), \quad . \quad . \quad . \quad . \quad (7)$$

that is, the sign in the bracket is now a plus sign.

As an illustration of calculating the constants of a wave, suppose a wave is represented by

$$y = a \sin\left(2000\pi t - \frac{\pi x}{17}\right),$$

where t is in seconds, y in cm. Then, comparing it with equation (5),

$$y = a \sin\frac{2\pi}{\lambda}(vt - x),$$

we have $\dfrac{2\pi v}{\lambda} = 2000\pi,$

and $\dfrac{2\pi}{\lambda} = \dfrac{\pi}{17}.$

$$\therefore \lambda = 2 \times 17 = 34 \text{ cm}$$

and $v = 1000\lambda = 1000 \times 34 = 34000 \text{ cm s}^{-1}$

$$\therefore \text{ frequency}, f, = \frac{v}{\lambda} = \frac{34000}{34} = 1000 \text{ Hz}$$

$$\therefore \text{ period}, T, = \frac{1}{f} = \frac{1}{1000} \text{ s.}$$

If two layers of the wave are 180 cm apart, they are separated by 180/34 wavelengths, or by $5\frac{10}{34}\lambda$. Their *phase difference* for a separation λ is 2π; and hence, for a separation $10\lambda/34$, omitting 5λ from consideration, we have:

$$\text{phase difference} = \frac{10}{34} \times 2\pi = \frac{10\pi}{17} \text{ radians.}$$

Principle of Superposition

When two waves travel through a medium, their combined effect at any point can be found by the *Principle of Superposition*. This states that *the resultant displacement at any point is the sum of the separate displacements due to the two waves*.

The principle can be illustrated by means of a long stretched spring ('Slinky'). If wave pulses are produced at each end simultaneously, the two waves pass through the wire. Fig. 25.13(a) shows the stages which occur as the two pulses pass each other. In Fig. 25.13(a)(i), they are some distance apart and are approaching each other, and in Fig. 25.13(a)(ii)

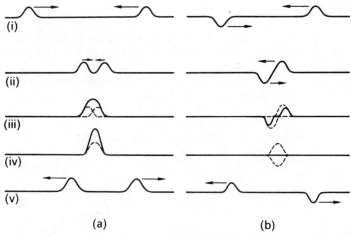

FIG. 25.13 Superposition of waves.

they are about to meet. In Fig. 25.13(a) (iii), the two pulses, each shown by broken lines, are partly overlapping. The resultant is the sum of the two curves. In Fig. 25.13(a) (iv), the two pulses exactly overlap and the greatest resultant is obtained. The last diagram shows the pulses receding from one another. The diagrams in Fig. 25.13(b) show the same sequence of events (i)–(v) but the pulses are equal and opposite. The Principle of Superposition is widely used in discussion of wave phenomena such as interference, as we shall see (p. 688).

Stationary Waves

We have already discussed progressive waves and their properties. Fig. 25.14 shows an apparatus which produces a different kind of wave (see also p. 662). If the weights on the scale-plan are suitably adjusted,

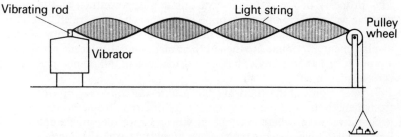

FIG. 25.14 Demonstration of stationary wave.

a number of *stationary vibrating loops* are seen on the string when one end is set vibrating. This time the wave-like profile on the string does *not* move along the medium, which is the string, and the wave is therefore called a *stationary* (or *standing*) wave.

The motion of the string when a stationary wave is produced can be seen by using a Xenon stroboscope (strobe). This instrument gives a flashing light whose frequency can be varied. The apparatus is set up in a darkened room and illuminated with the strobe. When the frequency of the strobe is nearly equal to that of the string, the string can be seen moving up and down slowly. Its observed frequency is equal to the difference between the frequency of the strobe and that of the string. Progressive stages in the motion of the string can now be seen and studied, and these are illustrated in Fig. 25.15.

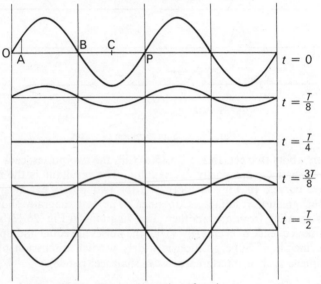

FIG. 25.15   Changes in motion of stationary wave.

The following points should be noted:

1. There are points such as B where the displacement is permanently zero. These points are called *nodes* of the stationary wave.

2. At points between successive nodes the vibrations *are in phase*. This property of the stationary wave is in sharp contrast to the progressive wave, where the phase of points near each other are all different (see p. 585). Thus when one point of a stationary wave is at its maximum displacement, *all* points are then at their maximum displacement. When a point (other than a node) has zero displacement, *all* points then have zero displacement.

3. Each point along the wave has a *different amplitude* of vibration from neighbouring points. Again this is different from the case of a progressive wave, where every point vibrates with the same amplitude. Points, e.g. C, which have the greatest amplitude are called *antinodes*.

4. The wavelength is equal to the distance OP, Fig. 25.15. Thus the wavelength λ, is twice the *distance between successive nodes or successive antinodes*. The distance between successive nodes or antinodes is $\lambda/2$; the distance between a node and a neighbouring antinode is $\lambda/4$.

Examples of stationary waves are discussed later in the book. The reader is referred to p. 641 for discussion of stationary waves in sound and to p. 768 for stationary electromagnetic waves.

Stationary Wave Equation

In deriving the wave equation of a progressive wave, we used the fact that the phase changes from point to point (p. 586). In the case of a stationary wave, we may find the equation of motion by considering the *amplitude* of vibration at each point because the amplitude varies while the phase remains constant.

Let ω be the angular frequency of the wave. The vibration of each particle may be represented by the equation

$$y = Y \sin \omega t, \qquad . \qquad . \qquad . \qquad . \qquad (8)$$

where Y is the amplitude of the vibration at the point considered. Y varies along the wave with the distance x from some origin. If we suppose the origin to be at an antinode, then the origin will have the greatest amplitude, A, say. Now the wave repeats at every distance λ, and it can be seen that the amplitudes at different points vary sinusoidally with their particular distance x. An equation representing the changing amplitude Y along the wave is thus:

$$Y = A \cos \frac{2\pi x}{\lambda} = A \cos kx, \qquad . \qquad . \qquad . \qquad (9)$$

where $k = 2\pi/\lambda$. When $x = 0$, $Y = A$; when $x = \lambda$, $Y = A$. When $x = \lambda/2$, $Y = -A$. This equation hence correctly describes the variation in amplitude along the wave, Fig. 25.15. Hence the equation of motion of a stationary wave is, with (8),

$$y = A \cos kx \cdot \sin \omega t. \qquad . \qquad . \qquad . \qquad (10)$$

From (10), $y = 0$ at all times when $\cos kx = 0$. Thus $kx = \pi/2$, $3\pi/2$, $5\pi/2$, ..., in this case. This gives values of x corresponding to $\lambda/4$, $3\lambda/4$, $5\lambda/4$, ... These points are *nodes* since the displacement at a node is always zero (p. 590). Thus equation (10) gives the correct distance, $\lambda/2$, between nodes.

A stationary wave can be considered as produced by the superposition of two progressive waves, of the same amplitude and frequency, travelling in opposite directions. This is shown mathematically on p. 643.

Wave Properties, Reflection and Refraction

Any wave motion can be *reflected*. The reflection of light and sound waves, for example, is discussed on pp. 677, 613 respectively.

Waves can also be *refracted*, that is, their direction changes when they enter a new medium. This is due to the change in velocity of the waves on entering a different medium. Refraction of light and sound, for example, is discussed later (pp. 679, 615).

Diffraction

Waves can also be 'diffracted'. *Diffraction* is the name given to the spreading of waves when they pass through apertures or around obstacles.

The general phenomenon of diffraction may be illustrated by using

water waves in a ripple tank, with which we assume the reader is
familiar. Fig. 25.16(i) shows the effect of widening the aperture and
Fig. 25.16(ii) the effect of shortening the wavelength and keeping the
same width of opening. In certain circumstances in diffraction, reinforcement of the waves, or complete cancellation occurs in particular
directions from the aperture, as shown in Fig. 25.16(i) and (ii). These
patterns are called 'diffraction bands' (p. 701).

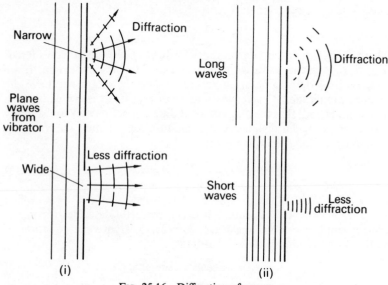

FIG. 25.16   Diffraction of waves.

Generally, the smaller the width of the aperture in relation to the
wavelength, the greater is the spreading or diffraction of the waves.
This explains why we cannot see round corners. The wavelength of *light
waves* is about 6×10^{-7} m (p. 690). This is so short that no appreciable
diffraction is obtained around obstacles of normal size. With very small
obstacles or narrow apertures, however, diffraction of light may be
appreciable (see p. 707). *Sound waves* have long wavelengths, for
example 50 cm, so that diffraction of sound waves occurs easily. For
this reason, it is possible to hear round corners. *Electromagnetic waves*
can be diffracted, as shown on p. 760.

Interference

When two or more waves of the same frequency overlap, the
phenomenon of *interference* occurs. Interference is easily demonstrated
in a ripple tank. Two sources, A and B, of the same frequency are used.
These produce circular waves which spread out and overlap, and the
pattern seen on the water surface is shown in Fig. 25.17.

The interference pattern can be explained from the Principle of
Superposition (p. 588). If the oscillations of A and B are in phase, crests
from A will arrive at the same time as crests from B at any point on
the line RS. Hence by the Principle of Superposition there will be

reinforcement or a large wave along RS. Along XY, however, crests from A will arrive before corresponding crests from B. In fact, every point on XY is half a wavelength, λ, nearer to A than to B, so that crests from A arrive at the same time as troughs from B. Thus, by the Principle of Superposition, the resultant is *zero*. At every point along PQ there is a $3\lambda/2$ path difference from A compared to that from B, so that the resultant is also zero along PQ.

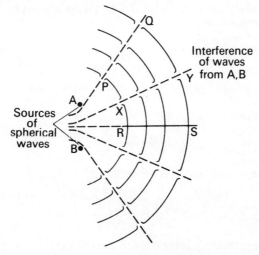

Fig. 25.17 Interference of waves.

Interference of *light waves* is discussed in detail on p. 687. An experiment to demonstrate the interference of *electromagnetic waves* (microwaves) is shown on p. 760. The interference of *sound waves* can be

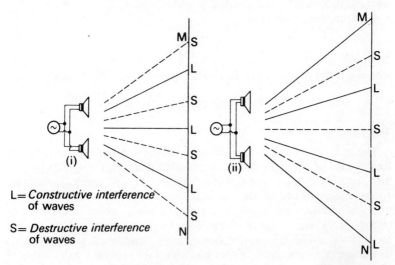

L = *Constructive interference* of waves

S = *Destructive interference* of waves

Fig. 25.18 Interference of sound waves.

demonstrated by connecting two loudspeakers in parallel to an audio-frequency oscillator, Fig. 25.18 (i). As the ear or microphone is moved along the line MN, alternate loud (L) and soft (S) sounds are heard according to whether the receiver of sound is on a line of reinforcement (constructive interference) or cancellation (destructive interference) of waves. Fig. 25.18 (i) indicates the positions of loud and soft sounds if the two speakers oscillate in phase. If the connections to *one* of the speakers is reversed, so that they oscillate out of phase, then the pattern is altered as shown in Fig. 25.18 (ii). The reader should try to account for this difference.

Velocity of Waves

We now list, for convenience, the velocity V of waves of various types, which are applied more fully in the appropriate sections of the book:

1. *Transverse wave on string*

$$V = \sqrt{\frac{T}{m}}, \qquad . \qquad . \qquad . \qquad . \quad (11)$$

where T is the tension and m is the mass per unit length.

2. *Sound waves in gas*

$$V = \sqrt{\frac{\gamma p}{\rho}}, \qquad . \qquad . \qquad . \qquad (12)$$

where p is the pressure, ρ is the density and γ is the ratio of the principal specific heat capacities of the gas.

3. *Longitudinal waves in solid*

$$V = \sqrt{\frac{E}{\rho}}, \qquad . \qquad . \qquad . \qquad (13)$$

where E is Young's modulus and ρ is the density.

4. *Electromagnetic waves*

$$V = \sqrt{\frac{1}{\mu \varepsilon}}, \qquad . \qquad . \qquad . \qquad (14)$$

where μ is the permeability and ε is the permittivity of the medium.

SOUND RECEPTION, REPRODUCTION, RECORDING

The Ear

The eye can detect colour changes, which are due to the different frequencies of the light waves; it can also detect variations in brightness, which are due to the different amounts of light energy it receives. In the sphere of sound, the ear is as sensitive as the eye; it can detect notes of different pitch, which are due to the different frequencies of sound waves, and it can also detect loud and soft notes, which are due to different amounts of sound energy falling on the ear per second.

We are not concerned in this book with the complete physiology of the ear. Among other features, it consists of the *outer ear*, A, a canal C

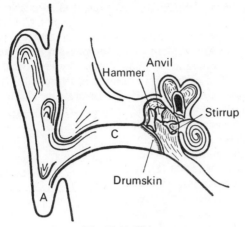

Fig. 25.19. The ear.

leading to the *drumskin*, and bones called the *ossicles*, Fig. 25.19. The ossicles consist of a bone called the 'hammer', fitting into another bone called the 'anvil', which is connected to the third bone called the 'stirrup'. When sound waves occur in the neighbourhood of the ear, they travel down the canal to the drumskin, which is also set into vibration. The part of the hammer in contact with the drumskin then vibrates, and strikes the anvil at the same rate. The motion is thus communicated to the stirrup, and from here it passes by a complicated mechanism to the auditory nerves, which set up the sensation of sound.

In 1843, OHM asserted that the ear perceives a simple harmonic vibration of the air as a simple or pure note. Although the waveforms of notes from instruments are far from being simple harmonic (see p. 610), the ear appears able to analyse a complicated waveform into the sum of a number of simple harmonic waves, which it then detects as separate notes.

Microphones

We now consider the principles of some *microphones*, instruments which convert sound energy to electrical energy. Details of microphones must be obtained from specialist works.

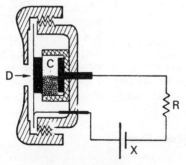

Fig. 25.20. Carbon microphone.

Carbon microphone, Fig. 25.20. This type of microphone is used in the hand set of a telephone. It contains carbon granules, C, whose electrical resistance decreases on compression and increases on release. Thus when sound waves are incident on the diaphragm D, a varying electric current of the same sound or audio frequency is produced along the telephone wires. The carbon microphone is a 'pressure' type since the magnitude of the current depends on the pressure changes in the air.

Ribbon microphone, Fig. 25.21. This is a sensitive microphone, with a uniform response over practically the whole of the audio-frequency (af) range from 40 to 15000 Hz. It has a corrugated aluminium ribbon R clamped between two pole-pieces N, S of a powerful magnet. When sound waves are incident on R, the ribbon vibrates perpendicular to the magnetic field. A varying induced emf of the same frequency is therefore obtained, as shown, and this is passed to an amplifier. The ribbon

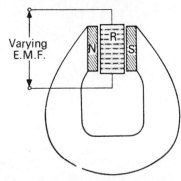

Fig. 25.21. Ribbon microphone.

microphone is a 'velocity' type, since the movement of R depends on the velocity changes in the air particles near it.

Moving-coil microphone, Fig. 25.22. This is also a sensitive microphone, widely used. A coil X, made from aluminium tape so that it is very light, is situated in the radial field between the pole-pieces N, S of a magnet M. When sound waves are incident on a diaphragm Y

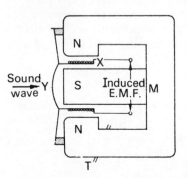

Fig. 25.22. Moving-coil microphone.

attached to X, the coil vibrates and 'cuts' magnetic flux. The varying induced emf obtained is passed to an amplifier. The moving-coil microphone is a pressure type. To avoid any effect of fluctuations in atmospheric pressure during use, the inside is kept at atmospheric pressure by means of a tube T.

The ribbon and moving-coil microphones are examples of *electrodynamic microphones*. They are widely used in the entertainment industry because their frequency response is extremely uniform. Unlike the carbon microphone, which has a relatively poor frequency response, no battery is used, and they are not subject to the background noise and 'hissing' obtained with a carbon microphone.

Loudspeaker. Telephone Earpiece

The *moving-coil loudspeaker* is used to reproduce sound energy from the electrical energy obtained with a microphone. It has a coil C or

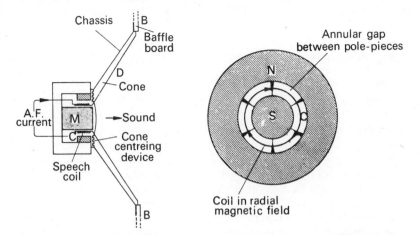

Fig. 25.23. Moving-coil loudspeaker.   Fig. 25.24. Speech coil in magnetic field.

speech coil, wound on a cylindrical former, which is positioned symmetrically in the radial field of a pot magnet M, Fig. 25.23. A thin cardboard cone D is rigidly attached to the former and loosely connected
to a large baffle-board B which surrounds it.

When C carries audio-frequency current, it vibrates at the same
frequency in the direction of its axis. This is due to the force on a current-
carrying conductor in a magnetic field, Fig. 25.24, whose direction is
given by Fleming's left-hand rule. Since the surface area of the cone is
large, the large mass of air in contact with it is disturbed and hence a
loud sound is produced.

When the cone moves forward, a
compression of air occurs in front of
the cone and simultaneously a rare-
faction behind it. The wave generated
behind the cone is then 180° out of
phase with that in front. If this wave
reaches the front of the cone quickly,
it will interfere appreciably with the
wave there. Hence the intensity of the
wave is diminished. This effect will be
more noticeable at low frequency, or
long wavelength, as the wave behind
then has time to reach the front before
the next vibration occurs. Generally,
then, the sound would lack low note

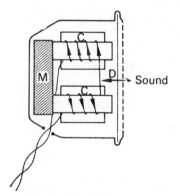

Fig. 25.25. Telephone earpiece.

or bass intensity. The large baffle reduces this effect appreciably. It
makes the path from the rear to the front so much longer that interference is negligible.

Telephone earpiece, Fig. 25.25. This has speech coils C wound round
soft-iron cores, which have a permanent magnet M between them. The
varying speech current produces a corresponding varying attraction on
the soft-iron diaphragm D, which thus generates sound waves of the

same frequency in the air. The sound is soft because the mass of air in contact with D is small. The permanent magnet is necessary to prevent distortion and to make the movement of D more sensitive to the varying attractive force.

Sound Recording and Reproduction

We now consider the principles of recording sound on tape and on film, and its reproduction. Details are beyond the scope of this book and must be obtained from manuals on the subject.

Tape recorder. Fig. 25.26 (i) illustrates the principle of tape recording in which flexible tape is used. It is coated with a fine uniform layer of a special form of ferric oxide which can be magnetised. The backing is a smooth plastic-base tape. When recording, the tape moves at a constant speed past the narrow gap between the poles of a ring of soft iron which has a coil round it. The coil carries the audio-frequency (af) current due to the sound recorded. On one half of a cycle, that part of the moving tape then in the gap is unmagnetised. On the other half of the same cycle, the next piece of tape in the gap is magnetised in the opposite direction, as shown. The rate at which pairs of such magnets is produced is equal to the frequency of the af current. The strength of the magnets is a measure of the magnetising current and hence of the intensity of the sound recorded.

In 'playback', the magnetised tape is now run at exactly the same speed past the same or another ring, the playback head, Fig. 25.26 (ii). As the small magnets pass the gap between the poles, the flux in the iron changes. An induced emf is thus obtained of the same frequency

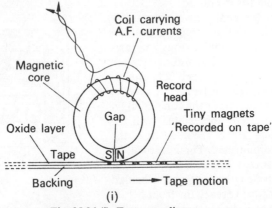

Fig. 25.26 (i). Tape recording.

and strength as that due to the original tape recording. This is amplified and passed to the loudspeaker, which reproduces the sound.

To obtain high-quality sound reproduction from tape, the output from a special high frequency *bias oscillator* is applied to the recording head in addition to the recording signal. This ensures that the magnets formed on the tape have strengths which are proportional to the recording signal. If the bias oscillator was not used, severe distortion due to

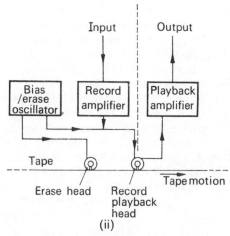

Fig. 25.26 (ii). Sound reproduction.

non-linearity would occur. The bias oscillator can also be used to *erase* the recording on the tape. This is done by applying its output to a coil in a special *erase head*. The erase head is similar to the record head but has a larger gap, so that the tape is in the magnetic field for a longer period. The bias signal takes the magnetic material on the tape through many thousands of hysteresis loops. These become progressively smaller until the magnetism disappears.

Sound Track

One method of recording sound on film, in the form of a *variable area* sound track, is illustrated in Fig. 25.27. A triangular aperture or mask T

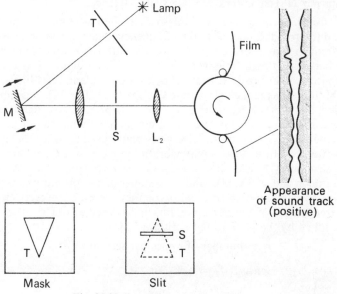

Fig. 25.27. Recording sound on film.

is brightly lit by a high-wattage lamp. After passing through T the light is reflected by a mirror M, and the rays are brought to a focus on to a slit S by a lens L_i. By rotating the mirror slightly, as shown, the image T′ of T, produced by the slit, can be moved up and down. This varies the length of slit illuminated. The light passing through the slit is collected by another lens system L_2 and focused on a strip of moving unexposed film. The mirror M is mounted on the moving system of a galvanometer, so that it moves at the same frequency as the audio-frequency currents passed through the galvanometer coil. The *area* of the film exposed thus varies as the audio-frequency signal. Hence when the film is developed, a permanent sound recording or *sound track* is obtained. A typical length of sound track is shown in Fig. 25.27.

Fig. 25.28 illustrates the principle of reproducing the sound. Light is focused on the sound track and passes through to a *photo-electric cell*. This contains a light-sensitive metal surface such as caesium, which then emits a number of electrons proportional to the light intensity. A current therefore flows in a resistor R. The sound track is coupled to the film, and as it moves, an audio-frequency current flows in R. The pd developed is amplified and passed to the loudspeaker.

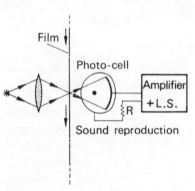

Fig. 25.28. Sound reproduction.

Frequency of Tuning-fork. Falling Plate Method

1. *Comparison method.* A tuning-fork is often used in experiments in sound to provide a note of known frequency. One method of measuring the frequency of a fork P is to compare it with the known frequency of another fork Q by a 'falling plate method'. One end of P is clamped in a vice; a light bristle B is attached to a prong, and rests lightly near the bottom of a smoked glass plate G, Fig. 25.29. The fork Q is similarly placed, and a bristle C, attached to one of its prongs, rests lightly on G. Both forks are sounded by drawing a bow across them, and the thread S suspending G is now burnt. The glass plate, usually in a groove, falls downward past the horizontally vibrating bristles, which then trace out two clear wavy 'tracks' BX, CY on G. Two horizontal lines, MN, are then drawn across BX, CY, and the number of complete waves between MN is counted on each trace. Suppose they are n_1, n_2 respectively, and the frequencies of the corresponding forks are f_1, f_2. Then if t is the time taken by the plate to fall a distance MN,

$$f_1 = \text{number of cycles per second} = \frac{n_1}{t}$$

and $f_2 = n_2/t$. Hence $f_1/f_2 = n_1/n_2$, or

$$f_1 = \frac{n_1}{n_2} \times f_2.$$

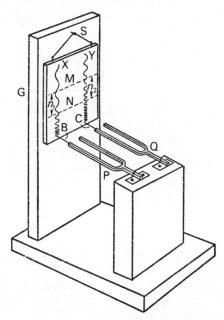

Fig. 25.29. Falling plate method.

Thus, knowing n_1, n_2, f_2, the unknown frequency f_1 can be calculated.

2. *Absolute method.* The unknown frequency f_1 of the tuning-fork P can also be calculated from its own trace. In this case the lengths s_1, s_2 corresponding to an *equal* number of consecutive waves are measured by a travelling microscope. Suppose there are n complete waves between LR, RS, Fig. 25.30. Then f_1 is given by

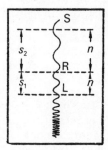

$$f_1 = n \sqrt{\frac{g}{s_2 - s_1}}, \qquad . \qquad . \qquad . \qquad (15)$$

where $g = 9.8$ m s^{-2} when s_1, s_2 are in metres.

To prove this formula, for f_1, let t be the time taken by the plate to fall a distance LR. Since the number of waves in LR is the same as in RS, the fork has also vibrated for a time t while the plate falls a distance RS. Thus if t is the time taken by the plate to fall a distance LR, $2t$ is the time it

Fig. 25.30.
(Not to Scale.)

takes to fall a distance LS. Suppose u is the velocity of the plate at the instant when the line L on it reaches the vibrating bristle on the fork.

Then $$s_1 = ut + \tfrac{1}{2}gt^2 \qquad . \qquad . \qquad . \qquad . \qquad (i)$$

from the dynamics equation $s = ut + \tfrac{1}{2}at^2$, since the acceleration, a, of the plate $= g$, the acceleration due to gravity. As the time taken by the plate to fall a distance LS, or $(s_1 + s_2)$, is $2t$, we have

$$s_1 + s_2 = u \cdot 2t + \tfrac{1}{2}g(2t)^2. \qquad . \qquad . \qquad (ii)$$

Multiplying equation (i) by 2, and subtracting from (ii) to eliminate u,

$$s_2 - s_1 = gt^2,$$

$$\therefore t = \sqrt{\frac{s_2 - s_1}{g}}$$

$$\therefore f_1 = \frac{n}{t} = n\sqrt{\frac{g}{s_2 - s_1}} \qquad . \qquad . \qquad . \qquad (16)$$

as given above in equation (15).

The falling-plate method gives only an approximate value of the frequency, as (a) it is difficult to determine an *exact* number of waves; (b) the attachment of a bristle to the tuning-fork prong lowers its frequency slightly (see p. 622); (c) there is friction between the style and the plate.

The Stroboscope

A *stroboscope* is an arrangement which can make a rotating object appear at rest when it is viewed, and thus enables a spinning wheel, for example, to be studied at leisure. The stroboscopic method can be used to determine the frequency of a tuning-fork, which is electrically maintained for the purpose.

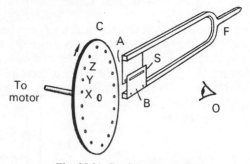

Fig. 25.31. Stroboscope method.

Two light metal plates, A, B, each with a slit S in them, are attached to prongs of the tuning-fork F so that the slits overlap each other when the fork is not sounded, Fig. 25.31. Behind the slit is a vertical circular white card C with black dots spaced at equal distances round the circumference, and the dots on the card can be seen through S. The tuning-fork is set into vibration, and the card is rotated by a motor about a horizontal axis through its centre with increasing speed. At first an observer O, viewing the dots through S, sees them moving round in an opposite direction to that in which the card is rotated. This is because the intermittent glimpses of the card through S occur quicker than the time taken for one dot to reach the place of the dot in front of it, with the result that the dots appear to be moving slowly back. As the speed of the card is increased further, a stage is reached when the dots appear perfectly stationary.

Through the slit S, glimpses of the dots are seen twice in every cycle of vibration of the tuning-fork. When the dots first appear stationary, a particular dot such as X moves to a neighbouring dot position Y at one glimpse of the wheel, then to the next dot position Z at the next glimpse, and so on. At the end of one second, $2f$ glimpses have occurred, where f is the frequency of the fork in Hz. If the wheel has m dots and is now rotating at n rev s^{-1}, there have been $m \times n$ dot successive movements in one second.

$$\therefore\ 2f = mn,\ \text{or}\ f = \frac{mn}{2}. \qquad . \qquad . \qquad . \qquad (17)$$

At twice the speed of revolution, $2n$ rev s^{-1}, the dots are again seen stationary. But this time only half their number is seen, as a particular dot moves through two dot places between successive glimpses. The dots may again be seen stationary at $3n$, $4n$, ... rev s^{-1} of the wheel if the stress on the wheel at higher speeds is below a dangerous level.

As an illustration, suppose a wheel has 40 equally spaced dots and is viewed stroboscopically by a fork of 300 Hz. If the dots are seen stationary at the lowest angular speed, the number of revs per sec, n, is given by

$$600 = n \times 40,\ \text{or}\ n = 15\ \text{rev s}^{-1}.$$

A neon lamp, providing intermittent flashes of light at a rate which can be varied by an electrical circuit, is used as a stroboscope in industry to adjust critically the speed of rotating wheels or machinery, which then appear stationary. The wear and tear with time of the moving parts of watches have been photographed with the aid of a stroboscope.

Relative movement of dots. It is sometimes impossible to keep the speed of rotation of the wheel in Fig. 25.31 constant, so that the dots appear to move slowly forward or back. Suppose, for example, that 2 dots per second cross the line of view in a backward direction in the case of the fork and wheel just discussed. Instead of 40×15 or 600 successive dot movements at 15 revs per second, when the wheel appears stationary, there are now $600 - 2$ or 598 successive dot movements. Since there are 40 dots round the wheel,

$$\therefore\ \text{new rate of revolution of wheel} = \frac{598}{40} = 14 \cdot 95\ \text{rev s}^{-1}.$$

Suppose that the 40 dots appear stationary again at a wheel rotation of 15 revs per sec when viewed stroboscopically with the fork of frequency 300 Hz, and the fork is now loaded with a small piece of plasticine. The fork frequency is then lowered to a value f'. In this case the time interval between successive glimpses is longer than before, so that the dots appear to move forward. Suppose the movement is 3 dots per 10 seconds across the field of view. The number of successive glimpses of the wheel is $10 \times 2f'$ in 10 seconds. In this time the number of successive dot movements is $10 \times 40 \times 15 - 3$, or 5997.

$$\therefore\ 20f' = 5997$$
$$\therefore\ f' = 299 \cdot 85.$$

Thus the frequency of the fork is lowered by $0 \cdot 15$ Hz.

EXERCISES 25

1. A tuning-fork is considered to produce a 'pure' note. (i) Write down an equation which represents the vibration of the prongs. (ii) Explain how an *exchange of energy* occurs during the motion of the prongs.

2. State, with reasons, whether the following waves are *longitudinal* (L) or *transverse* (T): *sound, waves on plucked string, water waves, light waves.* Draw sketches to illustrate your answer.

3. If the velocity of sound in air is 340 metres per second, calculate (i) the wavelength in cm when the frequency is 256 Hz, (ii) the frequency when the wavelength is 85 cm.

4. A plane-progressive wave is represented by the equation

$$y = 0 \cdot 1 \sin (200\pi t - 20\pi x/17),$$

where y is the displacement in millimetres, t is in seconds and x is the distance from a fixed origin O in metres (m).

Find (i) the frequency of the wave, (ii) its wavelength, (iii) its speed, (iv) the phase difference in radians between a point $0 \cdot 25$ m from O and a point $1 \cdot 10$ m from O, (v) the equation of a wave with double the amplitude and double the frequency but travelling exactly in the opposite direction.

5. Describe how a *sound wave* passes through air, using graphs which illustrate and compare the variation of (i) the *displacement* of the air particles, (ii) the *pressure changes*, while the wave travels.

6. In the *falling-plate* experiment of measuring the frequency of a tuning-fork, the successive lengths occupied by 25 complete waves were 9·85 and 14·1 cm. When the experiment was repeated, the successive lengths occupied by 25 complete waves were 9·0 and 13·0 cm. Calculate the frequency of the fork for each experiment, deriving any formula you employ.

7. Describe an absolute method for determining the frequency of a tuning-fork.

Two forks A and B vibrate in unison but when two slits are fixed to the prongs of A, so that they are in line when the prongs are at rest, 9 beats in 10 sec. are heard when the forks are sounded together. A is then made to vibrate in front of a stroboscopic disc, on which are marked 50 equally spaced radial lines. The disc is viewed through the slits and the lines appear at rest when the disc rotates at 25 rev s^{-1}. What is the frequency of B?

Describe and explain what would be seen if the speed of rotation of the disc were slightly decreased. (*L*).

8. Describe the nature of the disturbance set up in air by a vibrating tuning-fork and show how the disturbance can be represented by a sine curve. Indicate on the curve the points of (*a*) maximum particle velocity, (*b*) maximum pressure.

What characteristics of the vibration determine the pitch, intensity, and quality respectively of the note? (*N*.)

9. Distinguish between *longitudinal* and *transverse* wave motions, giving examples of each type. Find a relationship between the frequency, wave-length and velocity of propagation of a wave motion.

Describe experiments to investigate quantitatively for sound waves the phenomena of (a) reflection, (b) refraction, (c) interference. (*C*.)

10. State and explain the differences between progressive and stationary waves.

A progressive and a stationary simple harmonic wave each have the same frequency of 250 Hz and the same velocity of 30 m s $^{-1}$. Calculate (i) the phase difference between two vibrating points on the progressive wave which are 10 cm apart, (ii) the equation of motion of the progressive wave if its amplitude is 0·03 metre, (iii) the distance between nodes in the stationary wave, (iv) the equation of motion of the stationary wave if its amplitude is 0·01 metre.

11. (i) Describe a stroboscope and an experiment to demonstrate one of its uses. Explain the calculation involved. (ii) Explain how sound is recorded and reproduced in a tape recorder. (*L.*)

12. Describe a *ribbon* and a *moving-coil* microphone, and explain with the aid of a diagram how each functions.

Draw a labelled diagram of the principal features of *a moving-coil loud-speaker*. Explain the purpose of the *baffle-board*.

13. Give a brief account of the principle of the stroboscope.

Describe how such a device may be used to determine the frequency of a tuning-fork. (You may, if necessary, suppose that two forms of equal frequency are available.)

Give a short account of any use made outside the laboratory of the strobo-scope principle.

When viewed stroboscopically at a frequency of 300 vibrations second $^{-1}$ a circular disc with 40 equally spaced dots appears to have a backwards rotation such that two dots cross the viewing line each second. What is the least rate of rotation of the disc? (*N.*)

14. Explain the terms *damped oscillation, forced oscillation* and *resonance*. Give one example of each.

Describe an experiment to illustrate the behaviour of a simple pendulum (or pendulums) undergoing forced oscillation. Indicate qualitatively the results you would expect to observe.

What factors determine (*a*) the period of free oscillations of a mechanical system, and (*b*) the amplitude of a system undergoing forced oscillation? (*O. & C.*)

chapter twenty-six

Characteristics, properties, and velocity of sound waves

CHARACTERISTICS OF NOTES

NOTES may be similar to or different from each other in three respects: (i) *pitch*, (ii) *loudness*, (iii) *quality*; so that if each of these three quantities of a particular note is known, the note is completely defined or "characterised".

Pitch

Pitch is analogous to colour in light, which is characterised by the wavelength, or by the frequency, of the electromagnetic vibrations (p. 690). Similarly, the pitch of a note depends only on the frequency of the vibrations, and a high frequency gives rise to a high-pitched note. A low frequency produces a low-pitched note. Thus the high-pitched whistle of a boy may have a frequency of several thousand Hz, whereas the low-pitched hum due to A.C. mains frequency when first switched on may be a hundred Hz. The range of sound frequencies is about 15 to 20000 Hz and depends on the observer.

Musical Intervals

If a note of frequency 300 Hz, and then a note of 600 Hz, are sounded by a siren, the pitch of the higher note is recognised to be an upper octave of the lower note. A note of frequency 1000 Hz is recognised to be an upper octave of a note of frequency 500 Hz, and thus the *musical interval* between two notes is an upper octave if the ratio of their frequencies is 2 : 1. It can be shown that the musical interval between two notes depends on the *ratio* of their frequencies, and not on the actual frequencies. The table below shows the various

Note	C doh	D ray	E me	F fah	G soh	A lah	B te	c doh
Natural (Diatonic) Scale Frequency	256	288	320	341	384	427	480	512
Intervals between notes		9/8	10/9	16/15	9/8	10/9	9/8	16/15
Intervals above C	1·000	1·125	1·250	1·333	1·500	1·667	1·875	2·000
Equal Temperament Scale intervals above C	1·000	1·122	1·260	1·335	1·498	1·682	1·888	2·000

Note 1.—There are 12 semitones to the octave in the scale of equal temperament; each semitone has a frequency ratio of $2^{1/12}$.

Note 2.—The frequency of C is 256 on the scale of Helmholtz above; in music it is 261·2.

musical intervals and the corresponding ratio of the frequencies of the notes.

Intensity and Loudness

The *intensity* of a sound at a place is defined as the energy per second flowing through one square metre held normally at that place to the direction along which the sound travels. As we go farther away from a source of sound the intensity diminishes, since the intensity decreases as the square of the distance from the source (see also p. 565).

Suppose the displacement y of a vibrating layer of air is given by $y = a \sin \omega t$, where $\omega = 2\pi/T$ and a is the amplitude of vibration, see equation (1), p. 577. The velocity, v, of the layer is given by

$$v = \frac{dy}{dt} = \omega a \cos \omega t,$$

and hence the kinetic energy, W, is given by

$$W = \tfrac{1}{2} mv^2 = \tfrac{1}{2} m\omega^2 a^2 \cos^2 \omega t \qquad . \qquad . \qquad \text{(i)}$$

where m is the mass of the layer. The layer also has potential energy as it vibrates. Its total energy, W_0, which is constant, is therefore equal to the maximum value of the kinetic energy. From (i), it follows that

$$W_0 = \tfrac{1}{2} m\omega^2 a^2 \qquad . \qquad . \qquad . \qquad . \qquad \text{(ii)}$$

In 1 second, the air is disturbed by the wave over a distance V cm., where V is the velocity of sound in m s^{-1}; and if the area of cross-section of the air is 1 m^2, the volume of air disturbed is V m^3. The mass of air disturbed per second is thus $V\rho$ kg, where ρ is the density of air in kg m^{-3}, and hence, from (ii),

$$W = \tfrac{1}{2} V\rho\omega^2 a^2 \qquad . \qquad . \qquad . \qquad . \qquad \text{(iii)}$$

It therefore follows that *the intensity of a sound due to a wave of given frequency is proportional to the square of its amplitude of vibration.*

It can be seen from (ii) that the greater the mass m of air in vibration, the greater is the intensity of the sound obtained. For this reason the sound set up by the vibration of the diaphragm of a telephone earpiece cannot be heard except with the ear close to the earpiece. On the other hand, the cone of a loudspeaker has a large surface area, and thus disturbs a large mass of air when it vibrates, giving rise to a sound of much larger intensity than the vibrating diaphragm of the telephone earpiece. It is difficult to hear a vibrating tuning-fork a small distance away from it because its prongs set such a small mass of air vibrating. If the fork is placed with its end on a table, however, a much louder sound is obtained, which is due to the large mass of air vibrating by contact with the table.

Loudness is a sensation, and hence, unlike intensity, it is difficult to measure because it depends on the individual observer. Normally, the greater the intensity, the greater is the loudness of the sound (see p. 607).

The Decibel

We are already familiar with the fact that when the frequency of a note is doubled its pitch rises by an octave. Thus the increase in pitch sounds the same to the ear when the frequency increases from 100 to 200 Hz as from 500 to 1000 Hz. Similarly, it is found that increases in loudness depend on the *ratio* of the intensities, and not on the absolute differences in intensity.

If the power of a source of sound increases from 0·1 watt to 0·2 watt, and then from 0·2 watt to 0·4 watt, the loudness of the source to the ear increases in equal steps. The equality is thus dependent on the equality of the *ratio* of the powers, not their difference, and in commercial practice the increase in loudness is calculated by taking the *logarithm of the ratio of the powers to the base* 10, which is $\log_{10} 2$, or 0·3, in this case.

Relative intensities or powers are expressed in *bels*, after Graham Bell, the inventor of the telephone. If the power of a source of sound changes from P_1 to P_2, then

$$\text{number of bels} = \log_{10} \left(\frac{P_2}{P_1} \right).$$

In practice the bel is too large a unit, and the **decibel (db)** is therefore adopted. This is defined as one-tenth of a bel, and hence in the above case

$$\text{number of decibels} = 10 \log_{10} \left(\frac{P_2}{P_1} \right).$$

The minimum change of power which the ear is able to detect is about 1 db, which corresponds to an increase in power of about 25 per cent.

Calculation of Decibels

Suppose the power of a sound from a loudspeaker of a radio receiver is 50 milliwatts, and the volume control is turned so that the power increases to 500 milliwatts. The increase in power is then given by

$$10 \log_{10} \left(\frac{P_2}{P_1} \right) = 10 \times \log \frac{500}{50} = 10 \text{ db}.$$

If the volume control is turned so that the power increases to 1000 milliwatts, the increase in power compared with the original sound

$$= 10 \log_{10} \left(\frac{1000}{50} \right) = 10 \log_{10} 20 = 13 \text{ db}.$$

If the volume control is turned down so that the power decreases from 1000 to 200 milliwatts, the change in power

$$= 10 \log_{10} \left(\frac{200}{1000} \right) = 10 \log_{10} 2 - 10 \log_{10} 10 = - 7 \text{ db}.$$

The minus indicates a decrease in power. Besides its use in acoustics the decibel is used by radio and electrical engineers in dealing with changes in electrical power.

Intensity Levels. Threshold of Hearing

Since the intensity of sound is defined as the energy per second crossing 1 metre^2 normal to the direction of the sound, the unit of intensity is "watt metre^{-2}", symbol "W m^{-2}". The *intensity level* of a source is its intensity relative to some agreed 'zero' intensity level. If the latter has an intensity of P_0 watt metre^{-2}, a sound of intensity P watt metre^{-2} has an intensity level defined as:

$$10 \log_{10} \left(\frac{P}{P_0} \right) \text{ db.}$$

The lowest audible sound at a frequency of 1000 Hz, which is called the *threshold of hearing*, corresponds to an intensity P_0 of 10^{-12} watt m^{-2} or 10^{-10} microwatt cm^{-2}. This is chosen as the 'zero' of sound intensity level. An intensity level of a low sound of + 60 db is 60 decibels or 6 bels higher than 10^{-12} watt metre^{-2}. The intensity is thus 10^6 times as great, and is therefore equal to $10^6 \times 10^{-12}$ or 10^{-6} W m^{-2}.

Calculation of intensity level. The difference in intensity levels of two sounds of intensities P_1, P_2 watt metre^{-2} respectively is $10 \log_{10}(P_2/P_1)$ db. Thus the difference in intensity levels of a sound of intensity 8×10^{-5} W m^{-2} due to a person talking, and one of an intensity 10^{-1} W m^{-2} due to an orchestra playing,

$$= 10 \log_{10} \left(\frac{8 \times 10^{-5}}{10^{-1}} \right) = -40 + 10 \log_{10} 8 = -31 \text{ db.}$$

The negative sign indicates a decrease in intensity level. Similarly, if two intensity levels differ by 20 decibels, the ratio P'/P of the two intensities is given by

$$10 \log_{10} \left(\frac{P'}{P} \right) = 20,$$

or
$$\frac{P'}{P} = 10^2 = 100.$$

A source of sound such as a small loudspeaker produces a sound intensity round it proportional to $1/d^2$, where d is the distance from the loudspeaker (p. 607). Suppose the intensity level is 10 db at a distance of 20 m from the speaker. At a distance of 40 m the intensity will be four times less than at 20 m, a reduction of $10 \log_{10} 4$ db or about 6 db.

∴ intensity level here = 10 db − 6 db = 4 db.

At a point 10 m from the speaker the intensity will be four times greater than at 20 m. The intensity level here is thus $10 + 6$ or 16 db.

If the electrical power supplied to the loudspeaker is doubled, the sound intensity at each point is doubled. Thus if the original intensity level was 16 db, the new intensity level is higher by $10 \log_{10} 2$ db or about 3 db. The new intensity level is hence 19 db.

Loudness. The Phon

The loudness of a sound is a sensation, and thus depends on the observer, whereas power, or intensity, of a sound is independent of the observer. Observations show that sounds which appear equally loud to a person have different intensities or powers, depending on the frequency, f, of the sound. The curves a, b, c represent respectively three values of *equal loudness*, and hence the intensity at X, when the frequency is 1000 Hz, is less than the intensity at Y, when the frequency is 500, although the loudness is the same, Fig. 26.1.

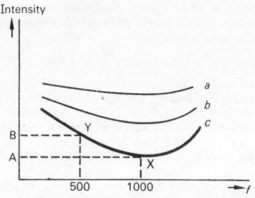

FIG. 26.1. Curves of equal loudness.

In order to measure loudness, therefore, scientists have adopted a "standard" source having a frequency of 1000 Hz, with which all other sounds are compared. The source H whose loudness is required is placed near the standard source, and the latter is then altered until the loudness is the same as H. The intensity or power level of the standard source is then measured, and if this is n decibels above the threshold value (10^{-10} microwatt per sq cm, p. 609) the loudness is said to be n *phons*. The phon, introduced in 1936, is thus a unit of loudness, whereas the decibel is a unit of intensity or power. *Noise meters*, containing a microphone, amplifier, and meter, are used to measure loudness, and are calibrated directly in phons. The "threshold of feeling", when sound produces a painful sensation to the ear, corresponds to a loudness of about 120 phons.

Quality or Timbre

If the same note is sounded on the violin and then on the piano, an untrained listener can tell which instrument is being used, without seeing it. We say that the *quality* or *timbre* of the note is different in each case.

The waveform of a note is never simple harmonic in practice; the nearest approach is that obtained by sounding a tuning-fork, which thus produces what may be called a "pure" note, Fig. 26.2 (i). If the same note is played on a violin and piano respectively, the waveforms produced might be represented by Fig. 26.2 (ii), (iii), which have the same frequency and amplitude as the waveform in Fig. 26.2 (i). Now curves of the shape of Fig. 26.2 (ii), (iii) can be analysed mathematically into the sum of a number of *simple harmonic* curves, whose frequencies are multiples of f_0, the frequency of the original waveform; the amplitudes of these curves diminish as the frequency increases. Fig. 26.2 (iv), for example, might be an analysis of a curve similar to Fig. 26.2 (iii), corresponding to a note on a piano. The ear is able to detect simple harmonic waves (p. 595), and thus registers the presence of notes of frequencies $2f_0$ and $3f_0$, in addition to f_0, when the note is sounded on the piano. The amplitude of the curve corresponding to f_0 is greatest,

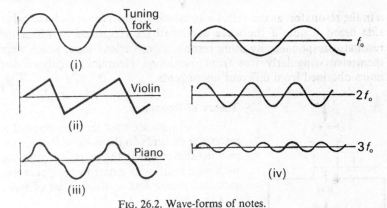

FIG. 26.2. Wave-forms of notes.

Fig. 26.2 (iv), and the note of frequency f_0 is heard predominantly because the intensity is proportional to the square of the amplitude (p. 607). In the background, however, are the notes of frequencies $2f_0$, $3f_0$, which are called the *overtones*. The frequency f_0 is called the *fundamental*.

As the waveform of the same note is different when it is obtained from different instruments, it follows that the analysis of each will differ; for example, the waveform of a note of frequency f_0 from a violin may contain overtones of frequencies $2f_0$, $4f_0$, $6f_0$. The musical "background" to the fundamental note is therefore different when it is sounded on different instruments, and hence *the overtones present in a note determine its quality or timbre*.

A *harmonic* is the name given to a note whose frequency is a simple multiple of the fundamental frequency f_0. The latter is thus termed the "first harmonic"; a note of frequency $2f_0$ is called the "second harmonic", and so on. Certain harmonics of a note may be absent from its overtones; for example, the only possible notes obtained from an organ-pipe closed at one end are f_0, $3f_0$, $5f_0$, $7f_0$, and so on (p. 647).

Helmholtz Resonators

HELMHOLTZ, one of the greatest scientists of the nineteenth century, devised a simple method of detecting the overtones accompanying the fundamental note. He used vessels, P, Q, of different sizes, containing air which "responded" or *resonated* (see p. 653) to a note of a particular frequency, Fig. 26.3. When a sound wave entered a small cavity or neck

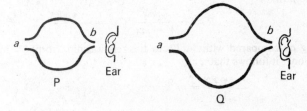

FIG. 26.3. Helmholtz resonators.

a in the resonator, as the vessel was called, an observer at *b* on the other side heard a note if the wave contained the frequency to which the resonator responded. By using resonators of various sizes, which were themselves singularly free from overtones, Helmholtz analysed the notes obtained from different instruments.

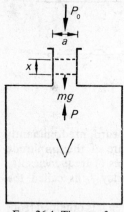

FIG. 26.4. Theory of a Resonator.

Theory of Resonator

We shall now see how the frequency of a resonator depends on the volume V of air inside it; and to define the situation, suppose we have a bottle with a narrow neck of cross-sectional area a and containing air of mass m, Fig. 26.4.

If the pressure outside is p_0, and the air-pressure inside is p, then, for equilibrium,

$$p_0 a + mg = pa \qquad . \qquad . \qquad \text{(i)}$$

When the air in the vessel is resonating to a particular note, the air in the neck moves up and down, acting like a damper or piston on the large mass of air of volume V beneath the neck. Suppose the air in the neck moves downward through a distance x at an instant. Then, assuming an adiabatic contraction, the increased pressure p_1 in the vessel is given by

$$p_1 (V - ax)^\gamma = pV^\gamma.$$

$$\therefore \quad p_1 = p \left[\frac{V}{V - ax} \right]^\gamma = p \left[1 + \frac{ax}{V - ax} \right]^\gamma$$

$$= p \left[1 + \frac{\gamma ax}{V - ax} \right],$$

by binomial expansion, assuming ax is small compared with $(V - ax)$; this is true for a narrow neck connected to a large volume V.

$$\therefore \quad p_1 - p = \frac{\gamma pax}{V - ax} \qquad . \qquad . \qquad . \qquad . \qquad . \qquad \text{(ii)}$$

The net downward force, P, on the air in the resonator

$$= p_0 a + mg - p_1 a = pa - p_1 a, \text{ from (i)}.$$

Hence, from (ii)

$$P = - \frac{\gamma pax}{V - ax} \times a = \frac{- \gamma pa^2 x}{V},$$

neglecting ax compared with V. From the relationship "force = mass × acceleration", it follows that

$$\frac{- \gamma pa^2 x}{V} = m \times \text{accn.},$$

or

$$\text{accn.} = - \frac{\gamma pa^2}{mV} \times x.$$

Thus the motion of the air in the neck is simple harmonic, and the period T is given by

$$T = 2\pi \sqrt{\frac{mV}{\gamma p a^2}}.$$

Hence the frequency, f, is given by

$$f = \frac{1}{T} = \frac{1}{2\pi} \sqrt{\frac{\gamma p a^2}{mV}} \qquad . \qquad . \qquad . \qquad \text{(iii)}$$

The velocity of sound, v, is given by $v = \sqrt{\gamma p/\rho}$, where ρ is the density of the air (p. 624), or $\gamma p = v^2 \rho$. If l is the length of the neck, the mass $m = al\rho$. Thus the frequency f can also be expressed by

$$f = \frac{1}{2\pi} \sqrt{\frac{v^2 \rho a^2}{al\rho V}} = \frac{v}{2\pi} \sqrt{\frac{a}{lV}} \qquad . \qquad . \qquad . \qquad \text{(iv)}$$

From these formulae for f, it follows that

$$f^2 V = constant.$$

The adiabatic changes at the neck are not perfect, and this result is thus only approximately true. In practice, the law more nearly obeyed is that given by

$$f^2 (V + c) = constant.$$

where c is a "correction" to V.

Experiment. In an experiment to verify the law, tuning-forks of known frequency, a bottle with a narrow neck, and a pipette and burette, are required. Water is run slowly into the bottle until resonance is obtained with the lowest note, for example. The volume of air V which is resonating is then found by subtracting the volume of the bottle below the neck, determined in a preliminary experiment, from the water run in. This is repeated for the various forks, and a graph of V is plotted against $1/f^2$. A straight line passing close to the origin is obtained, thus showing that $V + c = d/f^2$, where d is a constant, or $f^2(V + c) =$ constant.

PROPERTIES OF SOUND WAVES

Reflection

Like light waves, sound waves are reflected from a plane surface so that the angle of incidence is equal to the angle of reflection. This can be demonstrated by placing a tube T_1 in front of a plane surface AB and blowing a whistle gently at S, Fig. 26.5. Another tube T_2, directed towards N, is placed on the other side of the normal NQ, and moved

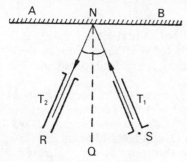

FIG. 26.5. Reflection of sound.

until a sensitive flame (see p. 650), or a microphone connected to a cathode-ray tube, is considerably affected at R, showing that the reflected wave passes along NR. It will then be found that angle RNQ = angle SNQ.

It can also be demonstrated that sound waves come to a focus when they are incident on a curved concave mirror. A surface shaped like a parabola reflects sound waves to long distances if the source of sound is placed at its focus (see also p. 404). The famous whispering gallery of St. Paul's is a circular-shaped chamber whose walls repeatedly reflect sound waves round the gallery, so that a person talking quietly at one end can be heard distinctly at the other end.

Acoustics of Rooms. Reverberation

A concert-hall, lecture-room, or a broadcasting studio requires special design to be acoustically effective. The technical problems concerned were first investigated in 1906 by SABINE in America, who was consulted about a hall in which it was difficult for an audience to hear the lecturer.

Generally, an audience in a hall hears sound from different directions at different times. They hear (a) sound *directly* from the speaker or orchestra, as the case may be, (b) sound from *echoes* produced by walls and ceilings, (c) sound *diffused* from the walls and ceilings and other objects present. The echoes are due to regular reflection at a plane surface (p. 391), but the diffused sound is scattered in different directions and reflection takes place repeatedly at other surfaces. When reflection occurs some energy is absorbed from the sound wave, and after a time the sound diminishes below the level at which it can be heard. The perseverance of the sound after the source ceases is known as **reverberation.** In the case of the hall investigated by Sabine the time of reverberation was about $5\frac{1}{2}$ seconds, and the sound due to the first syllable of a speaker thus overlapped the sound due to the next dozen or so syllables, making the speech difficult to comprehend. The quality of a sound depends on the time of reverberation. If the time is very short, for example 0·5 second, the music from an orchestra sounds thin or lifeless; if the time too long the music sounds muffled. The reverberation time at a B.B.C. concert-hall used for orchestral performances is about $1\frac{3}{4}$ seconds, whereas the reverberation time for a dance-band studio is about 1 second.

Sabine's Investigations. Absorptive Power

Sabine found that the time T of reverberation depended on the volume V of the room, its surface area A, and the *absorptive power*, a, of the surfaces. The time T is given approximately by

$$T = \frac{kV}{aA},$$

where k is a constant. In general some sound is absorbed and the rest is reflected; if too much sound is reflected T is large. If many thick curtains are present in the room too much sound is absorbed and T is small.

Sabine chose the absorptive power of unit area of an open window as the unit, since this is a perfect absorber. On this basis the absorptive power of a person in an audience, or of thick carpets and rugs, is 0·5, linoleum has an absorptive power of 0·12, and polished wood and glass have an absorptive power of 0·01. The absorptive power of a material depends on its pores to a large extent; this is shown by the fact that an unpainted brick has a high absorptive power, whereas the painted brick has a low absorptive power.

From Sabine's formula for T it follows that the time of reverberation can be shortened by having more spectators in the hall concerned, or by using felt materials to line some of the walls or ceiling. The seats in an acoustically-designed lecture-room have plush cushions at their backs to act as an absorbent of sound when the room is not full. B.B.C. studios used for plays or news talks should have zero reverberation time, as clarity is all-important, and the studios are built from special plaster or cork panels which absorb the sound completely. The structure of a room also affects the acoustics. Rooms with large curved surfaces tend to focus echoes at certain places, which is unpleasant aurally to the audience, and a huge curtain was formerly hung from the roof of the Albert Hall to obscure the dome at orchestral concerts.

Refraction

Sound waves can be refracted as well as reflected. TYNDALL placed a watch in front of a balloon filled with carbon dioxide, which is heavier than air, and found that the sound was heard at a definite place on the other side of the balloon. The sound waves thus converged to a focus on the other side of the balloon, which therefore has the same effect on sound waves as a convex lens has on light waves (see Fig. 28.12, p. 684). If the balloon is filled with hydrogen, which is lighter than air, the sound waves diverge on passing through the balloon. The latter thus acts similarly to a concave lens when light waves are incident on it (see p. 683).

The refraction of sound explains why sounds are easier to hear at night than during day-time. In the latter case the upper layers of air are colder than the layers near the earth. Now sound travels faster the higher the temperature (see 624), and sound waves are hence refracted in a direction away from the earth. The intensity of the sound waves thus diminish. At night-time, however, the layers of air near the earth are colder than those higher up, and hence sound waves are now refracted towards the earth, with a consequent increase in intensity.

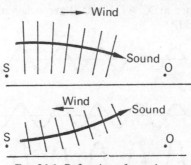

FIG. 26.6. Refraction of sound.

For a similar reason, a distant observer O hears a sound from a source S more easily when the wind is blowing towards him than away from him, Fig. 26.6. When the wind is blowing towards O, the bottom of the sound wavefront is moving slower than the upper part, and hence the wavefronts veer towards the observer, who therefore hears the sound easily. When the wind is blowing in the opposite direction the reverse is the case, and the wavefronts veer upwards away from the ground and O. The sound intensity thus diminishes. This phenomenon is hence another example of the refraction of sound.

Interference of Sound Waves

Besides reflection and refraction, sound waves can also exhibit the phenomenon of *interference*, whose principles we shall now discuss.

Suppose two sources of sound, A, B, have exactly the same frequency and amplitude of vibration, and that their vibrations are always in phase with each other, Fig. 26.7. Such sources are called "coherent" sources. *Their combined effect at a point is obtained by adding algebraically the displacements at the point due to the sources individually;* this is known as the *Principle of Superposition*. Thus their resultant effect at X, for example, is the algebraic sum of the vibrations at X due to the source A alone and the vibrations at X due to the source B alone. If X is equidistant from A and B, the vibrations at X due to the two sources are *always* in phase as (i) the distance AX travelled by the wave originating at A is equal to the distance BX travelled by the wave originating at B, (ii) the sources A, B are assumed to have the same frequency and to be always in phase with each other. Fig. 26.8 (i), (ii) illustrate the vibrations at X due to A, B, which have the same amplitude. The resultant vibration at X is obtained by adding the two curves, and has

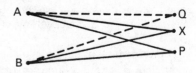

FIG. 26.7. Interference of sound.

an amplitude double that of either curve and a frequency the same as either, Fig. 26.8 (iii). Now the energy of a vibrating source is proportional to the square of its amplitude (p. 607). Consequently the sound energy at X is four times that due to A or B alone, and a loud sound is thus heard at X. As A and B are coherent sources, the loud sound is *permanent*.

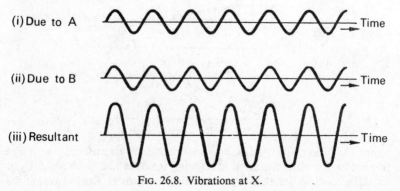

FIG. 26.8. Vibrations at X.

If Q is a point such that BQ is greater than AQ by a whole number of wavelengths (Fig. 26.7), the vibration at Q due to A is in phase with the vibration there due to B (see p. 593). A permanent loud sound is then obtained at Q. Thus a permanent loud sound is obtained at any point Y if the *path difference*, BY − AY, is given by

$$BY - AY = m\lambda,$$

where λ is the wavelength of the sources A, B, and m is an integer.

Destructive Interference

Consider now a point P in Fig. 26.7 whose distance from B is half a wavelength longer than its distance from A, i.e., $AP - BP = \lambda/2$. The vibration at P due to B will then be 180° out of phase with the vibration there to A (see p. 586), Fig. 26.9 (i), (ii). The resultant effect at P is thus zero, as the displacements at any instant are equal and opposite to each other, Fig. 26.9 (iii). No sound is therefore heard at P, and the permanent silence is said to be due to "destructive interference" between the sound waves from A and B.

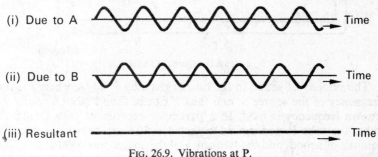

FIG. 26.9. Vibrations at P.

If the path difference, $AP - BP$, were $3\lambda/2$ or $5\lambda/2$, instead of $\lambda/2$, permanent silence would also exist at P as the vibrations there due to A, B would again be 180° out of phase. Summarising, then,

silence occurs if the path-difference is an odd number of half wavelengths, and

a loud sound occurs if the path-difference is a whole number of wavelengths

The total sound energy in all the positions of loud sound discussed above is equal to the total sound energy of the two sources A, B, from the principle of the conservation of energy. The extra sound at the positions of loud sound thus makes up for the absent sound in the positions of silence.

Quincke's Tube. Measurement of Velocity of Sound in a Tube

QUINCKE devised a simple method of obtaining permanent interference between two sound waves. He used a closed tube SAEB which had openings at S, E, and placed a source of sound at S, Fig. 26.10. A wave then travelled in the direction SAE round the tube, while another wave travelled in the opposite direction SBE; and since these waves are due to the same source, S, they always set out in phase, i.e., they are coherent.

Like a trombone, one side, B, of the tube can be pulled out, thus making SAE, SBE of different lengths. When SAE and SBE are equal in length an observer at E hears a loud sound, since the paths of the two waves are then equal. As B is pulled out the sound dies away and becomes a minimum when the path difference, SBE − SAE, is $\lambda/2$, where λ is the wavelength. In this case the two waves arrive 180° out of phase (p. 617). If the tube is pulled out farther, the sound increases in loudness to a maximum; the path difference is then λ. If k is the distance moved from one position of minimum sound, MN say, to the next position of minimum sound, PQ say, then $2k = \lambda$, Fig. 26.10. Thus the wavelength of the sound can be simply obtained by measuring k.

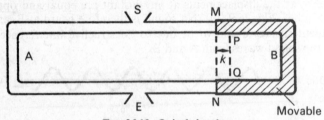

FIG. 26.10. Quincke's tube.

The velocity of sound in the tube is given by $V = f\lambda$, where f is the frequency of the source S, and thus V can be found when a source of known frequency is used. In a particular experiment with Quincke's tube, the tube B was moved a distance 4·28 cm between successive minima of sound, and the frequency of the source was 4000 Hz.

Thus $\lambda = 2 \times 4\cdot28$ cm,

and $V = f\lambda = 4000 \times 2 \times 4\cdot28 = 34240$ cm s^{-1} = 342·4 m s^{-1}

It can be seen that, unlike reflection and refraction, the phenomenon of interference can be utilised to measure the wavelength of sound waves. We shall see later that interference is also utilised to measure the wavelength of light waves (p. 689).

Velocity of Sound in Free Air. Hebb's Method

In 1905 HEBB performed an accurate experiment to measure the velocity of sound in free air which utilised a method of interference. He carried out his experiment in a large hall to eliminate the effect of wind, and obtained the temperature of the air by placing thermometers at different parts of the room. Two parabolic reflectors, R_1, R_2, are placed at each end of the hall, and microphones, M_1, M_2, are positioned at the respective foci, S_1, S_2 to receive sound reflected from R_1, R_2, Fig. 26.11. By means of a transformer, the currents in the microphones are induced into a telephone earpiece P, so that the *resultant* effect of the sound waves received by M_1, M_2 respectively can be heard.

A source of sound of known constant frequency is placed at the focus S_1. The sound waves are reflected from R_1 in a parallel direction (p. 613), and travel to R_2 where they are reflected to the focus S_2 and received by M_2. The microphone M_1 receives sound waves directly from the source, and hence the sound heard in the telephone earpiece is due to the resultant effect of two coherent sources. With the source

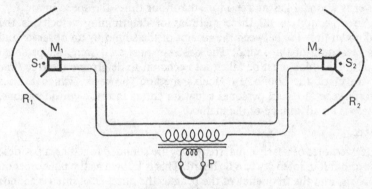

FIG. 26.11. Hebb's method.

and microphone maintained at its focus S_1, R_1 is moved along its axis in one direction. The positions of R_1 are noted when minima of sound are heard; and since the distance between successive minima corresponds to one wavelength, λ, the velocity of sound can be calculated from the relation $V = f\lambda$, as f and λ are known.

Other Velocity of Sound Determinations

The velocity of sound in *air* has been determined by many scientists. One of the first accurate determinations was carried out in 1738 by French scientists, who observed the time between the flash and the hearing of a cannon report about 30 km away. Their results confirmed that the velocity of sound increased as the temperature of the air increased (p. 624), and they obtained the result of 362 metres per second for the velocity at 0° C. Similar experiments were carried out by French scientists in 1822. In 1844 experiments carried out in the Tyrol district, several thousand metres above sea-level, showed that the velocity of sound was independent of the pressure of the air (p. 624).

REGNAULT, the eminent French experimental scientist of the nineteenth century, carried out an accurate series of measurements on the velocity of sound in 1864. Guns were fired at one place, breaking an electrical circuit automatically, and the arrival of the sound at a distant place was recorded by a second electrical circuit. Both circuits actuated a pen or style pressing against a drum rotating at a steady speed round its axis, which is known as a *chronograph*. Thus marks were made on the drum at the instant the sound occurred and the instant it was received. The small interval corresponding to the distance between the marks was determined from a wavy trace made on the drum by a style attached to an electrically-maintained tuning-fork whose frequency was known, and the speed of sound was thus calculated.

The velocity of sound in *water* was first accurately determined in 1826. The experiment was carried out by immersing a bell in the Lake of Geneva, and arranging to fire gunpowder at the instant the bell was struck. Miles away, the interval was recorded between the flash and the later arrival of the sound in the water, and the velocity was then calculated. This and other experiments have shown that the velocity in water is about 1435 m s^{-1}, more than four times the speed in air.

An objection to all these methods of determining velocity is the unknown time lag between the receipt of the sound by an observer and his recording of the sound. The observer has, as it were, a "personal equation" which must be taken into account to determine the true time of travel of the sound. In Hebb's method, however, which utilises interference, no such personal equation enters into the considerations, which is an advantage of the method.

Beats

If two notes of nearly equal frequency are sounded together, a periodic rise and fall in intensity can be heard. This is known as the phenomenon of *beats*, and the frequency of the beats is the number of intense sounds heard per second.

Consider a layer of air some distance away from two pure notes of nearly equal frequency, say 48 and 56 Hz respectively, which are sounding. The variation of the displacement, y_1, of the layer due to one fork alone is shown in Fig. 26.12 (i): the variation of the displacement y_2,

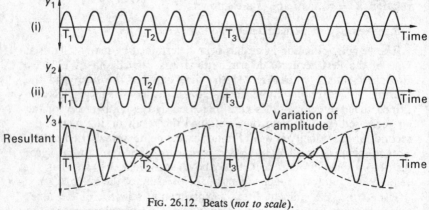

FIG. 26.12. Beats (*not to scale*).

of the layer due to the second fork alone is shown in Fig. 26.12 (ii). According to the Principle of Superposition (p. 588), the variation of the resultant displacement, y, of the layer is the algebraic sum of the two curves, which varies in amplitude in the way shown in Fig. 26.12 (iii). To understand the variation of y, suppose that the displacements y_1, y_2 are in phase at some instant T_1, Fig. 26.12. Since the frequency of the curve in Fig. 26.12 (i) is 48 cycles per sec the variation y_1 undergoes 3 complete cycles in $\frac{1}{16}$th second; in the same time, the variation y_2 undergoes $3\frac{1}{2}$ cycles, since its frequency is 56 cycles per second. Thus y_1 and y_2 are 180° out of phase with each other at this instant, and their resultant y is then a minimum at some instant T_2. Thus T_1T_2 represents $\frac{1}{16}$th of a second in Fig. 26.12 (iii). In $\frac{1}{8}$th of a second from T_1, y_1 has undergone 6 complete cycles and y_2 has undergone 7 complete cycles. The two waves are hence in phase again at T_3, where $T_1 T_3$ represents $\frac{1}{8}$th of a second, and their resultant at their instant is again a maximum, Fig. 26.12 (iii). In this way it can be seen that a loud sound is heard after every $\frac{1}{8}$ second, and thus the beat frequency is 8 cycles per second. This is the difference between the frequencies, 48, 56, of the two notes, and it is shown soon that *the beat frequency is always equal to the difference of the two nearly equal frequencies.*

It can now be seen that beats are a phenomenon of repeated interference. Unlike the cases in sound previously considered, however, the two sources are not coherent ones.

Beat Frequency Formula

Suppose two sounding tuning-forks have frequencies f_1, f_2 cycles per second close to each other. At some instant of time the displacement of a particular layer of air near the ear due to each fork will be a maximum to the right. The resultant displacement is then a maximum, and a loud sound or beat is heard. After this, the vibrations of air due to each fork go out of phase, and t seconds later the displacement due to each fork is again a maximum to the right, so that a loud sound or beat is heard again. One fork has then made exactly one cycle more than the other. But the number of cycles made by each fork in t seconds is $f_1 t$ and $f_2 t$ respectively. Assuming f_1 is greater than f_2,

$$\therefore \quad f_1 t - f_2 t = 1$$

$$\therefore \quad f_1 - f_2 = \frac{1}{t}$$

Now 1 beat has been made in t seconds, so that $1/t$ is the number of beats per second or beat frequency.

$$\therefore \quad f_1 - f_2 = beat \, frequency.$$

Mathematical derivation of beat frequency. Suppose y_1, y_2 are the displacements of a given layer of air due to two tuning-forks of frequencies f_1, f_2 respectively. If the amplitudes of each vibration are equal to a, then $y_1 = a \sin \omega_1 t$, $y_2 = a \sin (\omega_2 t + \theta)$, where $\omega_1 = 2\pi f_1$, $\omega_2 = 2\pi f_2$, and θ is the constant phase angle between the two variations.

$$\therefore \quad y = y_1 + y_2 = a \left[\sin \omega_1 t + \sin (\omega_2 t + \theta) \right]$$

$$\therefore \quad y = 2a \sin \left(\frac{\omega_1 + \omega_2}{2} t + \frac{\theta}{2} \right) \cdot \cos \left(\frac{\omega_1 - \omega_2}{2} t - \frac{\theta}{2} \right)$$

$$\therefore \quad y = A \sin \left(\frac{\omega_1 + \omega_2}{2} t + \frac{\theta}{2} \right)$$

where $A = 2a \cos \left(\frac{\omega_1 - \omega_2}{2} t - \frac{\theta}{2} \right)$.

We can regard A as the *amplitude* of the variation of y. The intensity of the resultant note is proportional to A^2, the square of the amplitude (p. 607) and

$$A^2 = 4a^2 \cos^2 \left(\frac{\omega_1 - \omega_2}{2} t - \frac{\theta}{2} \right) = 2a^2 \left[1 + \cos \overline{(\omega_1 - \omega_2) t - \theta} \right]$$

since $2 \cos^2 a = 1 + \cos 2a$. It then follows that the intensity varies at a frequency f given by $\omega = \omega_1 - \omega_2$,

i.e. $2\pi f = \omega_1 - \omega_2.$

But $\omega_1 = 2\pi f_1, \omega_2 = 2\pi f_2$.

$$\therefore \quad 2\pi f = 2\pi f_1 - 2\pi f_2$$

$$\therefore \quad f = f_1 - f_2$$

The frequency f of the beats is thus equal to the difference of the frequencies.

Uses of Beats

The phenomenon of beats can be used to measure the unknown frequency, f_1, of a note. For this purpose a note of known frequency f_2 is used to provide beats with the unknown note, and the frequency f of the beats is obtained by counting the number made in a given time. Since f is the difference between f_2 and f_1, it follows that $f_1 = f_2 - f$, or $f_1 = f_2 + f$. Thus suppose $f_2 = 1000$ Hz, and the number of beats per second made with a tuning-fork of unknown frequency f_1 is 4. Then $f_1 = 1004$ or 996 Hz.

To decide which value of f_1 is correct, the end of the tuning-fork prong is loaded with a small piece of plasticine, which diminishes the frequency a little, and the two notes are sounded again. If the beats are *increased*, a little thought indicates that the frequency of the note must have been originally 996 Hz. If the beats are decreased, the frequency of the note must have been originally 1004 Hz. The tuning-fork must not be overloaded, as the frequency may decrease, if it was 1004 Hz, to a frequency such as 995 Hz, in which case the significance of the beats can be wrongly interpreted.

Beats are also used to "tune" an instrument to a given note. As the instrument note approaches the given note, beats are heard, and the instrument can be regarded as "tuned" when the beats are occurring at a very slow rate.

Velocity of Sound in a Medium

When a sound wave travels in a medium, such as a gas, a liquid, or a solid, the particles in the medium are subjected to varying stresses, with resulting strains (p. 585). The velocity of a sound wave is thus partly governed by the *modulus of elasticity*, E, of the medium, which is defined by the relation

$$E = \frac{\text{stress}}{\text{strain}} = \frac{\text{force per unit area}}{\text{change in length (or volume)/ original length}}$$

$$\text{(or volume) (i)}$$

The velocity, V, also depends on the density, ρ, of the medium, and it can be shown that

$$V = \sqrt{\frac{E}{\rho}} \qquad . \qquad . \qquad . \qquad (1)$$

When E is in newton per metre^2 (N m^{-2}) and ρ in kg m^{-3}, then V is in metre per second (m s^{-1}). The relation (1) was first derived by Newton.

For a solid, E is Young's modulus of elasticity. The magnitude of E

for steel is about 2×10^{11} N m^{-2}, and the density ρ of steel is 7800 kg m^{-3}. Thus the velocity of sound in steel is given by

$$V = \sqrt{\frac{E}{\rho}} = \sqrt{\frac{2 \times 10^{11}}{7800}} = 5060 \text{ m s}^{-1}$$

For a liquid, E is the bulk modulus of elasticity. Water has a bulk modulus of $2 \cdot 04 \times 10^9$ N m^{-2}, and a density of 1000 kg m^{-3}. The calculated velocity of sound in water is thus given by

$$V = \sqrt{\frac{2 \cdot 04 \times 10^9}{1000}} = 1430 \text{ m s}^{-1}$$

The proof of the velocity formula requires advanced mathematics, and is beyond the scope of this book. It can partly be verified by the method of dimensions, however. Thus since density, ρ, = mass/volume, the dimensions of ρ are given by ML^{-3}. The dimensions of force (mass $\times$ acceleration) are MLT^{-2}, the dimensions of area are L^2; and the denominator in (i) has zero dimensions since it is the ratio of two similar quantities. Thus the dimensions of modulus of elasticity, E, are given by

$$\frac{ML}{T^2L^2} \text{ or } ML^{-1}T^{-2}$$

Suppose the velocity, V, $= kE^x\rho^y$, where k is a constant. The dimensions of V are LT^{-1}

$$\therefore \quad LT^{-1} = (ML^{-1}T^{-2})^x \times (ML^{-3})^y$$

using the dimensions of E and ρ obtained above. Equating the respective indices of M, L, T on both sides, then

$$x + y = 0 \quad . \quad . \quad . \quad . \quad . \quad . \quad \text{(ii)}$$
$$-x - 3y = 1 \quad . \quad . \quad . \quad . \quad . \quad . \quad \text{(iii)}$$
$$-2x = -1 \quad . \quad . \quad . \quad . \quad . \quad \text{(iv)}$$

From (iv), $x = 1/2$, from (ii), $y = -1/2$. Thus, as $V = kE^x\rho^y$,

$$V = kE^{\frac{1}{2}}\rho^{-\frac{1}{2}}$$

$$\therefore \quad V = k\sqrt{\frac{E}{\rho}}$$

It is not possible to find the magnitude of k by the method of dimensions, but a rigid proof of the formula by calculus shows that $k = 1$ since $V = \sqrt{\frac{E}{\rho}}$.

Velocity of Sound in a Gas. Laplace's Correction

The velocity of sound in a *gas* is also given by $V = \sqrt{\frac{E}{\rho}}$ where E is the *bulk modulus* of the gas and ρ is its density. Now it is shown on p. 162 that $E = p$, the pressure of the gas, if the stresses and strains in the gas take place isothermally. The formula for the velocity then becomes

$V = \sqrt{\frac{p}{\rho}}$ and as the density, ρ, of air is $1 \cdot 29$ kg per m^3 at S.T.P. and

$$p = 0\cdot76 \times 13600 \times 9\cdot8 \text{ N m}^{-2};$$

$$V = \sqrt{\frac{0\cdot76 \times 13600 \times 9\cdot8}{1\cdot29}} = 280 \text{ m s}^{-1} \text{ (approx.).}$$

This calculation for V was first performed by Newton, who saw that the above theoretical value was well below the experimental value of about 330 m s^{-1}. The discrepancy remained unexplained for more than a century, when LAPLACE suggested in 1816 that E should be the *adiabatic* bulk modulus of a gas, not its isothermal bulk modulus as Newton had assumed. Alexander Wood in his book *Acoustics* (Blackie) points out that adiabatic conditions are maintained in a gas because of the relative slowness of sound wave oscillations.* It is shown later that the adiabatic bulk modulus of a gas is γp where γ is the ratio of the principal specific heats of a gas (i.e., $\gamma = c_p/c_V$). The formula for the velocity of sound in a gas thus becomes

$$V = \sqrt{\frac{\gamma p}{\rho}} \qquad . \qquad . \qquad . \qquad . \qquad (2)$$

The magnitude of γ for air is 1·40, and *Laplace's correction*, as it is known, then amends the value of the velocity in air at 0° C to

$$V = \sqrt{\frac{1\cdot40 \times 0\cdot76 \times 13600 \times 9\cdot8}{1\cdot29}} = 331 \text{ m s}^{-1}$$

This is in good agreement with the experimental value.

Effect of Pressure and Temperature on Velocity of Sound in a Gas

Suppose that the mass of a gas is m, and its volume is v. Its density, ρ, is then m/v, and hence the velocity of sound, V, is given by

$$V = \sqrt{\frac{\gamma p}{\rho}} = \sqrt{\frac{\gamma p v}{m}}.$$

But $pv = mRT$, where R is the gas constant for unit mass of the gas and T is its absolute temperature. Thus $pv/m = RT$, and hence

$$V = \sqrt{\gamma RT} \qquad . \qquad . \qquad . \qquad . \qquad \text{(i)}$$

Since γ and R are constants for a given gas, it follows that *the velocity of sound in a gas is independent of the pressure* if the temperature remains constant. This has been verified by experiments which showed that the velocity of sound at the top of a mountain is about the same as at the bottom, p. 619. It also follows from (i) that *the velocity of sound is proportional to the square root of its absolute temperature*. Thus if the velocity in air at 16° C is 338 m s^{-1} by experiment, the velocity, V, at 0° C is calculated from

$$\frac{V}{338} = \sqrt{\frac{273}{289}},$$

from which $\qquad\qquad V = 338\sqrt{\frac{273}{289}} = 328\cdot5 \text{ m s}^{-1}$

---

* It was supposed for many years that the changes are so rapid that there is no time for transfer of heat to occur. The reverse appears to be the case. At ultrasonic (very high) frequencies adiabatic conditions no longer hold.

Ultrasonics

There are sound waves of higher frequency than 20000 Hz, which are inaudible to a human being. These are known as *ultrasonics*; and since velocity = wavelength × frequency, ultrasonics have short wavelengths compared with sound waves in the audio-frequency range.

In recent years ultrasonics have been utilised for a variety of industrial purposes. They are used on board coasting vessels for depth sounding, the time taken by the wave to reach the bottom of the sea from the surface and back being determined. Ultrasonics are also used to kill bacteria in liquids, and they are used extensively to locate faults and cracks in metal castings, following a method similar to that of radar. Ultrasonic waves are sent into the metal under investigation, and the beam reflected from the fault is picked up on a cathode-ray tube screen together with the reflection from the other end of the metal. The position of the fault can then easily be located.

Production of Ultrasonics

In 1881 CURIE discovered that a thin plate of quartz increased or decreased in length if an electrical battery was connected to its opposite faces. By correctly cutting the plate, the expansion or contraction could be made to occur along the axis of the faces to which the battery was applied. When an alternating voltage of ultrasonic frequency was connected to the faces of such a crystal the faces vibrated at the same frequency, and thus ultrasonic sound waves were produced.

Another method of producing ultrasonics is to place an iron or nickel rod inside a solenoid carrying an alternating current of ultrasonic frequency. Since the length of a magnetic specimen increases slightly when it is magnetised, ultrasonic sound waves are produced by the vibrations of the rod.

EXAMPLES

1. How does the velocity of sound in a medium depend upon the elasticity and density? Illustrate your answer by reference to the case of air and of a long metal rod. The velocity of sound in air being 330·0 m s^{-1} at 0° C and the coefficient of expansion 1/273 per degree, find the change in velocity per degree Centigrade rise of temperature. (*L.*)

First part. The velocity of sound, V, is given by $V = \sqrt{E/\rho}$, where E is the modulus of elasticity of the medium and ρ is its density. In the case of air, a gas, E represents the bulk of modulus of the air under adiabatic conditions, and $E = \gamma p$ (see p. 624). Thus $V = \sqrt{\gamma p/\rho}$ for air.

For a long metal rod, E is Young's modulus for the metal, assuming the sound travels along the length of the rod.

Second part. Since the coefficient of expansion is 1/273 per degree Centigrade, the absolute temperature corresponding to $t°$ C is given by $(273 + t)$. The velocity of sound in a gas is proportional to the square root of its absolute temperature, and hence

$$\frac{V}{V_0} = \sqrt{\frac{274}{273}},$$

where V is the velocity at $1°$ C and V_0 is the velocity at $0°$ C.

$$\therefore V = V_0 \sqrt{\frac{274}{273}} = 330 \times \sqrt{\frac{274}{273}} = 330·6 \text{ m s}^{-1}$$

$$\therefore \quad \text{change in velocity} = 0·6 \text{ m s}^{-1}$$

2. How would you find by experiment the velocity of sound in air? Calculate the velocity of sound in air in metre second^{-1} at $100°$ C if the density of air at S.T.P. is $1·29$ kg m^{-3}, the density of mercury at $0°$ C 13600 kg m^{-3}, the specific heat capacity of air at constant pressure $1·02$, and the specific heat capacity of air at constant volume $0·72$, in kJ kg^{-1}K^{-1}. (*L*.)

First part. See Hebb's method, p. 618 or p. 587.

Second part. The velocity of sound in air is given by

$$V = \sqrt{\frac{\gamma p}{\rho}}$$

with the usual notation. The density, ρ, of air is $1·29$ kg m^{-3}. The pressure p is given by

$$p = h\rho g$$
$$= 0·76 \times 13600 \times 9·8 \text{ N m}^{-2}$$

since S.T.P. denotes 76 cm mercury pressure and $0°$ C. Also,

$$\gamma = \frac{C_p}{C_v} = \frac{1·02}{0·72}$$

$$\therefore V = \sqrt{\frac{1·02 \times 0·76 \times 13600 \times 9·8}{0·72 \times 1·293}},$$

where V is the velocity at $0°$ C.

But $\quad\quad\quad\quad\quad\quad\quad\quad\quad$ velocity $\propto \sqrt{T}$,

where T is the absolute temperature of the air. Thus if V' is the velocity at $100°$ C,

$$\frac{V'}{V} = \sqrt{\frac{273 + 100}{273}} = \sqrt{\frac{373}{273}}$$

$$\therefore V' = \sqrt{\frac{373}{273}} V = \sqrt{\frac{373 \times 1·02 \times 0·76 \times 13600 \times 9·8}{273 \times 0·72 \times 1·293}}$$

$$\therefore V' = 388 \text{ m s}^{-1}$$

3. State briefly how you would show by experiment that the characteristics of the transmission of sound are such that (*a*) a finite time is necessary for transmission, (*b*) a material medium is necessary for propagation, (*c*) the disturbance may be reflected and refracted. The wavelength of the note emitted by a tuning-fork, frequency 512 Hz, in air at $17°$ C is $66·5$ cm. If the density of air at S.T.P. is $1·293$ kg m^{-3}, calculate the ratio of the two specific heat capacities of air. Assume that the density of mercury is 13600 kg m^{-3}. (*N*.)

First part. See text.

Second part. Since $V = f\lambda$, the velocity of sound at $17°$ C. is given by

$$V = 512 \times 0·665 \text{ m s}^{-1} \quad\quad . \quad\quad . \quad\quad . \quad\quad \text{(i)}$$

Now
$$\frac{V_\circ}{V} = \sqrt{\frac{273}{290}},$$

where V_0 is the velocity at $0°$ C, since the velocity is proportional to the square root of the absolute temperature.

$$\therefore \quad V_\circ = \sqrt{\frac{273}{290}} \times V = \sqrt{\frac{273}{290}} \times 512 \times 0\cdot665 \quad . \qquad . \qquad \text{(ii)}$$

But
$$V_\circ = \sqrt{\frac{\gamma p}{\rho}},$$

where $p = 0\cdot76$ m of mercury $= 0\cdot76 \times 13600 \times 9\cdot8$ N m^{-2}, and $\rho = 1\cdot293$ kg m^{-3}.

$$\begin{aligned}
\therefore \quad \gamma &= \frac{V_\circ{}^2 \times \rho}{p} \\
&= \frac{272 \times 512^2 \times 0\cdot665 \times 1\cdot293}{290 \times 0\cdot76 \times 13600 \times 9\cdot8} \\
&= 1\cdot39
\end{aligned}$$

Doppler Effect

The whistle of a train or a jet aeroplane appears to increase in pitch as it approaches a stationary observer; as the moving object passes the observer, the pitch changes and becomes lowered. The apparent alteration in frequency was first predicted by DOPPLER in 1845, who stated that a change of frequency of the wave-motion should be observed when a source of sound or light was moving, and it is accordingly known as the *Doppler effect*.

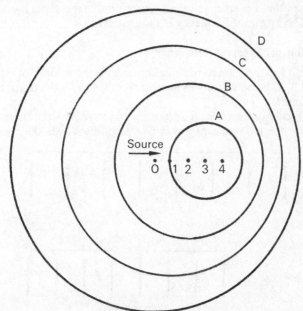

FIG. 26.13. Doppler effect.

The Doppler effect occurs whenever there is a *relative velocity* between the source of sound or light and an observer. In light, this effect was observed when measurements were taken of the wavelength of the colour of a moving star; they showed a marked variation. In sound, the Doppler effect can be demonstrated by placing a whistle in the end of a long piece of rubber tubing, and whirling the tube in a horizontal circle above the head while blowing the whistle. The open end of the tube acts as a moving source of sound, and an observer hears a rise and fall in pitch as the end approaches and recedes from him.

A complete calculation of the apparent frequency in particular cases is given shortly, but Fig. 26.13 illustrates how the change of wavelengths, and hence frequency, occurs when a source of sound is moving towards a stationary observer. At a certain instant the position of the moving source is at 4. At four successive seconds *before* this instant the source had been at the positions 3, 2, 1, 0 respectively. If V is the velocity of sound, the wavefront from the source when in the position 3 reaches the surface A of a sphere of radius V and centre 3 when the source is just at 4. In the same way, the wavefront from the source when it was in the position 2 reaches the surface B of a sphere of radius $2V$ and centre 2. The wavefront C corresponds to the source when it was in the position 1, and the wavefront D to the source when it was in the position O. Thus if the observer is on the right of the source S, he receives wavefronts which are relatively more crowded together than if S were stationary; the frequency of S thus appears to increase. When the observer is on the left of S, in which case the source is moving away from him, the wavefronts are farther apart than if S were stationary, and hence the observer receives correspondingly fewer waves per second. The apparent frequency is thus lowered.

Calculation of Apparent Frequency

Suppose V is the velocity of sound in air, u_s is the velocity of the source of sound S, u_o is the velocity of an observer O, and f is the true frequency of the source.

(i) *Source moving towards stationary observer.* If the source S were stationary, the f waves sent out in one second towards the observer O

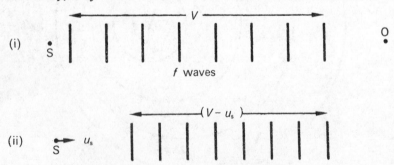

FIG. 26.14. Source moving towards stationary observer.

would occupy a distance V, and the wave length would be V/f, Fig. 26.14. (i). If S moves with a velocity u_s towards O, however, the f waves sent out occupy a distance $(V - u_s)$, because S has moved a distance u_s towards O in 1 sec, Fig. 26.14 (ii). Thus the wavelength λ' of the waves reaching O is now $(V - u_s)/f$.

But   velocity of sound waves = V.

$$\therefore \quad \text{apparent frequency,} \quad f' = \frac{V}{\lambda'} = \frac{V}{(V - u_s)/f}$$

$$\therefore \quad f' = \frac{V}{V - u_s} f \quad . \quad . \quad . \quad (3)$$

Since $(V - u_s)$ is less than V, f' is greater than f; the apparent frequency thus appears to increase when a source is moving towards an observer.

(ii) *Source moving away from stationary observer.* In this case the f waves sent out towards O in 1 sec occupy a distance $(V + u_s)$, Fig. 26.15.

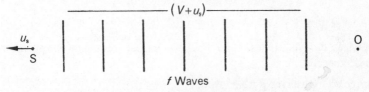

FIG. 26.15. Source moving away from stationary observer.

The wavelength λ' of the waves reaching O is thus $(V + u_s)/f$, and hence the apparent frequency f' is given by

$$f' = \frac{V}{\lambda'} = \frac{V}{(V + u_s)/f}$$

$$\therefore \quad f' = \frac{V}{V + u_s} \cdot f \quad . \quad . \quad . \quad . \quad (4)$$

Since $(V + u_s)$ is greater than V, f' is less than f, and hence the apparent frequency decreases when a source moves away from an observer.

(iii) *Source stationary, and observer moving towards it.* Since the source is stationary, the f waves sent out by S towards the moving observer O occupies a distance V, Fig. 26.16. The wavelength of the waves reaching O is hence V/f, and thus unlike the cases already considered, the wavelength is unaltered.

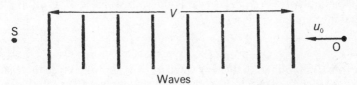

FIG. 26.16. Observer moving towards stationary source.

The velocity of the sound waves relative to O is not V, however, as O is moving relative to the source. The velocity of the sound waves relative to O is given by $(V + u_0)$ in this case, and hence the apparent frequency f' is given by

$$f' = \frac{(V + u_0)}{\text{wavelength}} = \frac{V + u_0}{V/f}$$

$$\therefore \quad f' = \frac{V + u_0}{V} \cdot f \qquad . \qquad . \qquad . \qquad . \qquad (5)$$

Since $(V + u_0)$ is greater than V, f' is greater than f; thus the apparent frequency is increased.

(iv) *Source stationary, and observer moving away from it*, Fig. 26.17. As in the case just considered, the wavelength of the waves reaching O is unaltered, and is given by V/f.

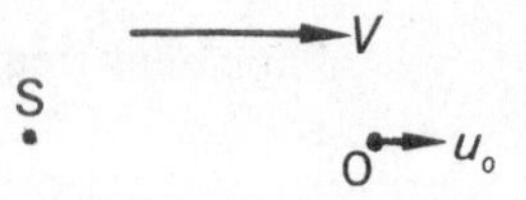

FIG. 26.17. Observer moving away from stationary source.

The velocity of the sound waves relative to $O = V - u_0$, and hence

$$\text{apparent frequency, } f', = \frac{V - u_0}{\text{wavelength}} = \frac{V - u_0}{V/f}$$

$$\therefore \quad f' = \frac{V - u_0}{V} \cdot f \qquad . \qquad . \qquad . \qquad (6)$$

Since $(V - u_0)$ is less than V, the apparent frequency f' appears to be decreased.

Source and Observer Both Moving

If the source and the observer are both moving, the apparent frequency f' can be found from the formula

$$f' = \frac{V'}{\lambda'}$$

where V' is the velocity of the sound waves relative to the observer, and λ' is the wavelength of the waves reaching the observer. This formula can also be used to find the apparent frequency in any of the cases considered before.

Suppose that the observer has a velocity u_0, the source a velocity u_s, and that both are moving in the *same* direction. Then

$$V' = V - u_0$$

and

$$\lambda' = (V - u_s)/f$$

as was deduced in case (i), p. 628.

$$\therefore \quad f' = \frac{V'}{\lambda'} = \frac{V - u_0}{(V - u_s)/f} = \frac{V - u_0}{V - u_s} \cdot f \quad . \qquad . \qquad (i)$$

If the observer is moving towards the source, $V' = V + u_0$, and the apparent frequency f' is given by

$$f' = \frac{V + u_0}{V - u_s} \cdot f \qquad . \qquad . \qquad . \qquad (ii)$$

From (i), it follows that $f' = f$ when $u_0 = u_s$, in which case there is no relative velocity between the source and the observer. It should also be noted that the motion of the observer affects only V', the velocity of the waves reaching the observer, while the motion of the source affects only λ', the wavelength of the waves reaching the observer.

The effect of the wind can also be taken into account in the Doppler effect. Suppose the velocity of the wind is u_w, in the direction of the line SO joining the source S to the observer O. Since the air has then a velocity u_w relative to the ground, and the velocity of the sound waves relative to the air is V, the velocity of the waves relative to ground is $(V + u_w)$ if the wind is blowing in the same direction as SO. All our previous expressions for f' can now be adjusted by replacing the velocity V in it by $(V + u_w)$. If the wind is blowing in the opposite direction to SO, the velocity V must be replaced by $(V - u_w)$.

When the source is moving at an angle to the line joining the source and observer, the apparent frequency changes continuously. Suppose the source is moving along AB with a velocity v, while the observer is stationary at O, Fig. 229. At S, the component velocity of v along OS is $v \cos \theta$, and is *towards* O. The observer thus hears a note of higher pitch whose frequency f' is given by

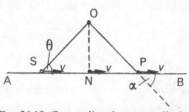

FIG. 26.18. Source direction perpendicular to observer.

$$f' = \frac{V}{V - v \cos \theta} f,$$

where V is the velocity of sound and f is the frequency of the source of sound. See equation (3), in which u_s now becomes $v \cos \theta$. When the source reaches P, Fig. 26.18, the component of v is $v \cos \alpha$ *away* from O, and the apparent frequency f'' is given by

$$f'' = \frac{V}{V + v \cos \alpha} f,$$

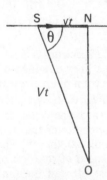

FIG. 26.19. Frequency heard when source at N.

from equation (4). The apparent frequency is thus lower than the frequency f of the source. When the source reaches N, the foot of the perpendicular from O to AB, the velocity v is perpendicular to ON and has thus no component towards the observer O. If the waves reach O shortly after, the observer hears a note of the same frequency f as the source.

Before the source S reaches N, however, it emits waves, travelling with a velocity V in air which reach O, Fig. 26.19. If S reaches N at the same instant as the waves reach O, the observer hears the note corresponding to the instant when the source was at S. In this case $SN = vt$ and $SO = Vt$, where t is the time-interval concerned. Thus:

$$\cos \theta = \frac{vt}{Vt} = \frac{v}{V}.$$

The frequency f' of the note heard by O when S just reaches N is hence given by

$$f' = \frac{V}{V - v \cos \theta} \cdot f = \frac{V}{V - v^2/V} \cdot f = \frac{V^2}{V^2 - v^2} \cdot f$$

Doppler's Principle in Light

The speed of distant stars and planets has been estimated from measurements of the wavelengths of the spectrum lines which they emit. Suppose a star or planet is moving with a velocity v away from the earth and emits light of wavelength λ. If the frequency of the vibrations is f cycles per second, then f waves are emitted in one second, where $c = f\lambda$ and c is the velocity of light *in vacuo*. Owing to the velocity v, the f waves occupy a distance $(c + v)$. Thus the *apparent wavelength* λ' to an observer on the earth in line with the star's motion is

$$\lambda' = \frac{c + v}{f} = \frac{c + v}{c} \cdot \lambda = \left(1 + \frac{v}{c}\right)\lambda$$

$$\therefore \quad \lambda' - \lambda = \text{``shift'' in wavelength} = \frac{v}{c}\lambda, \qquad . \qquad . \qquad \text{(i)}$$

and hence $\quad \dfrac{\lambda' - \lambda}{\lambda} = \text{fractional change in wavelength} = \dfrac{v}{c}.$ \qquad (ii)

From (i), it follows that λ' is greater than λ when the star or planet is moving away from the earth, that is, there is a "shift" or displacement *towards the red*. The position of a particular wavelength in the spectrum of the star is compared with that obtained in the laboratory, and the difference in the wavelengths, $\lambda' - \lambda$, is measured. From (i), knowing λ and c, the velocity v can be calculated.

If the star is moving *towards* the earth with a velocity u, the apparent wavelength λ'' is given by

$$\lambda'' = \frac{c - u}{f} = \frac{c - u}{c} \cdot \lambda = \left(1 - \frac{u}{c}\right)\lambda.$$

$$\therefore \quad \lambda - \lambda'' = \frac{u}{c}\lambda. \qquad . \qquad . \qquad . \qquad \text{(iii)}$$

Since λ'' is less than λ, there is a displacement towards the blue in this case.*

In measuring the speed of a star, a photograph of its spectrum is taken. The spectral lines are then compared with the same lines obtained by photographing in the laboratory an arc or spark spectrum of an element present in the star. If the former are displaced towards the red, the star is receding from the earth; if it is displaced towards the violet, the star is approaching the earth. By this method the velocities of the stars have been found to be between about 10 km s^{-1} and 300 km s^{-1}. The Doppler effect has also been used to measure the speed of rotation of the sun. Photographs are taken of the east and west edges of the sun; each contains absorption lines due to elements such as iron vaporised in the sun, and also some absorption lines due to oxygen in the earth's atmosphere. When the two photographs are put together so that the oxygen lines coincide, the iron lines in the two photographs are displaced relative to each other. In one case the edge of the sun approaches the earth, and in the other the opposite edge recedes from the earth. Measurements show a rotational speed of about 2 km s^{-1}.

---

* Equations (i)–(iii) apply for velocities much less than c, otherwise relativistic corrections are required. See *Introduction to Relativity*, Rosser (Butterworth).

Measurement of Plasma Temperature

In very hot gases or plasma, used in thermonuclear fusion experiments, the temperature is of the order of millions of degrees Celsius. At these high temperatures molecules of the glowing gas are moving away and towards the observer with very high speeds and, owing to the Doppler effect, the wavelength λ of a particular spectral line is apparently changed. One edge of the line now corresponds to an apparently increased wavelength λ_1 due to molecules moving directly away from the observer, and the other edge to an apparent decreased wavelength λ_2 due to molecules moving directly towards the observer. The line is thus observed to be *broadened*.

From our previous discussion, if v is the velocity of the molecules,

$$\lambda_1 = \frac{c + v}{c} \cdot \lambda$$

and

$$\lambda_2 = \frac{c - v}{c} \cdot \lambda$$

$$\therefore \quad \text{breadth of line, } \lambda_1 - \lambda_2 = \frac{2v}{c} \cdot \lambda \qquad . \qquad . \qquad \text{(i)}$$

The breadth of the line can be measured by a diffraction grating, and as λ and c are known, the velocity v can be calculated. By the kinetic theory of gases, the velocity v of the molecules is roughly the root-mean-square velocity, or $\sqrt{3RT}$, where T is the absolute temperature and R is the gas constant per gram of the gas. Consequently T can be found.

Doppler Effect and Radio Waves

A radio wave is an electromagnetic wave, like light, and travels with the same velocity, c, in free space of $3 \cdot 0 \times 10^5$ km s^{-1}. The Doppler effect with radio waves can be utilised for finding the speed of aeroplanes and satellites.

As an illustration, suppose an aircraft C sends out two radio beams at a frequency of 10^{10} Hz; one in a forward direction, and the other in a backward direction, each beam being inclined downward at an angle of 30° to the horizontal, Fig. 26.20. A Doppler effect is obtained when the radio waves are scattered at the ground at A, B, and when the

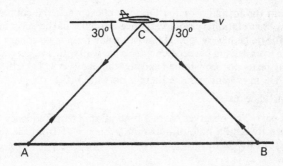

FIG. 26.20. Doppler effect and radio waves.

returning waves to C are combined, a beat frequency equal to their difference is measured. Suppose the beat frequency is 3×10^4 Hz.

If the velocity of the aircraft C is v, the velocity of radio waves is c and the frequency of the emitted beams is f, the apparent frequency f' of the waves reaching A is given by

$$f' = \frac{c}{c + v \cos \theta} \cdot f, \qquad \qquad \text{(i)}$$

where θ is 30°. The frequency f_1 of the wave received back at C from A is given by

$$f_1 = \frac{V'}{\lambda'}$$

where V' is the velocity of the wave relative to C and λ' is the wavelength of the waves reaching C. Since $V' = c - v \cos \theta$ and $\lambda' = c/f'$,

$$\therefore \quad f_1 = \frac{c - v \cos \theta}{c} f' = \frac{c - v \cos \theta}{c + v \cos \theta} \cdot f, \qquad \text{(ii)}$$

from (i). Similarly, the frequency f_2 of the waves received back at C from B is given by

$$f_2 = \frac{c + v \cos \theta}{c - v \cos \theta} \cdot f \qquad \qquad \text{(iii)}$$

$$\therefore \quad \text{beat frequency at C} = f_2 - f_1 = \frac{4 \, cv \cos \theta}{c^2 - v^2 \cos^2 \theta} \cdot f$$

Now $c = 3 \times 10^5$ km s^{-1}, $\theta = 30°$, $f_2 - f_1 = 3 \times 10^4$ Hz, $f = 10^{10}$ Hz, and $v^2 \cos^2 \theta$ is negligible compared with c^2.

$$\therefore \quad 3 \times 10^4 = \frac{4 \, cv \cos \theta}{c^2} \cdot f = \frac{4 \, v \cos \theta}{c} \cdot f$$

$$\therefore \quad v = \frac{3 \times 10^4 \times 3 \times 10^5}{4 \cos 30° \times 10^{10}}$$

$$= 0 \cdot 26 \text{ km s}^{-1}$$

$$= 936 \text{ km h}^{-1} \text{ (approx.)}$$

The speed of the aircraft relative to the ground is thus nearly 940 km h^{-1}

EXAMPLES

1. Obtain the formula for the Doppler effect when the source is moving with respect to a stationary observer. Give examples of the effect in sound and light. A whistle giving out 500 Hz moves away from a stationary observer in a direction towards and perpendicular to a flat wall with a velocity of $1 \cdot 5$ m s^{-1}. How many beats per sec will be heard by the observer? [Take the velocity of sound as 336 m s^{-1} and assume there is no wind.] (*C.*)

First part. See text.

Second part. The observer hears a note of apparent frequency f' from the whistle directly, and a note of apparent frequency f'' from the sound waves reflected from the wall.

Now

$$f' = \frac{V'}{\lambda'}$$

where V' is the velocity of sound in air relative to the observer and λ' is the wavelength of the waves reaching the observer. Since $V' = 336$ m s^{-1} and

$$\lambda' = \frac{336 + 1\cdot5}{500} \text{ m}$$

$$\therefore f' = \frac{336 \times 500}{337\cdot5} = 497\cdot8 \text{ Hz}$$

The note of apparent frequency f'' is due to sound waves moving towards the observer with a velocity of $1\cdot5$ m s^{-1}

$$\therefore \quad f'' = \frac{V'}{\lambda'} = \frac{336}{(336 - 1\cdot5)/500}$$

$$= \frac{336 \times 500}{334\cdot5} = 502\cdot2 \text{ Hz}$$

$$\therefore \text{ beats per second} = f'' - f' = 502\cdot2 - 497\cdot8 = 4\cdot4$$

2. Two observers A and B are provided with sources of sound of frequency 500. A remains stationary and B moves away from him at a velocity of $1\cdot8$m s^{-1} How many beats per sec are observed by A and by B, the velocity of sound being 330 m s^{-1}? Explain the principles involved in the solution of this problem. (*L.*)

Beats observed by A. A hears a note of frequency 500 due to its own source of sound. He also hears a note of apparent frequency f' due to the moving source B. With the usual notation,

$$f' = \frac{V'}{\lambda'} = \frac{330}{(330 + 1\cdot8)/500}$$

since the velocity of sound, V', relative to A is 330 m s^{-1} and the wavelength λ' of the waves reaching him is $(330 + 1\cdot8)/500$ m.

$$\therefore \quad f' = \frac{330 \times 500}{331\cdot8} = 497\cdot3$$

$$\therefore \text{ beats observed by A} = 500 - 497\cdot29 = 2\cdot71 \text{ Hz.}$$

Beats observed by B. The apparent frequency f' of the sound from A is given by

$$f' = \frac{V'}{\lambda'} \cdot$$

In this case $V' = $ velocity of sound relative to B $= 330 - 1\cdot8 = 328\cdot2$ m s^{-1} and the wavelength λ' of the waves reaching B is unaltered. Since $\lambda' = 330/500$ m, it follows that

$$f' = \frac{328\cdot2}{330/500} = \frac{328\cdot2 \times 500}{330} = 497\cdot27$$

$$\therefore \text{ beats heard by B} = 500 - 497\cdot27 = 2\cdot73 \text{ Hz}$$

EXERCISES 26

1. If the velocity of sound in air at $15°$C is 342 metres per second calculate the velocity at (*a*) $0°$C, (*b*) $47°$C. What is the velocity if the pressure of the air changes from 76 cm to 75 cm mercury, the temperature remaining constant at $15°$C?

2. Describe a determination (other than resonance) of the velocity of sound in air. How is the velocity dependent upon atmospheric conditions? Give Newton's expression for the velocity of sound in a gas, and Laplace's correction. Hence calculate the velocity of sound in air at 27°C. (Density of air at S.T.P. = 1·29 kg m^{-3}; C_p = 1·02 kJ kg^{-1} K^{-1}; C_v = 0·72 kJ kg^{-1} K^{-1}.) (L.)

3. Describe the factors on which the velocity of sound in a gas depends. A man standing at one end of a closed corridor 57 m long blew a short blast on a whistle. He found that the time from the blast to the sixth echo was two seconds. If the temperature was 17°C, what was the velocity of sound at 0°C? (C.)

4. Describe an experiment to find the velocity of sound in air at room temperature.

A ship at sea sends out simultaneously a wireless signal above the water and a sound signal through the water, the temperature of the water being 4°C. These signals are received by two stations, A and B, 40 km apart, the intervals between the arrival of the two signals being 16½ s at A and 22 s at B. Find the bearing from A of the ship relative to AB. The velocity of sound in water at t° C m s^{-1} = 1427 + 3·3t. (N.)

5. Write down an expression for the speed of sound in an ideal gas. Give a consistent set of units for the quantities involved.

Discuss the effect of changes of pressure and temperature on the speed of sound in air.

Describe an experimental method for finding a *reliable* value for the speed of sound in free air. (N.)

6. Describe an experiment to measure the velocity of sound in the open air. What factors may affect the value obtained and in what way may they do so?

It is noticed that a sharp tap made in front of a flight of stone steps gives rise to a ringing sound. Explain this and, assuming that each step is 0·25 m deep, estimate the frequency of the sound. (The velocity of sound may be taken to be 340 m s^{-1}.) (L.)

7. Explain why sounds are heard very clearly at great distances from the source (a) on still mornings after a clear night, and (b) when the wind is blowing from the source to the observer. (W.)

8. Describe one or two experiments to test each of the following statements: (a) If two notes are recognised by ear to be of the same pitch their sources are making the same number of vibrations per sec. (b) The musical interval between two notes is determined by the ratio of the frequencies of the vibrating sources of the notes. (L.)

9. Give a brief account of any important and characteristic wave phenomena which occur in sound. Why are sound waves in air regarded as longitudinal and not transverse?

An observer looking due north sees the flash of a gun 4 seconds before he records the arrival of the sound. If the temperature is 20°C and the wind is blowing from east to west with a velocity of 48 km per hour, calculate the distance between the observer and the gun. The velocity of sound in air at 0°C is 330 m s^{-1}. Why does the velocity of sound in air depend upon the temperature but not upon the pressure? (N.)

10. Explain upon what properties and conditions of a gas the velocity of sound through it depends.

Describe, and explain in detail, a laboratory method of measuring the velocity of sound in air. (*L.*)

Beats

11. Explain how *beats* are produced by two notes sounding together and obtain an expression for the number of beats heard per second.

A whistle of frequency 1000 Hz is sounded on a car travelling towards a cliff with a velocity of 18 m s^{-1}, normal to the cliff. Find the apparent frequency of the echo as heard by the car driver. Derive any relations used. (Assume velocity of sound in air to be 330 m s^{-1}.) (*L.*)

12. What is meant by (*a*) the *amplitude*, (*b*) the *frequency* of a vibration in the atmosphere? What are the corresponding characteristics of the musical sound associated with the vibration? How would you account for the difference in quality between two notes of the same pitch produced by two different instruments, e.g., by a violin and by an organ pipe?

What are 'beats'? Given a set of standard forks of frequencies 256, 264, 272, 280, and 288, and a tuning-fork whose frequency is known to be between 256 and 288, how would you determine its frequency to four significant figures? (*W.*)

13. Explain the origin of the beats heard when two tuning-forks of slightly different frequency are sounded together. Deduce the relation between the frequency of the beats and the difference in frequency of the forks. How would you determine which fork had the higher frequency?

A simple pendulum set up to swing in front of the 'seconds' pendulum ($T = 2$ s) of a clock is seen to gain so that the two swing in phase at intervals of 21 s. What is the time of swing of the simple pendulum? (*L.*)

Doppler's Principle

14. An observer beside a railway line determines the speed of a train by observing the change in frequency of the note of its whistle as it passes him. Explain why a change of frequency occurs and derive the relation from which the speed may be calculated. Describe an example of the same principle in another branch of physics.

Find the lowest velocity that can be measured in this way, if the true frequency of the whistle is 1000 Hz and the observer is unable to detect departures from this frequency of less than 20 Hz. (Assume the velocity of sound to be 340 m s^{-1}.) (*L.*)

15. Explain what is meant by the *Doppler effect* in sound. Does an observer hear the same pitch from a given source of sound irrespective of whether the source approaches the stationary observer at a certain velocity or the observer approaches the stationary source at the same velocity? Explain how you arrived at your answer.

The light of the H (calcium) line of the spectrum is deviated through an angle of 45° 12′ by a certain prism. When observed in the light of a distant nebula, the deviation is 44° 15′. Calculate the velocity of the nebula in the line of sight, taking the velocity of light in vacuo to be 3.00×10^8 m s^{-1} and the deviation to be inversely proportional to the wavelength of the light over the range of values to be considered. (*L.*)

16. Explain in each case the change in the apparent frequency of a note brought about by the motion of (i) the source, (ii) the observer, relative to the transmitting medium.

Derive expressions for the ratio of the apparent to the real frequency in the cases where (a) the source, (b) the observer, is at rest, while the other is moving along the line joining them.

The locomotive of a train approaching a tunnel in a cliff face at 95 km.p.h. is sounding a whistle of frequency 1000 Hz. What will be the apparent frequency of the echo from the cliff face heard by the driver? What would be the apparent frequency of the echo if the train were emerging from the tunnel at the same speed? (Take the velocity of sound in air as 330 m s^{-1}.) (L.)

17. (a) State the conditions necessary for 'beats' to be heard and derive an expression for their frequency.

(b) A fixed source generates sound waves which travel with a speed of 330 m s^{-1}. They are found by a distant stationary observer to have a frequency of 500 Hz. What is the wavelength of the waves? From first principles find (i) the wavelength of the waves in the direction of the observer, and (ii) the frequency of the sound heard if (1) the source is moving towards the stationary observer with a speed of 30 m s^{-1}, (2) the observer is moving towards the stationary source with a speed of 30 m s^{-1}, (3) both source and observer move with a speed of 30 m s^{-1} and approach one another. (N.)

18. What is the *Doppler effect*? Find an expression for it when the observer is at rest and there is no wind.

A whistle is whirled in a circle of 100 cm radius and traverses the circular path twice per second. An observer is situated outside the circle but in its plane. What is the musical interval between the highest and lowest pitch observed if the velocity of sound is 332 m s^{-1}? (L.)

19. Explain why the frequency of a wave motion appears, to a stationary observer, to change as the component of the velocity of the source along the line joining the source and observer changes. Describe two illustrations of this effect, one with sound and one with light.

A stationary observer is standing at a distance l from a straight railway track and a train passes with uniform velocity v sounding a whistle with frequency n_0. Taking the velocity of sound as V, derive a formula giving the observed frequency n as a function of the time. At which position of the train will $n = n_0$? Give a physical interpretation of the result. (C.)

Sound Intensity. Acoustics

20. Explain what is meant by (a) an intensity level in sound, (b) the statement that two intensity levels differ by 5 decibels. What considerations have determined the choice of a zero level in connection with the specification of *loudness*?

A loudspeaker produces a sound intensity level of 8 decibels above a certain reference level at a point P, 40 m from it. Find (a) the intensity level at a point 30 m from the loudspeaker, (b) the intensity level at P if the electrical power to the loudspeaker is halved. (L.)

21. Distinguish between the *intensity* and *loudness* of a sound. In what units would the intensity be measured? Define in each instance a unit employed to compare (a) the intensity and (b) the loudness of two sounds.

A source of sound is situated midway, between an observer and a flat wall. If the absorption coefficient of the wall is 0·25 find the ratio of the intensities of sound heard by the observer directly and by reflexion. Give the answer in decibels. (L.)

22. Describe a method for the accurate measurement of the velocity of sound in *free* air.

Indicate the factors which influence the velocity and how they are allowed for or eliminated in the experiment you describe.

At a point 20 m from a small source of sound the intensity is 0·5 microwatt cm^{-2}. Find a value for the rate of emission of sound energy from the source, and state the assumptions you make in your calculation. (*N.*)

23. Distinguish between *intensity* and *intensity level* of a sound.

The time taken for a sound to decay to one-millionth of its previous intensity after the source has been cut off is called the reverberation time. For a pure tone which gives an intensity level of 83 decibels in an empty lecture theatre the reverberation time was found to be 3·8 seconds. Calculate the sound intensity 7·6 seconds after the note was switched off. (Assume that the reference zero of intensity was 10^{-12} watt m^{-2}.)

Explain what was meant by a listener who stated that the note had a loudness of 70 phons.

Discuss how the acoustic properties of this lecture theatre might be improved. (*N.*)

24. (*a*) Discuss the relation between the *intensity level* and the *loudness* of a sound. Define suitable units in which each may be expressed.

(*b*) Give an account of the effect on the acoustics of a concert hall of such factors as: the design and material of the walls: the size of the audience; the frequency of the note. (*L.*)

25. A hall is 25 m long, 8 m wide and has walls 8 m high. The ceiling is a barrel vault of radius 5 m and the ends of the hall are plane. The floor is wood block and the walls are hard plaster, wood panelling and glass. The ceiling is also of hard plaster.

Indicate and give reasons for *three* defects of this hall as an auditorium and show how you would attempt to correct them.

Diagrams are essential in the answer to this question. (*N.*)

chapter twenty-seven

Vibrations in Pipes, Strings, Rods

Introduction

THE music from an organ, a violin, or a xylophone is due to vibrations in the air set up by oscillations in these instruments. In the organ, air is blown into a pipe, which sounds its characteristic note as the air inside it vibrates; in the violin, the strings are bowed so that they oscillate; and in a xylophone a row of metallic rods are struck in the middle with a hammer, which sets them into vibration.

Before considering each of the above cases in more detail, it would be best to consider the feature common to all of them. A violin string is fixed at both ends, A, B, and waves travel along m, n to each end of the string when it is bowed and are there reflected, Fig. 27.1 (i).

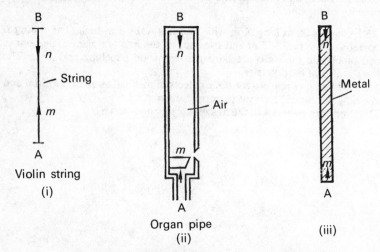

FIG. 27.1. Reflection of waves in instruments.

The vibrations of the particles of the string are hence due to *two waves of the same frequency and amplitude travelling in opposite directions*. A similar effect is obtained with an organ pipe closed at one end B, Fig. 27.1 (ii). If air is blown into the pipe at A, a wave travels along the direction m and is reflected at B in the opposite direction n. The vibrations of the air in the pipe are thus due to two waves travelling in opposite directions. If a metal rod is fixed at its middle in a vice and stroked at one end A, a wave travels along the rod in the direction m and is reflected at the other end B in the direction n, Fig. 27.1 (iii). The vibrations of the rod, which produce a high-pitched note, are thus due to two waves travelling in opposite directions.

The resultant effect of two waves travelling in opposite directions with equal amplitude and frequency can easily be demonstrated. A light string, or thread, is tied to the end of a clapper, P, of an electric bell, and the other end of the string is passed round a grooved wheel, Fig. 27.2.

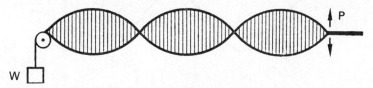

FIG. 27.2. Demonstration of stationary wave.

When the clapper vibrates, and a suitable weight W is attached to the string, a number of *stationary loops* is observed along the vibrating string, somewhat as shown in Fig. 27.2. By altering W a different number of stationary loops can be obtained. The wave along the string is known as a *stationary wave*, and we shall now discuss the formation of a stationary wave in detail.

Stationary Waves

Consider a plane-progressive wave *a* travelling in air along OA, Fig. 27.3. If it meets a wall at W a reflected wave *b* is obtained, and the condition of the air along W is due to the combined effects of *a*, *b*.

The layer of air at W must always be at rest since it is in contact with a fixed wall. For convenience, suppose that the displacements of the layers of air due to *a* at the instant shown are those represented by the sine wave in Fig. 27.3 (i), so that the displacement of the layer at W due to the incident wave is a maximum. Since the layer at W is always at rest, the displacements of the layers due to the wave *reflected* from the wall must be represented by the curve *b* at the same instant; otherwise the net displacement at W, which is the algebraic sum of WR, WH, will not be zero. From the curves *a*, *b* shown in Fig. 27.3 (i), it follows that the wave *b* reflected by the wall is 180° out of phase with the incident wave *a*.

At the instant *t* represented in Fig. 27.3 (i), the algebraic sum, S, of the displacements of the layers *everywhere* along OW is zero if the amplitudes of *a*, *b* are equal and the curves have the same wavelength. At an instant $T/4$ later, where T is the period of vibration of the layers the displacements of the layer due to the incident and reflected waves are those shown in Fig. 27.3 (ii). This can best be understood by imagining the incident wave *a* to have advanced to the right by ¼-wavelength, and the reflected wave to have advanced to the left by ¼-wavelength, which implies that the vibrating layers have now reached a displacement corresponding to a time $T/4$ later than *t*. The algebraic sums S of the displacement is then represented by the curve S in Fig. 27.3 (ii). At the end of a further time $T/4$, the displacements due to *a*, *b* are those shown

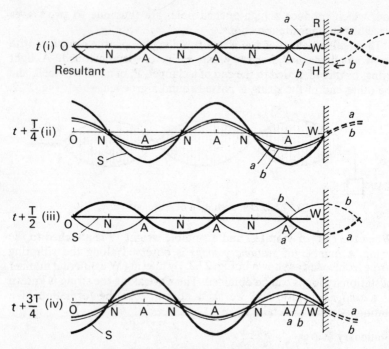

FIG. 27.3. Formation of stationary waves.

in Fig. 27.3 (iii); the waves have now advanced another ¼-wavelength in opposite directions. The algebraic sum S of the displacements is again zero everywhere along OW at this instant. After a further time $T/4$, the displacements of the layers, and the resultant displacement S, are those shown in Fig. 27.3 (iv). The wave in the air represented by S is called a stationary wave.

Nodes and Antinodes

We have now sufficient information to deduce the conditions of the layers of air along OW when a stationary wave is obtained. From the curves showing the resultant displacement, S, in Fig. 27.3, it can be seen that some layers, marked N, are *permanently* at rest; these are known as **nodes**. The layers marked A, however, are vibrating through an amplitude twice as big as the incident or reflected waves, see Fig. 27.3 (ii) and (iv), and these are known as **antinodes**. Layers between consecutive nodes are vibrating in phase with each other, but the amplitude of vibration varies from zero at a node N to a maximum

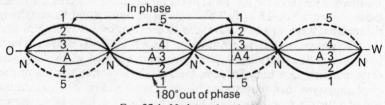

FIG. 27.4. Nodes and antinodes.

at the antinode A. Fig. 27.4 represents the displacement of the layers along OW at five different instants 1, 2, 3, 4, 5. It follows that

$$\textit{the distance between consecutive nodes, NN} = \frac{\lambda}{2} . \qquad . \qquad \text{(i)}$$

where λ is the wavelength of the stationary wave;

$$\textit{the distance between consecutive antinodes, AA} = \frac{\lambda}{2}, \qquad . \qquad \text{(ii)}$$

and

$$\textit{the distance from a node to the next antinode, NA} = \frac{\lambda}{4} . \qquad . \qquad \text{(iii)}$$

The importance of the nodes and antinodes in a stationary wave lies in their simple connection with the wavelength.

Differences Between Plane-Progressive and Stationary Waves

At the beginning of Chapter 25, we considered in detail the plane-progressive wave and its effect on the medium (pp. 584 and 587). It was then shown that each layer vibrates with constant amplitude at the same frequency, and that each layer is out of phase with others near to it. When a stationary wave is present in a medium, however, some layers (nodes) are permanently at rest; others between the nodes are vibrating in phase with different amplitudes, increasing to a maximum at the antinodes. A stationary wave is always set up when two plane-progressive waves of equal amplitude and frequency travel in opposite directions in the same medium.

Mathematical proof of stationary wave properties. The properties of the stationary wave, already deduced, can be obtained easily from a mathematical treatment. Suppose $y_1 = a \sin 2\pi \left(\dfrac{t}{T} - \dfrac{x}{\lambda} \right)$ is a plane-progressive wave travel-

ling in one direction along the x-axis (p. 587). Then $y_2 = a \sin 2\pi \left(\dfrac{t}{T} + \dfrac{x}{\lambda} \right)$

represents a wave of the same amplitude and frequency travelling in the opposite direction. The resultant displacement, y, is hence given by

$$y = y_1 + y_2 = a \left[\sin 2\pi \left(\frac{t}{T} - \frac{x}{\lambda} \right) + \sin 2\pi \left(\frac{t}{T} + \frac{x}{\lambda} \right) \right]$$

from which
$$y = 2a \sin \frac{2\pi t}{T} . \cos \frac{2\pi x}{\lambda} \qquad . \qquad . \qquad . \qquad \text{(i)}$$

using the transformation of the sum of two sine functions to a product.

$$\therefore \quad y = B \sin \frac{2\pi t}{T} \qquad . \qquad . \qquad . \qquad . \qquad . \qquad \text{(ii)}$$

where
$$B = 2a \cos \frac{2\pi x}{\lambda} \qquad . \qquad . \qquad . \qquad . \qquad \text{(iii)}$$

From (ii), B is the magnitude of the *amplitude* of vibration of the various layers; and from (iii) it also follows that the amplitude is a maximum and equal to $2a$ at $x = 0$, $x = \lambda/2$, $x = \lambda$, and so on. These points are thus antinodes, and consecutive antinodes are hence separated by a distance $\lambda/2$. The amplitude B is zero when $x = \lambda/4$, $x = 3\lambda/4$, $x = 5\lambda/4$, and so on. These points are thus nodes, and they are hence midway between consecutive antinodes.

The particle velocity in a stationary wave is the rate of change of the displacement (y) of the particle with respect to time (t). The velocity at the nodes is always zero since the particles there are permanently at rest. The velocity at an antinode increases from zero (when the particle is at the end of its oscillation) to a maximum (when the particle passes through its mean or original position). The corresponding displacement and velocity curves in the latter case are illustrated in Fig. 27.5 by P, M

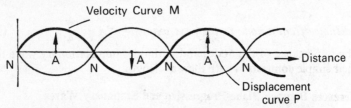

FIG. 27.5. Particle velocity due to stationary wave.

respectively. It will be noted that particles at neighbouring antinodes are moving in opposite directions at any instant.

Variation of Pressure in the Stationary Wave

Having considered the variation of the displacements and the particle velocities when a stationary wave travels in air, we must now turn our attention to the variation of *pressure* in the air.

Suppose that curve 1 represents the displacements at the antinodes and other points at an instant when they are a maximum, Fig. 27.6 (i). The layer of air immediately to the left of the node at a is then displaced towards a, since the displacement is positive from curve 1, and the layer immediately to the right of a also displaced towards a. The air at a is thus compressed, and the pressure is thus greater than normal, as represented by curve 1 in Fig. 27.6 (ii). The displacements of the layers on

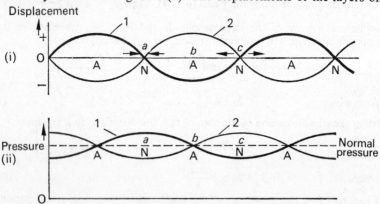

FIG. 27.6. Pressure variation due to stationary wave.

either side of the antinode at b are each a maximum to the left, and hence the pressure of the air is normal. The air on the left of the node c is displaced away from *a*, and the air on the right of *c* is also displaced away from *c*. The air is thus rarified here, and hence the pressure is less normal. By carrying out the same procedure at other points in the air, it can be seen that the pressure variation corresponds to the curve 1 in Fig. 27.6 (ii).

When the displacements change to those represented by curve 2 in Fig. 27.6 (i), the variation of pressure at the same instant is shown by curve 2 in Fig. 27.6 (ii). We can now see that *the pressure variation is always a maximum at a node of the stationary wave, and is always zero at an antinode of the stationary wave.* In a plane-progressive wave, however, the pressure variation is the same at every point in a medium (p. 585).

EXAMPLE

Distinguish between progressive and stationary wave motion. Describe and illustrate with an example how stationary wave motion is produced. Plane sound waves of frequency 100 Hz fall normally on a smooth wall. At what distances from the wall will the air particles have (*a*) maximum, (*b*) minimum amplitude of vibration? Give reasons for your answer. (The velocity of sound in air may be taken as 340 m s^{-1}.) (*L.*)

First part. See p. 587 and p. 643.

Second part. A stationary wave is set up between the source and the wall, due to the production of a reflected wave. The wall is a node, since the air in contact with it cannot move; and other nodes are at equal distances, *d*, from the wall. But if the wavelength is λ,

$$d = \frac{\lambda}{2} \text{ (p. 643)}.$$

Also

$$\lambda = \frac{V}{f} = \frac{340}{100} = 3 \cdot 4 \text{ m}$$

$$\therefore \quad d = \frac{3 \cdot 4}{2} = 1 \cdot 7 \text{ m}.$$

Thus minimum amplitude of vibration is obtained 1·7, 3·4, 5·1 m . . . from the wall.

The antinodes are midway between the nodes. Thus maximum amplitude of vibration is obtained 0·85, 2·55, 4·25m, . . . from the wall.

VIBRATIONS OF AIR IN PIPES

Closed Pipe

A *closed* or *stopped organ pipe* consists essentially of a metal pipe closed at one end Q, and a blast of air is blown into it at the other end P, Fig. 27.7 (i). A wave thus travels up the pipe to Q, and is reflected at this end down the pipe, so that a *stationary wave* is obtained. The end Q of the closed pipe must be a node N, since the layer in contact with Q must be permanently at rest, and the open end A, where the air is free to vibrate, must be an antinode A. The simplest stationary wave in the air in the pipe is hence represented by g in Fig. 27.7 (ii), where the pipe is positioned horizontally to show the relative displacement, y, of the layers at different distances, x, from the closed end Q; the axis of the stationary wave is Qx.

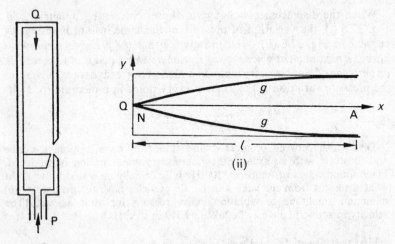

FIG. 27.7. (i). Closed (stopped) pipe. (ii). Fundamental of closed (stopped) pipe.

It can now be seen that the length l of the pipe is equal to the distance between a node N and a consecutive antinode A of the stationary wave. But NA $= \lambda/4$, where λ is the wavelength (p. 643).

$$\therefore \quad \frac{\lambda}{4} = l$$

$$\therefore \quad \lambda = 4l$$

But the frequency, f, of the note is given by $f = V/\lambda$, where V is the velocity of sound in air.

$$\therefore \quad f = \frac{V}{4l}$$

This is the frequency of the lowest note obtainable from the pipe, and it is known as its *fundamental*. We shall denote the fundamental frequency by f_0, so that

$$f_0 = \frac{V}{4l} \quad . \quad \quad . \quad \quad . \quad \quad . \quad \quad . \quad \quad . \quad (1)$$

Overtones of Closed Pipe

If a stronger blast of air is blown into the pipes, notes of higher frequency can be obtained which are simple multiples of the fundamental frequency f_0. Two possible cases of stationary waves are shown in Fig. 27.8. In each, the closed end of the pipe is a node, and the open end is an antinode. In Fig. 27.8 (i), however, the length l of the pipe is

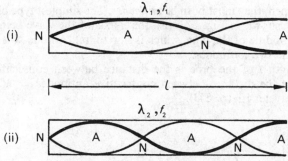

FIG. 27.8. Overtones in closed pipe.

related to the wavelength λ_1 of the wave by

$$l = \tfrac{3}{4} \lambda_1$$

$$\therefore \qquad \lambda_1 = \frac{4l}{3}$$

The frequency f_1 of the note is thus given by

$$f_1 = \frac{V}{\lambda_1} = \frac{3V}{4l} \qquad \cdot \quad \cdot \quad \cdot \quad \cdot \quad \text{(i)}$$

But $$f_0 = \frac{V}{4l}$$

$$\therefore \qquad f_1 = 3f_0 \qquad \cdot \quad \cdot \quad \cdot \quad \cdot \quad \text{(ii)}$$

In Fig. 27.8 (ii), when a note of frequency f_2 is obtained, the length l of the pipe is related to the wavelength λ_2 by

$$l = \frac{5\lambda_2}{4}$$

$$\therefore \qquad \lambda_2 = \frac{4l}{5}$$

$$\therefore \qquad f_2 = \frac{V}{\lambda_2} = \frac{5V}{4l} \qquad \cdot \quad \cdot \quad \cdot \quad \cdot \quad \text{(iii)}$$

$$\therefore \qquad f_2 = 5f_0 \qquad \cdot \quad \cdot \quad \cdot \quad \cdot \quad \text{(iv)}$$

By drawing other sketches of stationary waves, with the closed end as a node and the open end as an antinode, it can be shown that higher frequencies can be obtained which have frequencies of $7f_0$, $9f_0$, and so on. They are produced by blowing harder at the open end of the pipe. The frequencies obtainable at a closed pipe are hence f_0, $3f_0$, $5f_0$, and so on, i.e., the closed pipe gives only odd harmonics, and hence the frequencies $3f_0$, $5f_0$, etc. are possible *overtones*.

Open Pipe

An "open" pipe is one which is open at both ends. When air is blown into it at P, a wave m travels to the open end Q, where it is reflected in the direction n on encountering the free air, Fig. 27.9 (i). A stationary wave is hence set up in the air in the pipe, and as the two ends of the pipe are open, they must both be *antinodes*. The simplest type of wave is hence that shown in Fig. 27.9 (ii), the x-axis of the wave being drawn along the middle of the pipe, which is horizontal. A node N is midway between the two antinodes.

The length l of the pipe is the distance between consecutive antinodes. But the distance between consecutive antinodes $= \lambda/2$, where λ is the wavelength (p. 643).

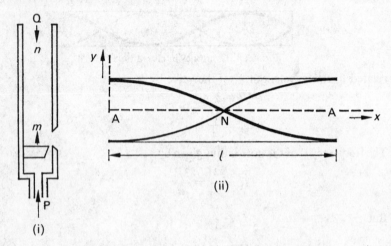

FIG. 27.9. (i). Open pipe.   (ii). Fundamental of open pipe.

$$\therefore \qquad \frac{\lambda}{2} = l$$

$$\therefore \qquad \lambda = 2l$$

Thus the frequency f_0 of the note obtained from the pipe is given by

$$f_0 = \frac{V}{\lambda} = \frac{V}{2l} \quad . \qquad . \qquad . \qquad . \qquad . \qquad (2)$$

This is the frequency of the fundamental note of the pipe.

Overtones of Open Pipe

Notes of higher frequencies than f_0 can be obtained from the pipe by blowing harder. The stationary wave in the pipe has always an antinode A at each end, and Fig. 27.10 (i) represents the case of a note of a frequency f_1.

The length l of the pipe is equal to the wavelength λ_1 of the wave in this case. Thus

$$f_1 = \frac{V}{\lambda_1} = \frac{V}{l}$$

But $\qquad\qquad\qquad f_0 = \frac{V}{2l}$, from (2) above.

$$\therefore \qquad f_1 = 2f_0 \qquad . \qquad . \qquad . \qquad . \qquad . \qquad (i)$$

In Fig. 27.10 (ii), the length $l = \frac{3}{2}\lambda_2$, where λ_2 is the wavelength in the pipe. The frequency f_2 is thus given by

$$f_2 = \frac{V}{\lambda_2} = \frac{3V}{2l}$$

as $\lambda_2 = \frac{2l}{3}$.

$$\therefore \quad f_2 = 3f_0 \qquad . \qquad . \qquad . \qquad . \qquad . \qquad \text{(ii)}$$

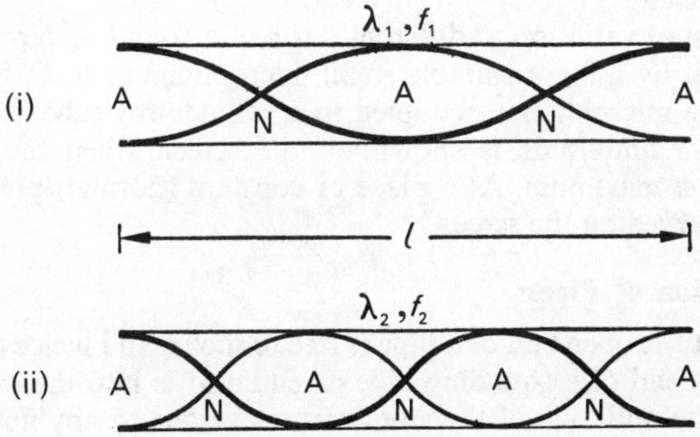

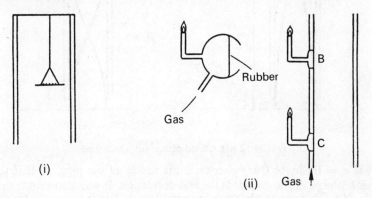

FIG. 27.10. Overtones of open pipes.

The frequencies of the overtones in the open pipe are thus $2f_0$, $3f_0$, $4f_0$, and so on, i.e., all harmonics are obtainable. The frequencies of the overtones in the closed pipe are $3f_0$, $5f_0$, $7f_0$, and so on, and hence the *quality* of the same note obtained from a closed and an open pipe is different (see p. 611).

Detection of Nodes and Antinodes, and Pressure Variation, in Pipes

The *nodes and antinodes* in a sounding pipe have been detected by suspending inside it a very thin piece of paper with lycopodium or fine sand particles on it, Fig. 27.11 (i). The particles are considerably agitated at the antinodes, but they are motionless at the nodes.

FIG. 27.11. (i). Detection of nodes and antinodes.
(ii). Detection of pressure.

The *pressure variation* in a sounding pipe has been examined by means of a sensitive flame, designed by Lord Rayleigh. The length of the flame can be made sensitive to the pressure of the gas supplied, so that if the pressure changes the length of flame is considerably affected. Several of the flames can be arranged at different parts of the pipe, with a thin rubber or mica diaphragm in the pipe, such as at B, C, Fig. 27.11 (ii). At a place of maximum pressure variation, which is a node (p. 645), the length of flame alters accordingly. At a place of constant (normal) pressure, which is an antinode, the length of flame remains constant.

The pressure variation at different parts of a sounding pipe can also be examined by using a suitable small microphone at B, C, instead of a flame. The microphone is coupled to a cathode-ray tube and a wave of maximum amplitude is shown on the screen when the pressure variation is a maximum. At a place of constant (normal) pressure, no wave is observed on the screen.

End-correction of Pipes

The air at the open end of a pipe is free to move, and hence the vibrations at this end of a sounding pipe extend a little into the air outside the pipe. The antinode of the stationary wave due to any note is thus a distance c from the open end in practice, known as the *end-correction*, and hence the wavelength λ in the case of a closed pipe is given by $\lambda/4 = l + c$, where l is the length of the pipe, Fig. 27.12 (i). In the case of an open pipe sounding its fundamental note, the wavelength λ is given by $\lambda/2 = l + c + c$, since *two* end-corrections are required, assuming the end-corrections are equal, Fig. 27.12 (ii). Thus $\lambda = 2(l + 2c)$. See also p. 648.

The mathematical theory of the end-correction was developed independently by Helmholtz and Rayleigh. It is now generally accepted

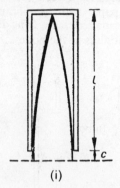
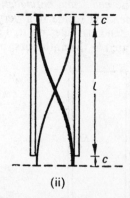

(i)   (ii)

FIG. 27.12. (i). Closed pipe.   (ii). Open pipe.

that $c = 0.58r$, or $0.6r$, where r is the radius of the pipe, so that the wider the pipe, the greater is the end-correction. It was also shown that the end-correction depends on the wavelength λ of the note, and tends to vanish for very short wavelengths.

Effect of Temperature, and End-correction, on the Pitch of Pipes

The frequency, f_0, of the fundamental note of a closed pipe of length l and end-correction c is given by

$$f_0 = \frac{V}{\lambda} = \frac{V}{4(l+c)} \qquad . \qquad . \qquad . \qquad \text{(i)}$$

with the usual notation, since $\lambda = 4(l+c)$. See p. 650. Now the velocity of sound, V, in air at $t°$ C is related to its velocity V_0 at $0°$ C by

$$\frac{V}{V_0} = \sqrt{\frac{273+t}{273}} = \sqrt{1+\frac{t}{273}} \qquad . \qquad . \qquad \text{(ii)}$$

since the velocity is proportional to the square root of the absolute temperature. Substituting for V in (i),

$$\therefore \ f_0 = \frac{V_0}{4(l+c)} \sqrt{1+\frac{t}{273}} \qquad . \qquad . \qquad . \qquad \text{(iii)}$$

From (iii), it follows that, with a given pipe, *the frequency of the fundamental increases as the temperature increases.* Also, for a given temperature and length of pipe, the frequency decreases as c increases. Now $c = 0.6r$, where r is the radius of the pipe. Thus *the frequency of the note from a pipe of given length is lower the wider the pipe,* the temperature being constant. The same results hold for an open pipe.

Resonance

If a diving springboard is bent and then allowed to vibrate freely, it oscillates with a frequency which is called its *natural frequency.* When a diver on the edge of the board begins to jump up and down repeatedly, the board is forced to vibrate at the frequency of the jumps; and at first, when the amplitude is small, the board is said to be undergoing *forced vibrations.* As the diver jumps up and down to gain increasing height for his dive, the frequency of the periodic downward force reaches a stage where it is practically the same as the natural frequency of the board. The amplitude of the board then becomes very large, and the periodic force is said to have set the board in *resonance.*

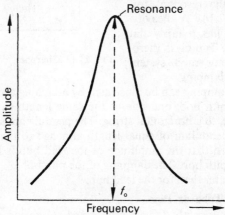

FIG. 27.13. Resonance curve.

A mechanical system which is free to move, like a wooden bridge or the air in pipes, has a natural frequency of vibration, f_0, which depends on its dimensions. When a periodic force of a frequency different from f_0 is applied to the system, the latter vibrates with a small amplitude and undergoes forced vibrations. When the periodic force has a frequency equal to the natural frequency f_0 of the system, the amplitude of vibration becomes a maximum, and the system is then set into resonance. Fig. 27.13 is a typical curve showing the variation of amplitude with frequency. Some time ago it was reported in the newspapers that a soprano who was broadcasting had broken a glass tumbler on the table of a listener when she had reached a high note. This is an example of resonance. The glass had a natural frequency equal to that of the note sung, and was thus set into a vibration sufficiently violent to break it.

The phenomenon of resonance occurs in other branches of Physics than Sound and Mechanics. When an electrical circuit containing a coil and capacitor is "tuned" to receive the radio waves from a distant transmitter, the frequency of the radio waves is equal to the natural frequency of the circuit and resonance is therefore obtained. A large current then flows in the electrical circuit. A dark line in a continuous spectrum, an absorption line, is an example of *optical resonance*. Thus some of the yellow wavelengths from the sun's spectrum are absorbed by molecules of sodium vapour in the cooler part of the sun's atmosphere, which are set into resonance (see p. 464).

Sharpness of resonance. As the resonance condition is approached, the effect of the damping forces on the amplitude increases. Damping prevents the amplitude from becoming infinitely large at resonance. The lighter the damping, the sharper is the resonance, that is, the amplitude diminishes considerably at a frequency slightly different from the resonant frequency, Fig. 27.14. A heavily-damped system has a fairly flat resonance curve. Tuning is therefore more difficult in a system which has light damping.

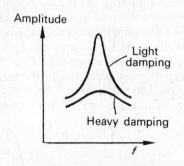

FIG. 27.14. Sharpness of resonance.

The effect of damping can be illustrated by attaching a simple pendulum carrying a pith bob, and one of the same length carrying a lead bob of equal size, to a horizontal string. The pendula are set into vibration by a third pendulum of equal length attached to the same string, and it is then seen that the amplitude of the lead bob is much greater than that of the pith bob. The damping of the pith bob due to air resistance is much greater than for the lead bob.

Resonance in a Tube or Pipe

If a person blows gently down a pipe closed at one end, the air inside vibrates freely, and a note is obtained from the pipe which is its funda-

mental (p. 646). A stationary wave then exists in the pipe, with a node
N at the closed end and an antinode A at the open end, as explained
on p. 645.

If the prongs of a tuning-fork are held over the top of
the pipe, the air inside it is set into vibration by the
periodic force exerted on it by the prongs. In general,
however, the vibrations are feeble, as they are *forced*
vibrations, and the intensity of the sound heard is correspondingly small. But when a tuning-fork of the same
frequency as the fundamental frequency of the pipe is
held over the latter, the air inside is set in *resonance* by
periodic force, and the amplitude of the vibrations is
large. A loud note, which has the same frequency as the
fork, is then heard coming from the pipe, and a stationary
wave is set up with the top of the pipe acting as an antinode

FIG. 27.15.
Resonance in
closed pipe.

and the fixed end as a node, Fig. 27.15. If a sounding tuning-fork is held over a pipe open at both ends, resonance
occurs when the stationary wave in the pipe has antinodes
at the two open ends, as shown by Fig. 27.9; the frequency of the
fork is then equal to the frequency of the fundamental of the open
pipe.

Resonance Tube Experiment. Measurement of Velocity of Sound and "End-Correction" of Tube

If a sounding tuning-fork is held over the open end of a tube T filled
with water, resonance is obtained at some position as the level of water
is gradually lowered, Fig. 27.16 (i). The stationary wave set up is then as

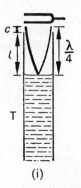
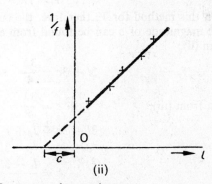

FIG. 27.16. Resonance tube experiment.

shown. If c is the end-correction of the tube (p. 650), and l is the length
from the water level to the top of the tube, then

$$l + c = \frac{\lambda}{4} \qquad . \qquad . \qquad . \qquad . \qquad . \qquad \text{(i)}$$

But
$$\lambda = \frac{V}{f},$$

where f is the frequency of the fork and V is the velocity of sound in air.

$$\therefore \quad l + c = \frac{V}{4f} \quad . \quad \quad . \quad \quad . \quad \quad . \quad \quad . \quad \text{(ii)}$$

If different tuning-forks of known frequency f are taken, and the corresponding values of l obtained when resonance occurs, it follows from equation (ii) that a graph of $1/f$ against l is a straight line, Fig. 27.16 (ii). Now from equation (ii), the gradient of the line is $4/V$; thus V can be determined. Also, the negative intercept of the line on the axis of l is c, from equation (ii); hence the end-correction can be found.

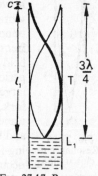

If only one fork is available, and the tube is sufficiently long, another method for V and c can be adopted. In this case the level of the water is lowered further from the position in Fig. 27.16 (i), until resonance is again obtained at a level L_1, Fig. 27.17. Since the stationary wave set up is that shown and the new length to the top from L_1 is l_1, it follows that

FIG. 27.17. Resonance at new water level.

$$l_1 + c = \frac{3\lambda}{4} \quad . \quad \quad . \quad \quad . \quad \quad . \quad \quad . \quad \text{(iii)}$$

But $\quad\quad\quad\quad\quad\quad\quad l + c = \dfrac{\lambda}{4}$, from (ii).

Subtracting, $\quad\quad\quad\quad\quad l_1 - l = \dfrac{\lambda}{2}$

$$\therefore \quad \lambda = 2(l_1 - l)$$

$$\therefore \quad V = f\lambda = 2f(l_1 - l) \quad . \quad \quad . \quad \text{(3)}$$

In this method for V, therefore, the end-correction c is eliminated. The magnitude of c can be found from equations (ii) and (iii). Thus, from (ii),

$$3l + 3c = \frac{3\lambda}{4}$$

But, from (iii), $\quad\quad\quad\quad l_1 + c = \dfrac{3\lambda}{4}$

$$\therefore \quad 3l + 3c = l_1 + c$$

$$\therefore \quad 2c = l_1 - 3l$$

$$\therefore \quad c = \frac{l_1 - 3l}{2} \quad . \quad \quad . \quad \quad . \quad \quad . \quad \text{(4)}$$

Hence c can be found from measurements of l_1 and l.

EXAMPLES

1. Describe the natural modes of vibration of the air in an organ pipe closed at one end, and explain what is meant by the term "end-correction". A cylindrical pipe of length 28 cm closed at one end is found to be at resonance when a tuning fork of frequency 864 Hz is sounded near the open end.

Determine the mode of vibration of the air in the pipe, and deduce the value of the end-correction. [Take the velocity of sound in air as 340 m s^{-1}.] (*L.*)

First part. See text.

Second part. Let λ = the wavelength of the sound in the pipe.

Then
$$\lambda = \frac{V}{f} = \frac{34000}{864} = 39 \cdot 35 \text{ cm}$$

If the pipe is resonating to its fundamental frequency f_0, the stationary wave in the pipe is that shown in Fig. 27.16 and the wavelength λ_0, is given by $\lambda_0/4 = 28$ cm. Thus $\lambda_0 = 112$ cm. Since $\lambda = 39 \cdot 35$ cm, the pipe cannot be sounding its resonant frequency. The first overtone of the pipe is $3f_0$, which corresponds to a wavelength λ_1 given by $3\lambda/4 = 28$ (see Fig. 27.8).

$$\therefore \quad \lambda_1 = \frac{112}{3} = 37\tfrac{1}{3} \text{ cm}$$

Consequently, allowing for the effect of an end-correction, the pipe is sounding its first overtone.

Let c = the end-correction in cm.

Then
$$28 + c = \frac{3\lambda_1}{4}$$

But, accurately,
$$\lambda_1 = \frac{V}{f} = \frac{34000}{864} = 39 \cdot 35$$

$$\therefore \quad 28 + c = \tfrac{3}{4} \times 39 \cdot 35$$

$$\therefore \quad c = 1 \cdot 5 \text{ cm.}$$

2. Explain the phenomenon of resonance, and illustrate your answer by reference to the resonance-tube experiment. In such an experiment with a resonance tube the first two successive positions of resonance occurred when the lengths of the air columns were 15·4 cm and 48·6 cm respectively. If the velocity of sound in air at the time of the experiment was 34000 cm s^{-1} calculate the frequency of the source employed and the value of the end-correction for the resonance tube. If the air column is further increased in length, what will be the length when the next resonance occurs? (*W.*)

First part. See text.

Second part. Suppose c is the end-correction in cm. Then, from p. 654,

$$48 \cdot 6 + c = \frac{3\lambda}{4} \qquad \cdot \qquad \cdot \qquad \cdot \qquad \text{(i)}$$

and
$$15 \cdot 4 + c = \frac{\lambda}{4} \qquad \cdot \qquad \cdot \qquad \cdot \qquad \text{(ii)}$$

Subtracting
$$\therefore \quad 33 \cdot 2 = \frac{\lambda}{2}$$

$$\therefore \quad 66 \cdot 4 = \lambda \qquad \cdot \qquad \cdot \qquad \cdot \qquad \text{(iii)}$$

$$\text{frequency, } f = \frac{V}{\lambda} = \frac{34000}{66 \cdot 4} = 512 \text{ Hz}$$

The end-correction, c, is given by substituting $\lambda = 66 \cdot 4$ in (i). Thus
$$48 \cdot 6 + c = \tfrac{3}{4} \times 66 \cdot 4$$

from which
$$c = 1 \cdot 2 \text{ cm.}$$

The next resonance occurs when the total length, a, of the stationary wave set up is $5\lambda/4$. From (iii), $a = \frac{5}{4} \times 66\cdot4 = 83\cdot0$ cm. Since the end-correction is $1\cdot2$ cm,

$$\therefore \quad \text{length of pipe} = 83\cdot0 - 1\cdot2 = 81\cdot8 \text{ cm.}$$

3. Explain, with diagrams, the possible states of vibration of a column of air in (a) an open pipe, (b) a closed pipe. An open pipe 30 cm long and a closed pipe 23 cm long, both of the same diameter, are each sounding its first over-tone, and these are in unison. What is the end-correction of these pipes? (L.)

First part. See text.

Second part. Suppose V is the velocity of sound in air, and f is the frequency of the note. The wavelength, λ, is thus V/f.

When the open pipe is sounding its first overtone, the length of the pipe plus end-corrections $= \lambda$.

$$\therefore \quad \frac{V}{f} = 30 + 2c \qquad . \quad . \quad . \quad . \quad \text{(i)}$$

since there are two end-corrections.

When the closed pipe is sounding its first overtone,

$$\frac{3\lambda}{4} = 23 + c$$

$$\therefore \quad \frac{3V}{4f} = 23 + c \qquad . \quad . \quad . \quad . \quad \text{(ii)}$$

From (i) and (ii), it follows that

$$23 + c = \tfrac{3}{4}(30 + 2c)$$

$$\therefore \quad 92 + 4c = 90 + 6c$$

$$\therefore \quad c = 1 \text{ cm.}$$

VIBRATIONS IN STRINGS

If a horizontal rope is fixed at one end, and the other end is moved up and down, a wave travels along the rope. The particles of the rope are then vibrating vertically, and since the wave travels horizontally, this is an example of a *transverse* wave (see p. 584). The waves propagated along the surface of the water when a stone is dropped into it are also transverse waves, as the particles of the water are moving up and down while the wave travels horizontally. A transverse wave is also obtained when a stretched string, such as a violin string, is plucked; and before we can study the vibrations in strings, we require to know the velocity of transverse waves along a string.

Velocity of Transverse Waves Along a Stretched String

Suppose that a transverse wave is travelling along a thin string of length l and mass s under a constant tension T. If we assume that the string has no "stiffness", i.e., that the string is perfectly flexible, the velocity V of the transverse wave along it depends only on the values of T, s, l. The velocity is given by

$$V = \sqrt{\frac{T}{s/l}},$$

$$\text{or} \quad V = \sqrt{\frac{T}{m}} \qquad . \qquad . \qquad . \qquad . \qquad . \qquad (5)$$

where m is the "mass per unit length" of the string.

When T is in *newtons* and m in *kilogramme per metre*, then V is in *metres per second*.

The formula for V may be partly deduced by the method of dimensions, in which all the quantities concerned are reduced to the fundamental units of mass, M, length, L, and time, T. Suppose that

$$V = kT^x s^y l^z \qquad . \qquad . \qquad . \qquad . \qquad (i)$$

where k, x, y, z, are numbers. The dimensions of velocity V are LT^{-1}, the dimensions of tension T, a force, are MLT^{-2}, the dimension of s is M, and the dimension of l is L. As the dimensions on both sides of (i) must be equal, it follows that

$$LT^{-1} = (MLT^{-2})^x (M^y)(L^z)$$

Equating the indices of M, L, T on both sides, we have

for M, $x + y = 0$
for L, $x + z = 1$
for T, $2x = 1$
$$\therefore \quad x = \tfrac{1}{2}, z = \tfrac{1}{2}, y = -\tfrac{1}{2}$$

Thus, from (i)

$$V = kT^{\frac{1}{2}} s^{-\frac{1}{2}} l^{\frac{1}{2}}$$

$$\therefore \quad V = k\sqrt{\frac{Tl}{s}} = k\sqrt{\frac{T}{s/l}}$$

A rigid mathematical treatment shows that $k = 1$, since $V = \sqrt{\dfrac{T}{s/l}}$. Since s/l is the "mass per unit length" of the string, it follows that

$$V = \sqrt{\frac{T}{m}},$$

where m is the mass per unit length.

Modes of Vibration of Stretched String

If a wire is stretched between two points N, N and is plucked in the middle, a transverse wave travels along the wire and is reflected at the fixed end. A *stationary wave* is thus set up in the wire, and the simplest mode of vibration is one in which the fixed ends of the wire are nodes, N, and the middle is an antinode, A, Fig. 27.18. Since the distance be-

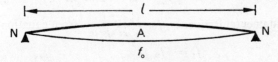

FIG. 27.18. Fundamental of stretched string.

tween consecutive nodes is $\lambda/2$, where λ is the wavelength of the transverse wave in the wire, it follows that

$$l = \frac{\lambda}{2},$$

where l is the length of the wire. Thus $\lambda = 2l$. The frequency f of the vibration is hence given by

$$f = \frac{V}{\lambda} = \frac{V}{2l},$$

where V is the velocity of the transverse wave. But $V = \sqrt{T/m}$, from previous.

$$\therefore \quad f = \frac{1}{2l}\sqrt{\frac{T}{m}}$$

This is the frequency of the *fundamental* note obtained from the string; and if we denote the frequency by the usual symbol f_0, we have

$$f_0 = \frac{1}{2l}\sqrt{\frac{T}{m}} \qquad . \qquad . \qquad . \qquad . \qquad (6)$$

Overtones of Stretched String

The first overtone f_1 of a string plucked in the middle corresponds to a stationary wave shown in Fig. 27.19, which has nodes at the fixed ends and an antinode in the middle. If λ_1 is the wavelength, it can be seen that

$$l = \frac{3}{2}\lambda_1,$$

$$\text{or } \lambda_1 = \frac{2l}{3}.$$

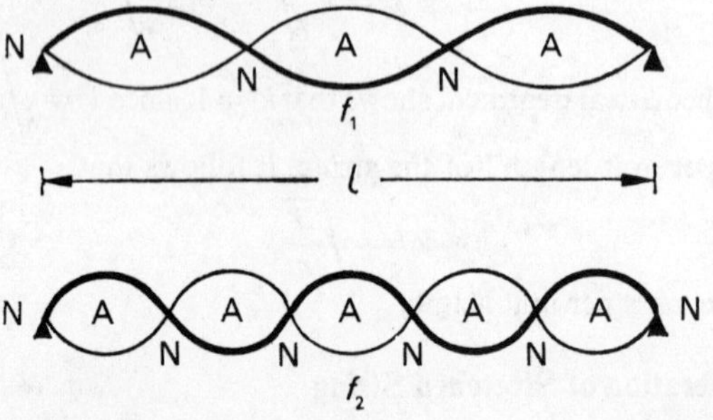

FIG. 27.19. Overtones of stretched string plucked in middle.

The frequency f_1 is thus given by

$$f_1 = \frac{V}{\lambda_1} = \frac{3V}{2l} = \frac{3}{2l}\sqrt{\frac{T}{m}} \qquad . \qquad . \qquad (i)$$

But the fundamental frequency, $f_0, = \frac{1}{2l}\sqrt{\frac{T}{m}}$, from equation (6).

$$\therefore \quad f_1 = 3f_0$$

The second overtone f_2 of the string when plucked in the middle corresponds to a stationary wave shown in Fig. 27.19. In this case $l = \frac{5}{2}\lambda_2$, where λ_2 is the wavelength.

$$\therefore \quad \lambda_2 = \frac{2l}{5}$$

$$\therefore \quad f_2 = \frac{V}{\lambda_2} = \frac{5V}{2l}$$

where f_2 is the frequency. But $V = \sqrt{T/m}$.

$$\therefore \quad f_2 = \frac{5}{2l}\sqrt{\frac{T}{m}} = 5f_0$$

The overtones are thus $3f_0$, $5f_0$, and so on.

Other notes than those considered above can be obtained by touching or "stopping" the string lightly at its midpoint, for example, so that the latter becomes a node in addition to those at the fixed ends. If the string is plucked one-quarter of the way along it from a fixed end, the simplest stationary wave set up is that illustrated in Fig. 27.20 (i). Thus the wavelength $\lambda = l$, and hence the frequency f is given by

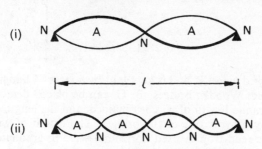

FIG. 27.20. Even harmonics in stretched string.

$$f = \frac{V}{\lambda} = \frac{V}{l} = \frac{1}{l}\sqrt{\frac{T}{m}}$$

$$\therefore \quad f = 2f_0, \text{ since } f_0 = \frac{1}{2l}\sqrt{\frac{T}{m}}.$$

If the string is plucked one-eighth of the way from a fixed end, a stationary wave similar to that in Fig. 27.20 (ii) may be set up. The wavelength, $\lambda' = l/2$, and hence the frequency $f' = \frac{V}{\lambda'} = \frac{2V}{l}$.

$$\therefore \quad f' = \frac{2}{l}\sqrt{\frac{T}{m}} = 4f_0$$

**Verification of the Laws of Vibration of a Fixed String.
The Sonometer**

As we have already shown (p. 658), the frequency of the fundamental of a stretched string is given by $f = \frac{1}{2l}\sqrt{\frac{T}{m}}$, writing f for f_0. It thus, follows that:

(1) $f \propto \dfrac{1}{l}$ *for a given tension* (*T*) *and string* (*m constant*).

(2) $f \propto \sqrt{T}$ *for a given length* (*l*) *and string* (*m constant*).

(3) $f \propto \dfrac{1}{\sqrt{m}}$ *for a given length* (*l*) *and tension* (*T*).

These are known as the "laws of vibration of a fixed string", first completely given by MERSENNE in 1636, and the *sonometer*, or *monochord*, was designed to verify them.

The sonometer consists of a hollow wooden box Q, with a thin horizontal wire attached to A on the top of it, Fig. 27.21. The wire passes

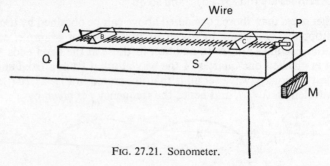

FIG. 27.21. Sonometer.

over a grooved wheel P, and is kept taut by a mass M hanging down at the other end. Wooden bridges, B, C, can be placed beneath the wire so that a definite length of wire is obtained, and the length of wire can be varied by moving one of the bridges. The length of wire between B, C can be read from a fixed horizontal scale S, graduated in centimetres, on the box below the wire.

(1) *To verify f* $\propto$ *1/l for a given tension* (*T*) *and mass per unit length* (*m*), the mass M is kept constant so that the tension, *T*, in the wire is constant. The length, *l*, of the wire between B, C is varied by moving C until the note obtained by plucking BC in the middle is the same as that produced by a sounding tuning-fork of known frequency *f*. If the observer lacks a musical ear, the "tuning" can be recognised by listening for beats when the wire and the tuning-fork are both sounding, as in this case the frequencies of the two notes are nearly equal (p. 620). Alternatively, a small piece of paper in the form of an inverted *V* can be placed on the middle of the wire, and the end of the sounding tuning-fork then placed on the sonometer box. The vibrations of the fork are then transmitted through the box to the wire, which vibrates in resonance with the fork if its length is "tuned" to the note. The paper will then vibrate considerably and may be thrown off the wire.

Different tuning-forks of known frequency *f* are taken, and the lengths, *l*, of the wire are observed when they are tuned to the corresponding note. A graph of *f* against 1/*l* is then plotted, and is found to be a straight line within the limits of experimental error. Thus *f* $\propto$ 1/*l* for a given tension and mass per unit length of wire.

(2) *To verify f ∝ √T for a given length and mass per unit length*, the
length BC between the bridges is kept fixed, so that the length of wire
is constant, and the mass M is varied to alter the tension. The experimental difficulty to be overcome is how to find the frequency *f* of the
note produced when the wire between B, C is plucked in the middle.
For this purpose a second wire, fixed to R, S on the sonometer, is utilised,
usually with a weight (not shown) attached to one end to keep the tension
constant, Fig. 27.22. This wire has bridges P, N beneath it, and N is
moved until the note from the wire between P, N is the same as the
note from the wire between B, C. Now the tension in PN is constant
as the wire is fixed to R, S. Thus, since frequency, *f*, ∝ 1/*l* for a given
tension and wire, the frequency of the note from BC is proportional to
1/*l*, where *l* is the length of PN.

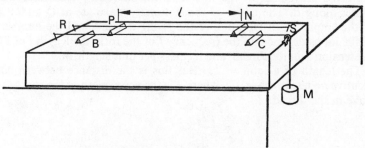

FIG. 27.22. Verification of *f* ∝ √*T*.

If a different mass is attached to the end of the wire BC, the tension
in the wire is altered. Keeping BC fixed, the bridge N is moved until
the note from PN is the same as that obtained from BC, and the length
PN (*l*) is again noted. Then, as before, the frequency of the new note in
BC is proportional to 1/*l*. By altering the mass M, and observing the
corresponding length *l*, a graph of 1/*l* can be plotted against √*T*, where
T is the weight of M. The graph is a straight line, within the limits of experimental error, and hence 1/*l* ∝ √*T*. Thus *f* ∝ √*T* for a given length of
wire and mass per unit length.

(3) *To verify f ∝ 1/√m for a given length and tension*, wires of different
material are connected to B, C, and the same mass M and the same
length BC are taken. The frequency, *f*, of the note obtained from BC
is again found by using the second wire RS in the way already described.
The mass per unit length, *m*, is the mass per metre length of wire, and
is given by $\pi r^2 \rho$ kg m⁻¹, where *r* is the radius of the wire in m and
ρ is its density in kg m⁻³, as ($\pi r^2 \times 1$) m³ is the volume of 1 m of
the wire. Since *f* ∝ 1/*l*, where *l* is the length on the second wire, a graph
of 1/*l* against 1/√*m* should be a straight line; thus a graph of *l* against
√*m* should be a straight line if *f* ∝ 1√*m* for a given length and mass
per unit length. Experiment shows this is the case.

Melde's Experiment on Vibrations in Stretched String

MELDE gave a striking demonstration of the stationary wave set up
in a vibrating string. He used a light thread with one end attached to

the prong P of an electrically-maintained tuning-fork, and the other end connected to a weight W after passing over a grooved wheel Q, Fig. 27.23. With the vibrations of the prong perpendicular to the length

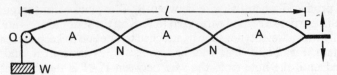

FIG. 27.23. Melde's experiment.

of the thread a number of "loops" can be observed, which are due to the rapid movement of the thread as the stationary transverse wave passes along it. The nodes of the stationary wave are at Q and at P, as the amplitude of vibration of the prong is small, and the number n of the loops depends on the frequency f of the fork, the length l of PQ, the tension T in the thread, and its mass per unit length, m.

The length of a loop $= l/n$. But this is the distance between consecutive nodes, which is $\lambda/2$, where λ is the wavelength of the stationary wave in the thread.

$$\therefore \quad \frac{\lambda}{2} = \frac{l}{n}$$

$$\therefore \quad \lambda = \frac{2l}{n}$$

$$\therefore \quad f = \frac{V}{\lambda} = \frac{n}{2l} V,$$

where V is the velocity of the transverse wave. The frequency of the transverse wave is the same as that of the tuning-fork.

But
$$V = \sqrt{\frac{T}{m}} \text{ (p. 657)}$$

$$\therefore \quad f = \frac{n}{2l} \sqrt{\frac{T}{m}}, \qquad . \quad . \quad . \quad . \quad (7)$$

$$\therefore \quad n\sqrt{T} = \text{constant},$$

if f, T, m are kept constant. In this case, therefore,

$$n = \frac{\text{constant}}{\sqrt{T}}$$

$$\text{or } n^2 \propto \frac{1}{T} \qquad . \quad . \quad . \quad . \quad . \quad (8)$$

This relation can be verified by varying the tension T in the thread, and obtaining a corresponding whole number, n, of loops. A graph of n^2 against $1/T$ is then plotted, and is a straight line passing through the origin.

When the prong P of the tuning-fork is vibrating in the same direction as the length of string, the latter moves from position 3 to position 2 as the prong moves from one end a of an oscillation to the other end b,

FIG. 27.24. Melde's experiment.

Fig. 27.24. As the string continues from position 2 to position 1, the prong moves back from b to a to complete 1 cycle of oscillation. The fork thus goes through one complete cycle in the same time as the particles of the string go through half a cycle, and hence the frequency of the transverse wave is *half* the frequency f of the fork, unlike the first case considered, Fig. 27.24. Instead of equation (7), we now have

$$\frac{f}{2} = \frac{n}{2l} \sqrt{\frac{T}{m}}$$

or $n = fl/\sqrt{T/m}$,

and hence with the same tension, thread, and length l, the number of loops n is half that obtained previously (p. 662).

Measurement of the Frequency of A.C. Mains

The frequency of the alternating current (A.C.) mains can be determined with the aid of a sonometer wire. The alternating current is passed into the wire MP, and the poles N, S of a powerful magnet are placed on either side of the wire so that the magnetic field due to it is perpendicular to the wire, Fig. 27.25. As a result of the magnetic effect of the current, a force acts on the wire which is perpendicular to the directions of both the magnetic field and the current, and hence the wire is subjected to a tranverse force. If the current is an alternating one of 50 Hz, the magnitude of the force varies at the rate of 50 Hz. By adjusting the tension in the sonometer wire, whose magnitude is read on the spring-balance A, a position can be reached when the wire is seen to be vibrating through a large amplitude; in this case the wire is *resonating* to the applied force, Fig. 27.25.

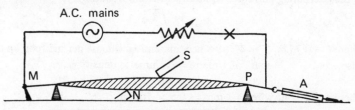

FIG. 27.25. Measurement of frequency of A.C. mains.

The length l of wire between the bridges is now measured, and the tension T and the mass per unit length, m, are also found. The frequency f of the alternating current is then given by

$$f = \frac{1}{2l} \sqrt{\frac{T}{m}} .$$

Velocity of Longitudinal Waves in Wires

If a sonometer wire is stroked along its length by a rosined cloth, a high-pitched note is obtained. This note is due to *longitudinal* vibrations in the wire, and must be clearly distinguished from the note produced when the wire is plucked, which sets up *transverse* vibrations of the wire and a corresponding transverse wave. As we saw on p. 594, the velocity V of a longitudinal wave in a medium is

$$V = \sqrt{\frac{E}{\rho}},$$

where E is Young's modulus for the wire and ρ is its density. The wavelength, λ, of the longitudinal wave is $2l$, where l is the length of the wire, since a stationary longitudinal wave is set up. Thus the frequency f of the note is given by

$$f = \frac{V}{\lambda} = \frac{1}{2l}\sqrt{\frac{E}{\rho}}.$$

The frequency of the note may be obtained with the aid of an audio oscillator, and thus the velocity of sound in the wire, or its Young's modulus, can be calculated.

EXAMPLES

1. Explain the meaning of the term *resonance*, giving in illustration two methods of obtaining resonance between the stretched string of a sonometer and a tuning-fork of fixed frequency. A sonometer wire of length 76 cm is maintained under a tension of value 4 kgf and an alternating current is passed through the wire. A horse-shoe magnet is placed with its poles above and below the wire at its midpoint, and the resulting forces set the wire in resonant vibration. If the density of the material of the wire is 8800 kg m⁻³ and the diameter of the wire is 1 mm, what is the frequency of the alternating current? (*L.*)

First part. See text.

Second part. The wire is set into resonant vibration when the frequency of the alternating current is equal to its natural frequency, f.

Now
$$f = \frac{1}{2l}\sqrt{\frac{T}{m}} \qquad . \qquad . \qquad . \qquad . \qquad . \qquad \text{(i)}$$

where $l = 0.76$ m, $T = 4 \times 9.8$ newton, and $m =$ mass per metre in kg m⁻¹.

Also,   mass of 1 metre = volume × density
$$= \pi r^2 \times 1 \times 8800 \text{ kg},$$

where radius r of wire $= \frac{1}{2}$ mm $= 0.5 \times 10^{-3}$ m

From (i), $\quad \therefore \quad f = \dfrac{1}{2 \times 0.76}\sqrt{\dfrac{4 \times 9.8}{\pi \times 0.5^2 \times 10^{-6} \times 1 \times 8800}}$

$$= 49.6 \text{ Hz}.$$

2. What data would be required in order to predict the frequency of the note emitted by a stretched wire (*a*) when it is plucked, (*b*) when it is stroked along its length? A weight is hung on the wire of a vertical sonometer. When the vibrating length of the wire is adjusted to 80 cm the note it emits when

plucked is in tune with a standard fork. On adding a further weight of 100 g the vibrating length has to be altered by 1 cm in order to restore the tuning. What is the initial weight on the wire? (*L.*)

First part. When the wire is plucked the vibrations of the particles produce transverse waves, and the frequency of the note is given by $f = \dfrac{1}{2l}\sqrt{\dfrac{T}{m}}$. When the wire is stroked along its length, the vibrations of the particles produce a *longitudinal* wave, and the velocity of the wave is given by $V = \sqrt{E/\rho}$, where E is Young's modulus for the wire and ρ is its density. The frequency in the latter case thus depends on the magnitudes of E and ρ, as well as on the length of the wire.

Second part.   Let W = the initial weight on the wire in kgf,

and f = the frequency of the fork

Since $= \dfrac{1}{2l}\sqrt{\dfrac{T}{m}}$

we have $f = \dfrac{1}{2 \times 0.80}\sqrt{\dfrac{Wg}{m}}$.   .   .   .   (i)

When a weight of 0·1 kgf is added, the frequency increases since the tension increases. The new length of the wire = 0·81 m.

$$\therefore\ \ f = \dfrac{1}{2 \times 0.81}\sqrt{\dfrac{(W + 0.1)g}{m}}$$.   .   (ii)

From (i) and (ii), it follows that

$$\dfrac{1}{1.60}\sqrt{\dfrac{Wg}{m}} = \dfrac{1}{1.62}\sqrt{\dfrac{(W + 0.1)g}{m}}$$

$$\therefore\quad 162^2\,W = 160^2\,(W + 0.1)$$

$$W = \dfrac{160^2 \times 0.1}{162^2 - 160^2} = 4 \text{ kgf (approx.)}$$

VIBRATIONS IN RODS

Sound waves travel through liquids and solids, as well as through gases, and in the nineteenth century an experiment to measure the velocity of sound in iron was carried out by tapping one end of a very long iron tube. The speed of sound in iron is much greater than in air, and the sound through the iron thus arrived at the other end of the pipe before the sound transmitted through the air. From a knowledge of the interval between the sounds, the length of the pipe, and the velocity of sound in air, the velocity of sound in iron was determined. More accurate methods were soon forthcoming for the velocity of sound in substances such as iron, wood, and glass, and they depend mainly on the formation of stationary waves in rods of these materials.

Consider a rod AA fixed by a vice B at its mid-point N, Fig. 27.26. If the rod is stroked along its length by a rosined cloth, a stationary longitudinal wave is set up in the rod on account of reflection at its ends, and a high-pitched note is obtained. Since the mid-point of the rod is fixed, this is a node, N, of the stationary wave; and since the ends

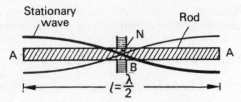

FIG. 27.26. Stationary wave in rod.

of the rod are free, these are antinodes, A. Thus the length l of the rod is equal to half the wavelength, $\lambda/2$, of the wave in the rod, and hence $\lambda = 2l$. Thus the velocity of the sound in the rod, $V = f\lambda = f \times 2l$, where f is the frequency of the note from the rod.

Kundt's Tube

About 1868, KUNDT devised a simple method of showing the stationary waves in air or any other gas. He used a closed tube T containing the gas, and sprinkled some dry lycopodium powder, or cork dust, along the entire length, Fig. 27.27. A rod AE, clamped at its mid-point, is placed with one end projecting into T, and a disc E is attached at this end so that it just clears the sides of the tube, Fig. 27.27. When the

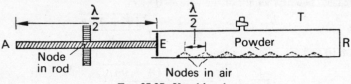

FIG. 27.27. Kundt's tube.

rod is stroked at A by a rosined cloth in the direction EA, the rod vibrates longitudinally and a high-pitched note can be heard. The end E acts as a vibrating source of the same frequency, and a sound wave thus travels through the air in T and is reflected at the fixed end R. If the rod is moved so that the position of E alters, a position can be found when the stationary wave in the air in T causes the lycopodium powder to become violently agitated. The powder then settles into definite small heaps at the nodes, which are the positions of permanent rest of the stationary wave, and the distance between consecutive nodes can best be found by measuring the distance between several of them and dividing by the appropriate number.

Determination of Velocity of Sound in a Rod

Kundt's tube can be used to determine the velocity of sound, V_r, in the rod. Suppose the length of the rod is l: then $\lambda/2 = l$, or $\lambda = 2l$, where λ is the wavelength of the sound wave *in the rod* (p. 643). Thus the frequency of the high-pitched note obtained from the rod is given by

$$f = \frac{V_r}{\lambda} = \frac{V_r}{2l} \qquad . \qquad . \qquad . \qquad . \qquad \text{(i)}$$

If l_1 is the distance between consecutive nodes of the stationary wave in the air, we have $\lambda_1/2 = l_1$, where λ_1 is the wavelength of the sound wave *in the air*. Thus $\lambda_1 = 2l_1$, and hence the frequency of the wave, which is also f, is given by

$$f = \frac{V_a}{\lambda} = \frac{V_a}{2l_1}, \qquad \cdot \qquad \cdot \qquad \cdot \qquad \cdot \qquad \text{(ii)}$$

where V_a is the velocity of sound in air. From (i) and (ii) it follows that

$$\frac{V_r}{2l} = \frac{V_a}{2l_1}$$

$$\therefore \qquad V_r = \frac{l}{l_1} V_a \qquad \cdot \qquad \cdot \qquad \cdot \qquad \cdot \qquad \text{(9)}$$

Thus knowing V_a, l, l_1, the velocity of sound in the rod, V_r, can be calculated. By using glass, brass, copper, steel and other substances in the form of a rod, the velocity of sound in these media have been determined. Kundt also used liquids in the tube T instead of air, and employed fine iron filings instead of lycopodium powder to detect the nodes in the liquid. In this way he determined the velocity of sound in liquids.

Determination of Young's Modulus of a Rod

On p. 622, it was shown that the velocity of sound, V, in a medium is always given by

$$V = \sqrt{\frac{E}{\rho}},$$

where E is the appropriate modulus of elasticity of the medium and ρ is its density. In the case of a rod undergoing longitudinal vibrations, as in Kundt's tube experiment, E is Young's modulus (see p. 594). Thus if V_r is the velocity of sound in the rod,

$$V_r = \sqrt{\frac{E}{\rho}},$$

and $\qquad \therefore \qquad E = V_r^2 \rho \qquad \cdot \qquad \cdot \qquad \cdot \qquad \cdot \qquad \text{(10)}$

Since V_r is obtained by the method explained above, and ρ can be obtained from tables, it follows that E can be calculated.

Determination of Velocity of Sound in a Gas

If the air in Kundt's tube T is replaced by some other gas, and the rod stroked, the average distance l' between the piles of dust in T is the distance between consecutive nodes of the stationary wave in the gas. The wavelength, λ_g, in the gas is thus $2l'$, and the frequency f is given by

$$f = \frac{V_g}{\lambda_g} = \frac{V_g}{2l'},$$

where V_g is the velocity of sound in the gas. But the wavelength, λ, of the wave in the rod $= 2l$, where l is the length of the rod (p. 666); hence f is also given by

$$f = \frac{V_r}{\lambda} = \frac{V_r}{2l}$$

$$\therefore \quad \frac{V_g}{2l'} = \frac{V_r}{2l}$$

$$\therefore \quad V_g = \frac{l'}{l} V_r \quad . \quad . \quad . \quad . \quad . \quad (11)$$

Knowing l', l, and V_r, the latter obtained from a previous experiment (p. 667), the velocity of sound in a gas, V_g, can be calculated. The velocity of sound in a gas can also be found by the more direct method described below.

Determination of Ratio of Specific Heat Capacities of a Gas, and its Molecular Structure

The velocity of sound in a gas, V_g, is given by

$$V_g = \sqrt{\frac{\gamma p}{\rho}},$$

where γ is the ratio (c_p/c_v) of the two principal specific heat capacities of the gas, p is its pressure, and ρ is its density. See p. 624. Thus

$$\gamma = \frac{V_g^2 \rho}{p} \quad . \quad . \quad . \quad . \quad . \quad (12)$$

Now it has already been shown that V_g can be found; and knowing ρ and p, γ can be calculated. The determination of γ is one of the most important applications of Kundt's tube, as kinetic theory shows that $\gamma = 1.66$ for a monatomic gas and 1.40 for a diatomic gas. Thus Kundt's tube provides valuable information about the molecular structure of a gas. When RAMSEY isolated the hitherto-unobtainable argon from the air, Lord Rayleigh in 1895 suggested a Kundt's tube experiment for finding the ratio γ, of its specific heats. It was then discovered that γ was about 1.65, showing that argon was a monatomic gas. The dissociation of the molecules of a gas at high temperatures has been investigated by containing it in Kundt's tube surrounded by a furnace, and measuring the magnitude of γ when the temperature was changed.

Comparison of Velocities of Sound in Gases by Kundt's Tube

The ratio of the velocities of sound in two gases can be found from a Kundt's tube experiment. The two gases, air and carbon dioxide for example, are contained in tubes A, B respectively, into which the ends of a metal rod R project, Fig. 27.28. The middle of the rod is clamped. By stroking the rod, and adjusting the positions of the movable discs Y, X in turn, lycopodium powder in each tube can be made to settle into heaps at the various nodes. The average distances, d_a, d_b, between successive nodes in A, B respectively are then measured.

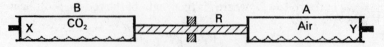

FIG. 27.28. Comparison of velocities of sound in gases.

The frequency f of the sound wave in A, B is the same, being the frequency of the note obtained from R. Since $f = V/\lambda$, it follows that

$$\frac{V_g}{\lambda_g} = \frac{V_a}{\lambda_a} , \qquad . \qquad . \qquad . \qquad . \qquad \text{(i)}$$

where V_g, V_a are the velocities of sound in carbon dioxide and air respectively, and λ_g, λ_a are the corresponding wavelengths.

Now

$$\frac{\lambda_g}{\lambda_a} = \frac{d_b}{d_a} \qquad . \qquad . \qquad . \qquad . \qquad \text{(ii)}$$

since the distance between successive nodes is half a wavelength. From (i),

$$\frac{V_g}{V_a} = \frac{\lambda_g}{\lambda_a}$$

$$\therefore \quad \frac{V_g}{V_a} = \frac{d_b}{d_a} \qquad . \qquad . \qquad . \qquad . \qquad \text{(13)}$$

The two velocities can thus be compared as d_b, d_a are known; and if the velocity of sound, V_a, in air is known, the velocity in carbon dioxide can be calculated.

Vibrations in Plates. Chladni's Figures

We have already studied the different modes of vibration of the air in a pipe, the particles of a string, and the particles of a rod. About 1790 CHLADNI examined the vibrations of a glass *plate* by sprinkling sand on it. If the plate on a stand is gripped firmly at the corner N and bowed in the middle A of one side, the particles arrange themselves into a symmetrical pattern which shows the nodes of the stationary wave in the plate, Fig. 27.29 (i). By gripping the plate firmly at other points N, thus making a node at these points, and bowing at A, a series

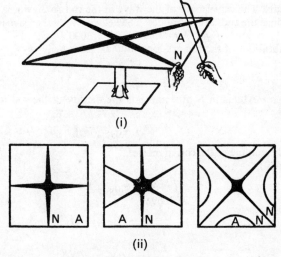

FIG. 27.29. Chladni's figures.

of different patterns can be obtained. These are known as *Chladni's figures*, Fig. 27.29 (ii). Each pattern corresponds to a particular mode of vibration. These modes are not harmonically related in frequency, unlike the case of the vibration of air in pipes and the vibration of strings.

EXAMPLES

1. Describe and explain the way in which a Kundt tube may be used to determine the ratio of the specific heats of a gas. A Kundt tube is excited by a brass rod 150 cm long and the distance between successive nodes in the tube is 13·6 cm; what is the ratio of the velocity of sound in brass to that in air? (*L.*)

First part. See text.

Second part. Since both ends of the rod are successive antinodes, the wavelength λ_1 in the rod $= 2 \times 150 = 300$ cm. The wavelength λ_2 in the air $= 2 \times 13\cdot6 = 27\cdot2$ cm.

The frequency f of the note in the rod and the air is the same.

$$\therefore \ f = \frac{V_1}{\lambda_1} = \frac{V_2}{\lambda_2}$$

where V_1, V_2 are the velocities of sound in the rod and in the air.

$$\therefore \ \frac{V_1}{V_2} = \frac{\lambda_1}{\lambda_2} = \frac{300}{27\cdot2} = 11\cdot0$$

2. Describe the dust tube experiment. How may it be used to compare the velocities of sound in different gases? The fundamental frequency of longitudinal vibration of a rod clamped at its centre is 1500 Hz. If the mass of the rod is 96·0 g, find the increase in its total length produced by a stretching force of 10 kgf (*L.*)

First part. The dust tube is Kundt's tube. See p. 666.

Second part. The wavelength of the wave in the rod $= 2l$, where l metre is its length, since the ends are antinodes. The velocity V, of the wave is given by

$$V = f\lambda = 1500 \times 2l = 3000\ l \quad . \quad\quad . \quad\quad . \quad\quad (i)$$

Since the vibrations of the rod are longitudinal,

$$V = \sqrt{\frac{E}{\rho}}.$$

E is Young's modulus in $N\ m^{-2}$ and ρ is the density of the rod in $kg\ m^{-3}$

$$\therefore \ V = \sqrt{\frac{E}{0\cdot096/v}} = \sqrt{\frac{Ev}{0\cdot096}} \quad\quad . \quad\quad . \quad\quad (ii)$$

where v is the volume of the rod in $metre^3$.
From (i) and (ii),

$$\sqrt{\frac{Ev}{0\cdot096}} = 3000\ l$$

$$\therefore \quad \frac{Ev}{l^2} = 0\cdot096 \times 3000^2$$

$$\therefore \quad \frac{EA}{l} = 0\cdot096 \times 3000^2 \quad\quad . \quad\quad . \quad\quad . \quad\quad (iii)$$

since $v = Al$, where A is the area of cross-section of the rod. Now if x is the increase in length produced by 10 kgf, it follows from the definition of E that

$$\text{force} = \frac{EAx}{l} = 10 \times 9 \cdot 8 \text{ newtons}$$

$$\therefore \quad \frac{EA}{l} = \frac{10 \times 9 \cdot 8}{x}$$

From (iii), $\therefore 0 \cdot 096 \times 3,000^2 = \dfrac{10 \times 9 \cdot 8}{x}$

$$\therefore \quad x = \frac{10 \times 9 \cdot 8}{0 \cdot 096 \times 3000^2} = 1 \cdot 1 \times 10^{-4} \text{ metre.}$$

EXERCISES 27

Pipes

1. Write down in terms of wavelength, λ, the distance between (i) consecutive nodes, (ii) a node and an adjacent antinode, (iii) consecutive antinodes. Find the frequency of the fundamental of a closed pipe 15 cm long if the velocity of sound in air is 340 m s^{-1}.

2. Discuss what is meant by the statement that *sound is a wave motion.* Use the example of the passage of a sound wave through air to explain the terms wavelength (λ), frequency (f), and velocity (v) of a wave. Show that $v = f\lambda$.

Explain the increase in loudness (or 'resonance') which occurs when a sounding tuning-fork is held near the open end of an organ pipe when the length of the pipe has certain values, the other end of the pipe being closed. Find the shortest length of such a pipe which resonates with a 440 Hz tuning-fork, neglecting end corrections. (Velocity of sound in air = 350 m s^{-1}.) (*O. & C.*)

3. What are the chief characteristics of a progressive wave motion? Give your reasons for believing that sound is propagated through the atmosphere as a longitudinal wave motion, and find an expression relating the velocity, the frequency, and the wavelength.

Neglecting end effects, find the lengths of (*a*) a closed organ pipe, and (*b*) an open organ pipe, each of which emits a fundamental note of frequency 256 Hz. (Take the speed of sound in air to be 330 m s^{-1}.) (*O.*)

4. (*a*) Explain in terms of the properties of a gas, but without attempting mathematical treatment, how the vibration of a sound source, such as a loudspeaker diaphragm, can be transmitted through the air around it.

Explain, also, the reflection which occurs when the vibration reaches a fixed barrier, such as a wall.

(*b*) Plane, simple harmonic, progressive sound waves of wavelength 1·2 m and speed 348 m s^{-1}, are incident normally on a plane surface which is a perfect reflector of sound. What statements can be made about the amplitude of vibration and about air pressure changes at points distant (i) 30 cm, (ii) 60 cm, (iii) 90 cm, (iv) 10 cm from the reflector? Justify your answers. (*O. & C.*)

5. Describe the motion of the air in a tube closed at one end and vibrating in its fundamental mode. An observer (a) holds a vibrating tuning-fork over the open end of a tube which resounds to it, (b) blows lightly across the mouth of the tube. Describe and explain the difference in the quality of the notes that he hears.

A uniform tube, 60·0 cm long, stands vertically with its lower end dipping into water. When the length above water is 14·8 cm, and again when it is 48·0 cm, the tube resounds to a vibrating tuning fork of frequency 512 Hz. Find the lowest frequency to which the tube will resound when it is open at both ends. (L.)

6. Discuss the factors which determine the pitch of the note given by a 'closed' pipe. Explain why the fundamental frequency and the quality of the note from a 'closed' pipe differ from those of the note given under similar conditions by a pipe of the same length which is open at both ends. (N.)

7. What is meant by (a) *a stationary wave motion* and (b) *a node*?

Describe how the phenomenon of resonance may be demonstrated using a loudspeaker, a source of alternating voltage of variable frequency and a suitable tube open at one end and closed at the other. Explain how resonance occurs in the arrangement you describe, draw a diagram showing the position of the nodes in the tube in a typical case of resonance and state clearly the meaning of the diagram. How would you demonstrate the position of the nodes experimentally? (O. & C.)

8. Explain the meaning of (a) the *end correction* of a resonance tube, (b) *beats*. Establish a formula for the frequency of beats in terms of the super-imposed frequencies.

A closed resonance tube with an end correction of 0·60 cm is made to sound its fundamental note on a day when the air temperature is 17°C. It is found to be in unison with a siren whose disc, which has 12 holes, is revolving at a rate of 43·0 rev s^{-1}. Calculate (i) the length of the tube, (ii) the frequency of the beats produced if the experiment is repeated on a day when the air temperature has fallen to 12°C, the rate of revolution of the siren's disc being unaltered. (The velocity of sound in air at 0°C may be taken as 331·5 m s^{-1}.) (L.)

9. Distinguish between the formation of an echo and the formation of a stationary sound wave by reflection, explaining the general circumstances in which each is produced.

Describe an experiment in which the velocity of sound in air may be determined by observations on stationary waves.

An organ pipe is sounded with a tuning-fork of frequency 256 Hz. When the air in the pipe is at a temperature of 15°C, 23 beats are heard in 10 seconds; when the tuning-fork is loaded with a small piece of wax, the beat frequency is found to decrease. What change of temperature of the air in the pipe is necessary to bring the pipe and the unloaded fork into unison? (C.)

10. Describe and give the theory of one experiment in each instance by which the velocity of sound may be determined, (a) in free air, (b) in the air in a resonance tube.

What effect, if any, do the following factors have on the velocity of sound in free air; frequency of the vibrations; temperature of the air; atmospheric pressure; humidity?

State the relationship between this velocity and temperature. (L.)

Strings. Rods

11. What is meant by a *wave motion*? Define the terms *wavelength* and *frequency* and derive the relationship between them.

Given that the velocity v of transverse waves along a stretched string is related to the tension F and the mass m per unit length by the equation

$$v = \sqrt{\frac{F}{m}},$$

derive an expression for the natural frequencies of a string of length l when fixed at both ends.

Explain how the vibration of a string in a musical instrument produces sound and how this sound reaches the ear. Discuss the factors which determine the quality of the sound heard by the listener. *(O. & C.)*

12. Distinguish between a *progressive* wave and a *stationary* wave. Explain in detail how you would use a sonometer to establish the relation between the fundamental frequency of a stretched wire and (a) its length, (b) its tension. You may assume a set of standard tuning-forks and a set of weights in steps of half a kilogram to be available.

A pianoforte wire having a diameter of 0·90 mm is replaced by another wire of the same material but with diameter 0·93 mm. If the tension of the wire is the same as before, what is the percentage change in the frequency of the fundamental note? What percentage change in the tension would be necessary to restore the original frequency? *(L.)*

13. What is meant by (a) a forced vibration, (b) resonance? Give an example of each from (i) mechanics, (ii) sound.

Using the same axes sketch graphs showing how the amplitude of a forced vibration depends upon the frequency of the applied force when the damping of the system is (a) light, (b) heavy. Point out any special features of the graphs.

A sonometer wire is stretched by hanging a metal cylinder of density 8000 kg m^{-3}. at the end of the wire. A fundamental note of frequency 256 Hz is sounded when the wire is plucked.

Calculate the frequency of vibration of the same length of wire when a vessel of water is placed so that the cylinder is totally immersed. *(N.)*

14. Describe an experiment to determine the velocity of sound in a gas, e.g. nitrogen. How would you expect the velocity to be affected by (a) temperature, (b) pressure and (c) humidity? Give reasons for your answers.

What information about the nature of a gas can be obtained from a measurement of the velocity of sound in that gas, the pressure and density being known? *(L.)*

15. Describe experiments to illustrate the differences between (a) *transverse* waves, (b) *longitudinal* waves, (c) *progressive* waves and (d) *stationary* waves? To which classes belong (i) the vibrations of a violin string, (ii) the sound waves emitted by the violin into the surrounding air?

A wire whose mass per unit length is 10^{-3} kg m^{-1} is stretched by a load of 4 kg over the two bridges of a sonometer 1 m apart. If it is struck at its middle point, what will be (a) the wavelength of its subsequent fundamental vibrations, (b) the fundamental frequency of the note emitted? If the wire were struck at a point near one bridge what further frequencies might be heard? (Do not derive standard formulae.) (Assume $g = 10$ m s^{-2}.) *(O. & C.)*

16. A uniform wire vibrates transversely in its fundamental mode. On what factors, other than the length does the frequency of vibration depend, and what is the form of the dependence for each factor?

Describe the experiment you would perform to verify the form of dependence for *one* factor.

A wire of diameter 0·040 cm and made of steel of density 8000 kg m^{-3} is under constant tension of 8·0 kgf. A fixed length of 50 cm is set in transverse vibration. How would you cause the vibration of frequency about 840 Hz to predominate in intensity? (*N.*)

17. (*a*) The velocity of sound in air being known, describe how Young's modulus for brass may be found using Kundt's tube. (*b*) Discuss how the frequency of a note heard by an observer is affected by movement of (i) the source, (ii) the observer along the line joining source and observer. (*L.*)

18. Give an expression for the velocity of a transverse wave along a thin flexible string and show that it is dimensionally correct. Explain how reflexion may give rise to transverse *standing waves* on a stretched string and use the expression for the velocity to derive the frequency of the fundamental mode of vibration.

A steel wire of length 40·0 cm and diameter 0·0250 cm vibrates transversely in unison with a tube, open at each end and of effective length 60·0 cm, when each is sounding its fundamental note. The air temperature is 27°C. Find in kilograms force the tension in the wire. (Assume that the velocity of sound in air at 0°C is 331 m s^{-1} and the density of steel is 7800 kg m^{-3}.) (*L.*)

19. It may be shown theoretically that the frequency f of the fundamental note emitted as the result of the transverse vibration of a stretched wire is given by $f = kT^{\frac{1}{2}}l^{-1}$, where T is the tension in the wire and l its length, k being a constant for a given wire. Describe the experiments you would perform to check this relation. assuming that tuning-forks of known frequencies covering a range of one octave are available.

A brass wire is tuned so that its fundamental frequency is 100 Hz, and a horse-shoe magnet is placed so that the mid-point of the wire lies between its poles. On passing an alternating current through the wire it vibrates, the amplitude of vibration depending upon the frequency of the current. Explain this, and show for what frequencies the vibration will be particularly strong. (*C.*)

20. Explain why the velocity of sound in a gas depends upon the ratio of its principal specific heats.

Give a detailed account of a method of determining this ratio for carbon dioxide gas at atmospheric temperature, assuming that the ratio for air is known. Give the theory of the experiment. (*W.*)

OPTICS

chapter twenty-eight

Wave theory of light

Historical

I T has already been mentioned that light is a form of energy which stimulates our sense of vision. One of the early theories of light, about 400 B.C., suggested that particles were emitted from the eye when an object was seen. It was realised, however, that something is *entering* the eye when a sense of vision is caused, and about 1660 the great Newton proposed that particles, or corpuscles, were emitted from a luminous object. The *corpuscular theory of light* was adopted by many scientists of the day owing to the authority of Newton, but HUYGENS, an eminent Dutch scientist, proposed about 1680 that light energy travelled from one place to another by means of a wave-motion. If the *wave theory of light* was correct, light should bend round a corner, just as sound travels round a corner. The experimental evidence for the wave theory in Huygens' time was very small, and the theory was dropped for more than a century. In 1801, however, THOMAS YOUNG obtained evidence that light could produce wave effects (p. 688), and he was among the first to see clearly the close analogy between sound and light waves. As the principles of the subject became understood other experiments were carried out which showed that light could spread round corners, and Huygens' wave-theory of light was revived. Newton's corpuscular theory was rejected since it was incompatible with experimental observations (see p. 679), and the wave theory of light has played, and is still playing, an important part in the development of the subject.

In 1905 the great mathematical physicist EINSTEIN suggested that the energy in light could be carried from place to place by particles whose energy depended on the wavelength of the light. This was a return to a corpuscular theory, though it was completely different from that of Newton, as we see later. Experiments carried out at his suggestion showed that the theory was true, and the particles carrying light energy are known as "photons." It is now considered that *either* the wave theory *or* the particle theory of light can be used in a problem on light, depending on the circumstances of the problem. In this book we shall consider Huygens' wave theory, which was the foundation of many notable advances in the subject.

Wavefront. Rays

We have already considered the topic of waves in the Sound section

(p. 583). As we shall presently see, close analogies exist between light and sound waves.

Consider a point source of light, S, in air, and suppose that a disturbance, or wave, originates at S as a result of vibrations occurring inside the atoms of the source, and travels outwards. After a time t the wave has travelled a distance ct, where c is the velocity of light in air, and the light energy has thus reached the surface of a sphere of centre S and radius ct, Fig. 28.1. The surface of the sphere is called the *wavefront* of the light at this instant, and every point on it is vibrating "in step" or *in phase* with every other point. As time goes on the wave travels further and new wavefronts are obtained which are the surfaces of spheres of centre S.

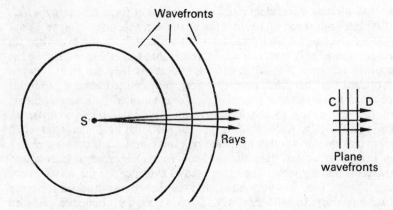

FIG. 28.1. Wavefronts and rays.

At points a long way from S, such as C or D, the wavefronts are portions of a sphere of very large radius, and the wavefronts are then substantially *plane*. Light from the sun reaches the earth in plane wavefronts because the sun is so far away; plane wavefronts also emerge from a convex lens when a point source of light is placed at its focus.

The significance of the wavefront, then, is that it shows how the light energy travels from one place in a medium to another. A *ray* is the name given to the direction along which the energy travels, and consequently a ray of light passing through a point is perpendicular to the wavefront at that point. The rays diverge near S, but they are approximately parallel a long way from S, as plane wavefronts are then obtained, Fig. 28.1.

Huygens' Construction for the New Wavefront

Suppose that the wavefront from a centre of disturbance S has reached the surface AB in a medium at some instant, Fig. 28.2. To obtain the position of the new wavefront after a further time t, Huygens postulated that *every point*, $A,\ldots, C,\ldots, E,\ldots, B$, *on AB becomes a new or* "*secondary*" *centre of disturbance*. The wavelet from A then reaches the surface M of a sphere of radius vt and centre A, where v is the velocity

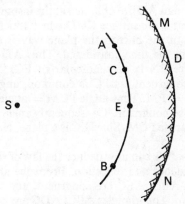

Fig. 28.2. Huygens' construction.

of light in the medium; the wavelet from C reaches the surface D of a sphere of radius *vt* and centre C; and so on for every point on AB. According to Huygens, *the new wavefront is the surface MN which touches all the wavelets from the secondary sources*; and in the case considered, it is the surface of a sphere of centre S.

In this simple example of obtaining the new wavefront, the light travels in the same medium. Huygens' construction, however, is especially valuable for deducing the new wavefront when the light travels from one medium to another, as we shall soon show.

Reflection at Plane Surface

Suppose that a beam of parallel rays between HA and LC is incident on a plane mirror, and imagine a plane wavefront AB which is normal to the rays, reaching the mirror surface, Fig. 28.3. At this instant the point A acts as a centre of disturbance. Suppose we require the new wavefront at a time corresponding to the instant when the disturbance at B reaches C. The wavelet from A reaches the surface of a sphere of radius AD at this instant; and as other points between AC on the mirror, such as P, are reached by the disturbances originating on AB, wavelets

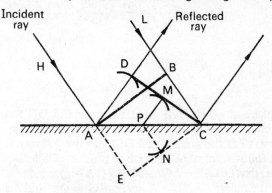

Fig. 28.3. Reflection at plane surface.

of smaller radius than AD are obtained at the instant we are considering. The new wavefront is the surface CMD which touches all the wavelets.

In the absence of the mirror, the plane wavefront AB would reach the position EC in the time considered. Thus AD = AE = BC, and PN = PM, where PN is perpendicular to EC. The triangles PMC, PNC are hence congruent, as PC is common, angles PMC, PNC are each 90°, and PN = PM. Thus angle PCM = angle PCN. But triangles ACD, AEC are also congruent. Consequently angle ACD = angle ACE = angle PCN = angle PCM, Since EC is a plane. Hence CMD is a *plane* surface.

Law of reflection. We can now deduce the law of reflection concerning the angles of incidence and reflection. From the above, it can be seen that the triangles ABC, AEC are congruent, and that triangles ADC, AEC are congruent. The triangles ABC, ADC are hence congruent, and therefore angle BAC = angle DCA. Now these are the angles made by the wavefront AB, CD respectively with the mirror surface AC. Since the incident and reflected rays, for example HA, AD, are normal to the wavefronts, these rays also make equal angles with AC. It now follows that the angles of incidence and reflection are equal.

Point Object

Consider now a point object O in front of a plane mirror M, Fig. 28.4. A spherical wave spreads out from O, and at some time the wavefront reaches ABC. In the absence of the mirror the wavefront would reach a position DEF in a time *t* thereafter, but every point between D and F on the mirror acts as a secondary centre of disturbance and wavelets are reflected back into the air. At the end of the time *t*, a surface DGC is drawn to touch all the wavelets, as shown. DGC is part of a spherical surface which advances into the air, and it appears to have come from a point I as a centre below the mirror, which is therefore a virtual image.

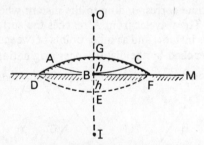

Fig. 28.4. Point object.

The sphere of which DGF is part has a chord DF. Suppose the distance from B, the midpoint of the chord, to G is h. The sphere of which DEF is part has the same chord DF, and the distance from B to E is also h. It follows, from the theorem of product of intersection of chords of a circle, that $DB.BF = h(2r - h) = h(2R - h)$, where r is the radius OE and R is the radius IG. Thus $R = r$, or IG = OE, and hence IB = OB. The image and object are thus equidistant from the mirror.

Refraction at Plane Surface

Consider a beam of parallel rays between LO and MD incident on the plane surface of a water medium from air in the direction shown, and suppose that a plane wavefront has reached the position OA at a certain

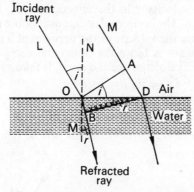

Fig. 28.5. Refraction at plane surface.

instant, Fig. 28.5. Each point between O, D becomes a new centre of disturbance as the wavefront advances to the surface of the water, and the wavefront changes in direction when the disturbance enters the liquid.

Suppose that t is the time taken by the light to travel from A to D. The disturbance from O travels a distance OB, or vt, in water in a time t, where v is the velocity of light in water. At the end of the time t, the wavefronts in the water from the other secondary centres between O, D reach the surfaces of spheres to each of which DB is a tangent. Thus DB is the new wavefront in the water, and the ray OB which is normal to the wavefront is consequently the refracted ray.

Since c is the velocity of light in air, $AD = ct$. Now

$$\frac{\sin i}{\sin r} = \frac{\sin \text{LON}}{\sin \text{BOM}} = \frac{\sin \text{AOD}}{\sin \text{ODB}}$$

$$\therefore \frac{\sin i}{\sin r} = \frac{AD/OD}{OB/OD} = \frac{AD}{OB} = \frac{ct}{vt} = \frac{c}{v} \qquad . \qquad . \qquad . \qquad \text{(i)}$$

But c, v are constants for the given media.

$$\therefore \quad \frac{\sin i}{\sin r} \text{ is a constant,}$$

which is Snell's law of refraction (p. 420).

It can now been seen from (i) that the refractive index, n, of a medium is given by $n = \dfrac{c}{v}$, where c is the velocity of light *in vacuo* and v is the velocity of light in the medium.

Newton's Corpuscular Theory of Light

Prior to the wave theory of light, Newton had proposed a corpuscular or particle theory of light. According to Newton, particles are emitted by a source of light, and they travel in a straight line until the boundary of a new medium is encountered.

In the case of *reflection at a plane surface*, Newton stated that at some very small distance from the surface M, represented by AB, the particles were acted upon by a repulsive force, which gradually diminished the component of the velocity v in the direction of the normal and then reversed it, Fig. 28.6. The *horizontal* component of the velocity remained unaltered, and hence the velocity of the particles of light as they moved away from M is again v. Since the horizontal components of the incident and reflected velocities are the same, it follows that

$$v \sin i = v \sin i' \qquad . \qquad . \qquad . \qquad \text{(i)}$$

where i' is the angle of reflection.

$$\therefore \quad \sin i = \sin i', \text{ or } i = i'$$

Thus the corpuscular theory explains the law of reflection at a plane surface.

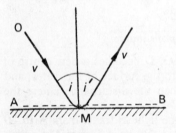

FIG. 28.6. Newton's corpuscular theory of reflection.

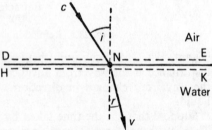

FIG. 28.7. Newton's corpuscular theory of refraction.

To explain *refraction at a plane surface* when light travels from air to a denser medium such as water, Newton stated that a force of attraction acted on the particles as they approached beyond a line DE very close to the boundary N, Fig. 28.7. The vertical component of the velocity of the particles was thus increased on entering the water, the horizontal component of the velocity remaining unaltered, and beyond a line HK close to the boundary the vertical component remained constant at its increased value. The resultant velocity, v, of the particles in the water is thus *greater* than its velocity, c, in air.

Suppose i, r are the angles of incidence and refraction respectively. Then, as the horizontal components of the velocity is unaltered.

$$c \sin i = v \sin r$$

$$\therefore \quad \frac{\sin i}{\sin r} = \frac{v}{c}$$

$$\therefore \quad n = \frac{v}{c} = \text{the refractive index}$$

Since n is greater than 1, the velocity of light (v) in water is greater than the velocity (c) in air, as was stated above. This is according to Newton's corpuscular theory. On the wave theory, however, $n = \frac{c}{v}$ (see p. 679);

and hence the velocity of light (v) in water is *less* than the velocity (c) in air according to the wave theory. The corpuscular theory and wave theory are thus in conflict, and Foucault's experimental results showed that the corpuscular theory, as enunciated by Newton, could not be true (p. 559).

Dispersion

The dispersion of colours produced by a medium such as glass is due to the difference in speeds of the various colours in the medium. Thus suppose a plane wavefront AC of white light is incident in air on a plane glass surface, Fig. 28.8. In the time the light takes to travel in air from C to D, the red light from the centre of disturbance A reaches a position shown by the wavelet at R. The blue light from A reaches another position shown by the wavelet at B, since the speed of blue light in glass is less than that of red light, so that AB is less than AR. On drawing the new wavefronts DB, DR, it can be seen that the blue wavefront BD is refracted more in the glass than the red wavefront DR. The refracted blue ray is AB and the refracted red ray is AR, and hence dispersion occurs.

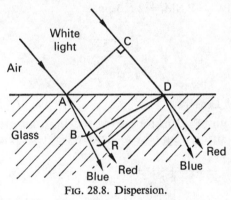

FIG. 28.8. Dispersion.

Refraction Through Prism at Minimum Deviation

Consider a wavefront HB incident on the face XB of a prism of angle A, Fig. 28.9. If the wavefront emerges along EC, then the light travels

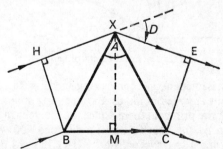

FIG. 28.9. Refraction at minimum deviation.

a distance HXE in air in the same time as the light travels a distance BC in glass.

$$\therefore \quad \frac{HX + XE}{c} = \frac{BC}{v},$$

where c is the velocity of light in air and v is the velocity in glass

$$\therefore \quad HX + XE = \frac{c}{v} BC = n\,BC \qquad . \qquad . \qquad (i)$$

At minimum deviation, the wavefront passes symmetrically through the prism (p. 443).

$$\therefore \quad HX = XE$$

From (i), $\therefore \quad 2\,HX = n\,BC$

$$\therefore \quad n = \frac{2\,HX}{BC} \qquad . \qquad . \qquad . \qquad (ii)$$

But $HX = XB \cos BXH = XB \cos \left[\dfrac{180° - (A + D)}{2} \right]$

$$= XB \sin \left(\frac{A + D}{2} \right),$$

and $BC = 2\,BM = 2\,XB \sin \dfrac{A}{2}$

From (ii), $\therefore \quad n = \sin \dfrac{\left(\dfrac{A + D}{2} \right)}{\sin \dfrac{A}{2}}$

Focal Length of Lens

The focal length of a lens (or curved spherical mirror) can also be found by wave theory. Suppose a plane wavefront AHB, parallel to

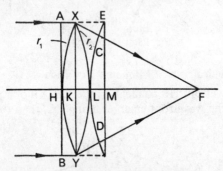

FIG. 28.10. Focal length of lens.

the principal axis, is incident on a converging lens, Fig. 28.10. After refraction the wavefront emerges in air as a converging spherical wavefront CLD of centre F, the principal focus, since the incident rays are parallel.

The time taken by the light to travel a distance AX + XE in air is equal to the time taken to travel a distance HKL in glass. Thus if c is the velocity in air and v is the velocity in glass,

$$\therefore \quad \frac{AX + XE}{c} = \frac{HKL}{v}$$

$$\therefore \quad AX + XE = \frac{c}{v} \cdot HKL = n \cdot HKL \qquad (i)$$

From the geometry of a circle, $AX = HK = \dfrac{h^2}{2r_1}$,

where XK is h and r_1 is the radius of curvature of the lens, assumed thin (see also p. 695).

Also, $\qquad XE = KM = KL + LM = \dfrac{h^2}{2r_2} + \dfrac{h^2}{2f}$,

where r_2 is the radius of curvature of the surface XLY and $FL = f$. Substituting in (i),

$$\therefore \quad \frac{h^2}{2r_1} + \frac{h^2}{2r_2} + \frac{h^2}{2f} = n \left(\frac{h^2}{2r_1} + \frac{h^2}{2r_2} \right)$$

Simplifying, $\qquad\qquad \therefore \quad \dfrac{1}{f} = (n-1) \left(\dfrac{1}{r_1} + \dfrac{1}{r_2} \right).$

Power of a Lens   ~~Notes to be made.~~

We have now to consider the effect of lenses on the *curvature* of wavefronts. A plane wavefront has obviously zero curvature and a spherical wavefront has a small curvature if the radius of the sphere is large. *The "curvature" of a spherical wavefront is defined as* $1/r$, where r is the radius of the surface which constitutes the wavefront, and hence the curvature is zero when r is infinitely large, as in the case of a plane wavefront.

When a plane wavefront is incident on a converging lens L, a spherical wavefront, S, of radius f emerges from L, where f is the focal length of the lens, Fig. 28.11 (i). Parallel rays, which are normal to the plane

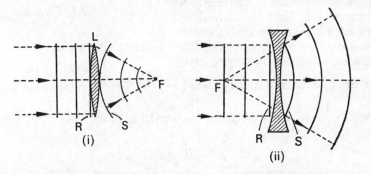

FIG. 28.11. (i). Converging lens.   (ii). Diverging lens.

wavefront, are thus refracted towards F, the focus of the lens. Now the curvature of a plane wavefront is zero, and the curvature of the spherical wavefront S is $1/f$. Thus the convex lens impresses a curvature of $1/f$ on a wavefront incident on it, and $1/f$ is accordingly defined as the *converging power* of the lens.

$$\therefore \quad \text{Power } P, = \frac{1}{f} . \qquad . \qquad . \qquad (91)$$

Fig. 28.11 (ii) illustrates the effect of a *diverging* lens on a plane wavefront R. The front S emerging from the lens has a curvature opposite to S in Fig. 28.11 (i), and it appears to be diverging from a point F behind the concave lens, which is its focus. The curvature of the emerging wavefront is thus $1/f$, where f is the focal length of the lens, and the powers of the convex and concave lens are opposite in sign.

The power of a converging lens is positive, since its focal length is positive, while the power of a diverging lens is negative. Opticians use a unit of power called the *dioptre*, D, which is defined as the power of a lens of 100 cm focal length. A lens of focal length f cm has thus a power P given by

$$P = \frac{1/f}{1/100} \text{ dioptres}$$

$$\text{or} \qquad P = \frac{100}{f} \text{ dioptres} \qquad . \qquad . \qquad . \qquad (92)$$

A lens of $+ 8$ dioptres, or $+ 8D$, is therefore a converging lens of focal length 12·5 cm, and a lens of $- 4D$ is a diverging lens of 25 cm focal length.

The Lens Equation

Suppose that an object O is placed a distance u from a converging lens, Fig. 28.12. The spherical wavefront. A from O which reaches the lens has a radius of curvature u, and hence a curvature $1/u$. Since the converging lens adds a curvature of $1/f$ to the wavefront (proved page 683), the spherical wavefront B emerging from

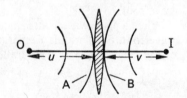

Fig. 28.12. Effect of lens on wavefront.

the lens into the air has a curvature $\left(\dfrac{1}{u} + \dfrac{1}{f}\right)$. But the curvature is also given by $\dfrac{1}{v}$, where v is the image distance IB from the lens.

$$\therefore \quad \frac{1}{v} = \frac{1}{u} + \frac{1}{f}$$

It can be seen that the curvature of A is of an opposite sign to that of B. Hence the usual lens equation $\dfrac{1}{v} + \dfrac{1}{u} = \dfrac{1}{f}$ is obtained. A similar method can be used for a diverging lens, which is left as an exercise for the student.

EXERCISES 28

1. A parallel beam of monochromatic radiation travelling through glass is incident on the plane boundary between the glass and air. Using Huygens' principle draw diagrams (one in each case) showing successive positions of the wave fronts when the angle of incidence is (a) 0°, (b) 30°, (c) 60°. Indicate clearly and explain the constructions used. (The refractive index of glass for the radiation used is 1·5.) (N.)

2. State Snell's law of refraction. How is the law explained in terms of the wave theory of light?

An equiangular glass prism is placed in a broad beam of parallel mono-chromatic light as shown, the face AB of the prism being perpendicular to the direction of the incident light, Fig. 28A. By sketching typical rays, show that most of the light which is refracted by the prism emerges as two beams of parallel light deviated respectively through $\pm \theta$, and calculate the value of θ if the refractive index of the glass is 1·5. Why does not all the light falling on the prism emerge in this way?

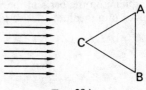

FIG. 28A.

The prism is turned round so that the light is incident normally on the face AB. Describe as fully as you can what happens to most of the light with the prism in this position. (O. & C.)

3. A plane wave-front of monochromatic light is incident normally on one face of a glass prism, of refracting angle 30°, and is transmitted. Using Huygens' construction trace the course of the wave-front. Explain your diagram and find the angle through which the wave-front is deviated. (Refractive index of glass = 1·5.) (N.)

4. State *Snell's law of refraction* and define *refractive index*.

Show how refraction of light at a plane interface can be explained on the basis of the wave theory of light.

Light travelling through a pool of water in a parallel beam is incident on the horizontal surface. Its speed in water is $2 \cdot 2 \times 10^8$ m s^{-1}. Calculate the maximum angle which the beam can make with the vertical if light is to escape into the air where its speed is $3 \cdot 0 \times 10^8$ m s^{-1}.

At this angle in water, how will the path of the beam be affected if a thick layer of oil, of refractive index 1·5, is floated on to the surface of the water? (O. & C.)

5. How did Huygens explain the reflection of light on the wave theory? Using Huygens' conceptions, show that a series of light waves diverging from a point source will, after reflection at a plane mirror, appear to be diverging from a second point, and calculate its position. (C.)

6. Explain how the *corpuscular theory* of Newton accounted for the laws of reflection and refraction. What experimental evidence showed that the theory was incorrect?

7. What is Huygens' principle?

Draw and explain diagrams which show the positions of a light wave-front at successive equal time intervals when (a) parallel light is reflected from a plane mirror, the angle of incidence being about 60°, (b) monochromatic light originating from a small source in water is transmitted through the surface of the water into the air.

Describe an experiment, and add the necessary theoretical explanation, to show that in air the wavelength of blue light is less than that of red light. (*N.*)

8. Using Huygens' concept of secondary wavelets show that a plane wave of monochromatic light incident obliquely on a plane surface separating air from glass may be refracted and proceed as a plane wave. Establish the physical significance of the refractive index of the glass.

In what circumstances does dispersion of light occur? How is it accounted for by the wave theory?

If the wavelength of yellow light in air is $6 \cdot 0 \times 10^{-7}$ m, what is its wavelength in glass of refractive index $1 \cdot 5$? (*N.*)

9. Describe fully a method for measuring the velocity of light in air. Explain, on the basis of the wave theory, the relation between the refractive index of a medium relative to air and the velocity of light. (*L.*)

chapter twenty-nine

Interference, diffraction, polarisation of light

INTERFERENCE OF LIGHT

THE beautiful colours seen in thin films of oil in the road, or in soap bubbles, are due to a phenomenon in light called *interference*. Newton discovered that circular coloured rings were obtained when white light illuminated a convex lens of large radius of curvature placed on a sheet of plane glass (p. 693), which is another example of interference. As we saw in Sound, interference can be used to measure the wavelength of sound waves (p. 617). By a similar method the phenomenon can be used to measure the wavelengths of different colours of light. Interference of light has also many applications in industry.

The essential conditions, and features, of interference phenomena have already been discussed in connection with sound waves. As there is an exact analogy between the interference of sound and light waves we can do no better than recapitulate here the results already obtained on pp. 616–617:

1. Permanent interference between two sources of light can only take place if they are *coherent* sources, i.e., they must have the same frequency and be always in phase with each other or have a constant phase difference. (This implies that the two sources of light must have the same colour.)

2. If the coherent monochromatic light sources are P, Q, a bright light is observed at B if the path-difference, QB—PB, is a whole number of wavelengths, Fig. 29.1. (This corresponds to the case of a loud sound

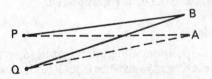

FIG. 29.1. Interference of light.

heard at B if P, Q were two coherent sources of sound.) A bright light is observed at A if PA = QA, in which case the path-difference is zero.

3. If the path-difference is an odd number of half wavelengths, darkness is observed at the point under consideration. (This corresponds to silence at the point in the case of two coherent sound sources.)

Young's Experiment

From the preceding, it can be understood that two conditions are essential to obtain an interference phenomenon. (i) Two coherent sources of light must be produced, (ii) the coherent sources must be very close to each other as the wavelength of light is very small, otherwise the bright and dark pattern in front of the sources tend to be too fine to see and no interference pattern is obtained.

One of the first demonstrations of the interference of light waves was given by YOUNG in 1801. He placed a source, S, of monochromatic light in front of a narrow slit C, and arranged two very narrow slits A, B, close to each other, in front of C. Much to his delight, Young observed bright and dark bands on either side of O on a screen T, where O is on the perpendicular bisector of AB, Fig. 29.2.

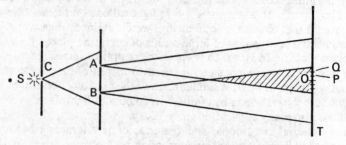

FIG. 29.2. Young's experiment.

Young's observations can be explained by considering the light from S illuminating the two slits A, B. Since the light diverging from A has exactly the same frequency as, and is always in phase with, the light diverging from B, A and B act as two close coherent sources. Interference thus takes place in the shaded region, where the light beams overlap, Fig. 29.2. As AO = OB, a bright band is obtained at O. At a point P close to O, such that BP − AP = $\lambda/2$, where λ is the wavelength of the light from S, a dark band is obtained. At a point Q such that BQ − AQ = λ, a bright band is obtained; and so on for either side of O. Young demonstrated that the bands were due to interference by covering A or B, when the bands disappeared.

Separation of Bands

Suppose P is the position of the mth bright band, so that BP − AP = $m\lambda$, Fig. 29.3. Let OP = x_m = distance from P to O, the centre of the band system, where MO is the perpendicular bisector of AB. If a length PN equal to PA is described on PB, then BN = BP − AP = $m\lambda$. Now in practice AB is very small, and PM is very much larger than AB. Thus AN meets PM practically at right angles. It then follows that

$$\text{angle PMO} = \text{angle BAN} = \theta \text{ say.}$$

From triangle BAN, $$\sin \theta = \frac{BN}{AB} = \frac{m\lambda}{a},$$

where $a = AB =$ the distance between the slits. From triangle PMO,

$$\tan \theta = \frac{PO}{MO} = \frac{x_m}{D},$$

where $D = MO =$ the distance from the screen to the slits. Since θ is very small, $\tan \theta = \sin \theta$.

$$\therefore \qquad \frac{x_m}{D} = \frac{m\lambda}{a}$$

$$\therefore \qquad x_m = \frac{mD\lambda}{a}$$

If Q is the neighbouring or $(m - 1)$th bright band, it follows that

$$OQ = x_{m-1} = \frac{(m-1)D\lambda}{a}$$

$\therefore$ separation y between successive bands $= x_m - x_{m-1} = \dfrac{\lambda D}{a}$ (i)

$$\therefore \qquad \lambda = \frac{ay}{D} \qquad . \qquad . \qquad . \qquad . \qquad \text{(ii)}$$

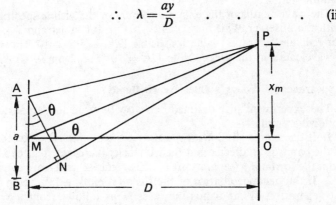

FIG. 29.3. Theory of Young's experiment (exaggerated).

Measurement of Wavelength by Young's Interference Bands

A laboratory experiment to measure wavelength by Young's inter-ference bands is shown in Fig. 29.4. Light from a small filament lamp is focused by a lens on to a narrow slit S, such as that in the collimator of a spectrometer. Two narrow slits A, B, about a millimetre apart, are placed a short distance in front of S, and the light coming from A, B is viewed in a low-powered microscope or eyepiece M about two metres away. Some coloured interference bands are then observed by M. A red and then a blue filter, F, placed in front of the slits, produces red and then blue bands. Observation shows that the separation of the red bands is more than that of the blue bands. Now $\lambda = ay/D$, from (ii), where y is the separation of the bands. It follows that the wavelength of red light is *longer* than that of blue light.

An approximate value of the wavelength of red or blue light can be found by placing a Perspex rule R in front of the eyepiece and moving it until the graduations are clearly seen, Fig. 29.4. The average distance, y, between the bands is then measured on R. The distance a between

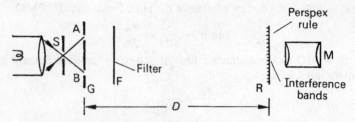

FIG. 29.4. Laboratory experiment on Young's interference bands.

the slits can be found by magnifying the distance by a convex lens, or by using a travelling microscope. The distance D from the slits to the Perspex rule, where the bands are formed, is measured with a metre rule. The wavelength λ can then be calculated from $\lambda = ay/D$, and is of the order 6×10^{-7} m. Further details of the experiment can be obtained from *Advanced Level Practical Physics* by Nelkon and Ogborn (Heinemann).

The wavelengths of the extreme colours of the visible spectrum vary with the observer. This may be 4×10^{-7} m for violet and 7×10^{-7} m for red; an "average" value for visible light is $5 \cdot 5 \times 10^{-7}$ m, which is a wavelength in the green. Note $1 \text{ Å} = 10^{-10}$ m, $1 \text{ nm} = 10^{-9}$ m.

Appearance of Young's Interference Bands

The experiment just outlined can also be used to demonstrate the following points:—

1. If the source slit S is moved nearer the double slits the separation of the bands is unaffected but their intensity increases. This can be seen from the formula y (separation) $= \lambda D/a$, since D and a are constant.

2. If the distance apart a of the slits is diminished, keeping S fixed, the separation of the bands increases. This follows from $y = \lambda D/a$.

3. If the source slit S is widened the bands gradually disappear. The slit S is then equivalent to a large number of narrow slits, each producing its own band system at different places. The bright and dark bands of different systems therefore overlap, giving rise to uniform illumination. It can be shown that, to produce interference bands which are recognisable, the slit width of S must be less than $\lambda D'/a$, where D' is the distance of S from the two slits A, B.

4. If one of the slits, A or B, is covered up, the bands disappear.

5. If white light is used the central band is white, and the bands either side are coloured. Blue is the colour nearer to the central band and red is farther away. The path difference to a point O on the perpendicular bisector of the two slits A, B is zero for all colours, and consequently each colour produces a bright band here. As they overlap, a white band is formed. Farther away from O, in a direction parallel to the slits, the shortest visible wavelengths, blue, produce a bright band first.

Fresnel's Biprism Experiment

Fresnel used a biprism R which had a very large angle of nearly 180°, and placed a narrow slit S, illuminated by monochromatic light, in

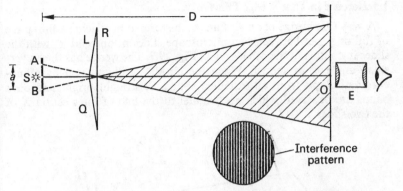

FIG. 29.5. Fresnel's biprism experiment (not to scale).

front of it so that the refracting edge was parallel to the slit, Fig. 29.5. The light emerging after refraction from the two halves, L, Q, of the prism can be considered to come from two sources, A, B, which are the virtual images of the slit S in L, Q respectively. Thus A, B are coherent sources; further, as R has a very large obtuse angle, A and B are close together. Thus an interference pattern is observed in the region of O where the emergent light from the two sources overlap, as shown by the shaded portion of Fig. 29.5, and bright and dark bands can be seen through an eyepiece E at O directed towards R, Fig. 29.6. By using cross-wires, and moving the eyepiece by a screw arrangement, the distance y between successive bright bands can be measured. Now it was shown on p. 689 that $\lambda = ay/D$, where a is the distance between A, B and D is the distance of the source slit from the eyepiece. The distance D is measured with a metre rule. The distance a can be found by moving a convex lens between the fixed biprism and eyepiece until a magnified image of the two slits A, B is seen clearly, and the magnified distance b between them is measured. The magnification m is (image distance $\div$ object distance) for the lens, and a can be calculated from $a = b/m$. Knowing a, y, D, the wavelength λ can be determined.

FIG. 29.6 Fresnel's biprism interference bands (magnified).

If A is the large angle, nearly 180°, of the biprism, each of the small base angles is $(180° - A)/2$, or $90° - A/2$. The small deviation d in radians of light from the slit S is $(n - 1)\,\theta$, where θ is the magnitude of the base angle in radians (p. 457), and hence the distance A, B between the virtual images of the slit $= 2td = 2t\,(n - 1)\,\theta$, where t is the distance from S to the biprism.

Interference in Thin Wedge Films

A very thin wedge of an air film can be formed by placing a thin piece of foil or paper between two microscope slides at one end Y, with the slides in contact at the other end X, Fig. 29.7. The wedge has then a very small angle θ, as shown. When the air-film is illuminated by monochromatic light from an extended source S, straight bright and dark bands are observed which are parallel to the line of intersection X of the two slides.

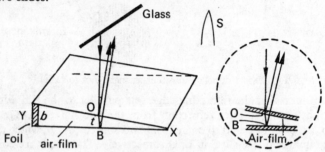

FIG. 29.7. Thin wedge film.

The light reflected down towards the wedge is partially reflected upwards from the lower surface O of the top slide. The remainder of the light passes through the slide and some is reflected upward from the top surface B of the lower slide. The two trains of waves are coherent, since both have originated from the same centre of disturbance at O, and they produce an interference phenomenon if brought together by the eye or in an eyepiece. Their path difference is $2t$, where t is the small thickness of the air-film at O. At X, where the path difference is apparently zero, we would expect a bright band. But a *dark* band is observed at X. This is due to a phase change of 180°, equivalent to an extra path difference of $\lambda/2$, which occurs when a wave is reflected at a denser medium. See pp. 641, 694. The optical path difference between the two coherent beams is thus actually $2t + \lambda/2$, and hence, if the beams are brought together to interfere, a bright band is obtained when $2t + \lambda/2 = m\lambda$, or $2t = (m - \frac{1}{2})\lambda$. A dark band is obtained at a thickness t given by $2t = m\lambda$.

The bands are located at the air-wedge film, and the eye or microscope must be focused here to see them. The appearance of a band is the contour of all points in the air-wedge film where the optical path difference is the same. If the wedge surfaces make perfect optical contact at one edge, the bands are straight lines parallel to the line of intersection of the surfaces. If the glass surfaces are uneven, and the contact at one edge is not regular, the bands are not perfectly straight. A particular band still shows the locus of all points in the air-wedge which have the same optical path difference in the air-film.

In *transmitted light*, the appearance of the bands are complementary to those seen by reflected light, from the law of conservation of energy. The bright bands thus correspond in position to the dark bands seen by reflected light, and the band where the surfaces touch is now bright instead of dark.

Thickness of Thin Foil. Expansion of Crystal

If there is a bright band at Y at the edge of the foil, Fig. 29.7, the thickness b of the foil is given by $2b = (m + \frac{1}{2})\lambda$, where m is the number of bright bands between X and Y. If there is a dark band at Y, then $2b = m\lambda$. Thus by counting m, the thickness b can be found. The small angle θ of the wedge is given by b/a, where a is the distance XY, and by measuring a with a travelling microscope focused on the air-film, θ can be found. If a liquid wedge is formed between the plates, the optical path difference becomes $2nt$, where the air thickness is t, n being the refractive index of the liquid. An optical path difference of λ now occurs for a change in t which is n times less than in the case of the air-wedge. The spacing of the bright and dark bands is thus n times closer than for air, and measurement of the relative spacing enables n to be found.

The coefficient of expansion of a crystal can be found by forming an air-wedge of small angle between a fixed horizontal glass plate and the upper surface of the crystal, and illuminating the wedge by monochromatic light. When the crystal is heated a number of bright bands, m say, cross the field of view in a microscope focused on the air-wedge. The increase in length of the crystal in an upward direction is $m\lambda/2$, since a change of λ represents a change in the thickness of the film of $\lambda/2$, and the coefficient of expansion can then be calculated.

Newton's Rings

Newton discovered an example of interference which is known as "Newton's rings". In this case a lens L is placed on a sheet of plane glass, L having a lower surface of very large radius of curvature, Fig. 29.8. By means of a sheet of glass G monochromatic light from a sodium

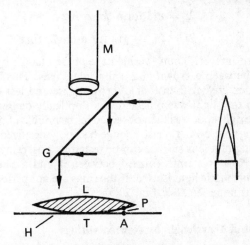

FIG. 29.8. Newton's rings.

flame, for example, is reflected downwards towards L; and when the light reflected upwards is observed through a microscope M focused on

H, a series of bright and dark rings is seen. The circles have increasing radius, and are concentric with the point of contact T of L with H.

Consider the air-film PA between A on the plate and P on the lower lens surface. Some of the incident light is reflected from P to the microscope, while the remainder of the light passes straight through to A, where it is also reflected to the microscope and brought to the same focus. The two rays of light have thus a net path difference of $2t$, where $t =$ PA. The same path difference is obtained at all points round T which are distant TA from T; and hence if $2t = m\lambda$, where m is an integer and λ is the wavelength, we might expect a bright *ring* with centre T. Similarly, if $2t = (m + \frac{1}{2})\lambda$, we might expect a dark ring.

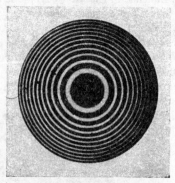

FIG. 29.9. Newton's rings, formed by interference of yellow light between a convex lens and a flat glass plate.

When a ray is reflected from an optically *denser* medium, however, a phase change of 180° occurs in the wave, which is equivalent to its acquiring an extra path difference of $\lambda/2$ (see also p. 692). The truth of this statement can be seen by the presence of the dark spot at the centre, T, of the rings. At this point there is no geometrical path difference between the rays reflected from the lower surface of the lens and H, so that they should be in phase when they are brought to a focus and should form a bright spot. The dark spot means, therefore, that one of the rays suffers a phase change of 180°. Taking the phase change into account, it follows that

$$2t = m\lambda \text{ for a } dark \text{ ring} \qquad . \qquad . \qquad . \qquad (1)$$

and

$$2t = (m + \frac{1}{2})\lambda \text{ for a } bright \text{ ring} \qquad . \qquad . \qquad (2)$$

where m is an integer. Young verified the phase change by placing oil of sassafras between a crown and a flint glass lens. This liquid had a refractive index greater than that of crown glass and less than that of flint glass, so that light was reflected at an optically denser medium at each lens. A bright spot was then observed in the middle of the Newton's rings, showing that no net phase change had now occurred.

The grinding of a lens surface can be tested by observing the appearance of the Newton's rings formed between it and a flat glass plate when monochromatic light is used. If the rings are not perfectly circular, the grinding is imperfect. See Fig. 29.9.

Measurement of Wavelength by Newton's Rings

The radius r of a ring can be expressed in terms of the thickness, t, of the corresponding layer of air by simple geometry. Suppose TO is produced to D to meet the completed circular section of the lower surface PQ of the lens, PO being perpendicular to the diameter TD through T,

Fig. 29.10. Then, from the well-known theorem concerning the segments of chords in a circle, TO. OD = QO. OP. But AT = r = PO, QO = OP = r, AP = t = TO, and OD = $2a$ − OT = $2a$ − t.

$$\therefore \quad t\,(2a - t) = r \times r = r^2$$

$$\therefore \quad 2at - t^2 = r^2$$

But t^2 is very small compared with $2at$, as a is large.

$$\therefore \quad 2at = r^2$$

$$\therefore \quad 2t = \frac{r^2}{a} \qquad \qquad \qquad . \qquad (i)$$

But
$$2t = (m + \tfrac{1}{2})\lambda \text{ for a bright ring.}$$

$$\therefore \quad \frac{r^2}{a} = (m + \tfrac{1}{2})\lambda \quad . \qquad \frac{d^2}{4} = R\left(n+\tfrac{1}{2}\right)\lambda \qquad (3)$$

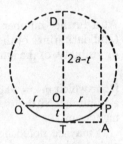

The first bright ring obviously corresponds to the case of $m = 0$ in equation (3); the second bright ring corresponds to the case of $m = 1$. Thus the radius of the 15th bright ring is given from (3) by $r^2/a = 14\tfrac{1}{2}\lambda$, from which $\lambda = 2r^2/29a$. Knowing r and a, therefore, the wavelength λ can be calculated. Experiment shows that the rings become narrower when blue or violet light is used in place of red light, which proves, from equation (3), that the wavelength of violet light is shorter than the wavelength of red light. Similarly it can be proved that the wavelength of yellow light is

FIG. 29.10.
Theory of radius of
Newton's rings.

shorter than that of red light and longer than the wavelength of violet light.

The radius r of a particular ring can be found by using a travelling microscope to measure its diameter. The radius of curvature, a, of the lower surface of the lens can be measured accurately by using light of known wavelength λ', such as the green in a mercury-vapour lamp or the yellow of a sodium flame; since $a = r^2/(m + \tfrac{1}{2}) \cdot \lambda'$ from (3), the radius of curvature a can be calculated from a knowledge of r, m, λ'.

Visibility of Newton's Rings

When white light is used in Newton's rings experiment the rings are coloured, generally with violet at the inner and red at the outer edge. This can be seen from the formula $r^2 = (m + \tfrac{1}{2})\lambda a$, (3), as $r^2 \propto \lambda$. Newton gave the following list of colours from the centre outwards:

First order: Black, blue, white, yellow, orange, red. *Second order:* Violet, blue, green, yellow, orange, red. *Third order:* Purple, blue, green, yellow, orange, red. *Fourth order:* Green, red. *Fifth order:* Greenish-blue, red. *Sixth order:* Green-blue, pale-red. *Seventh order:* Greenish-blue, reddish-white. Beyond the seventh order the colours overlap and hence white light is obtained. The list is known generally as "Newton's scale of colours". Newton left a detailed description of the colours obtained with different thicknesses of air.

When Newton's rings are formed by sodium light, close examination shows that the clarity, or visibility, of the rings gradually diminishes as one moves outwards from the central spot, after which the visibility improves again. The variation in clarity is due to the fact that sodium light is not monochromatic but consists of *two* wavelengths, λ_2, λ_1, close to one another. These are (i) $\lambda_2 = 5 \cdot 890 \times 10^{-7}$ m (D_2), (ii) $\lambda_1 = 5 \cdot 896 \times 10^{-7}$ m (D_1). Each wavelength produces its own pattern of rings, and the ring patterns gradually separate as m, the number of the ring, increases. When $m\lambda_1 = (m + \frac{1}{2})\lambda_2$, the bright rings of one wavelength fall in the dark spaces of the other and the visibility is a minimum. In this case

$$5896m = 5890 \, (m + \tfrac{1}{2}).$$
$$\therefore \quad m = \frac{5890}{12} = 490 \text{ (approx.)}$$

At a further number of ring m_1, when $m_1\lambda_1 = (m_1 + 1)\lambda_2$, the bright (and dark) rings of the two ring patterns coincide again, and the clarity, or visibility, of the interference pattern is restored. In this case

$$5896m_1 = 5890 \, (m_1 + 1),$$

from which $m_1 = 980$ (approx.). Thus at about the 500th ring there is a minimum visibility, and at about the 1000th ring the visibility is a maximum.

It may be noted here that the bands in films of varying thickness, such as Newton's rings and the air-wedge bands, p. 692, appear to be formed in the film itself, and the eye must be focused on the film to see them. We say that the bands are "localised" at the film. With a thin film of uniform thickness, however, bands are formed by parallel rays which enter the eye, and these bands are therefore localised at infinity.

"Blooming" of Lenses

Whenever lenses are used, a small percentage of the incident light is reflected from each surface. In compound lens systems, as in telescopes and microscopes, this produces a background of unfocused light, which results in a reduction in the clarity of the final image. There is also a reduction in the intensity of the image, since less light is transmitted through the lenses.

The amount of reflected light can be considerably reduced by evaporating a thin coating of a fluoride salt such as magnesium fluoride on to the surfaces, Fig. 29.11. Some of the light, of average

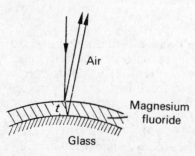

Fig. 29.11. Blooming of lens.

wavelength λ, is then reflected from the air-fluoride surface and the remainder penetrates the coating and is partially reflected from the fluoride-glass surface. Destructive interference occurs between the two reflected beams when there is a phase difference of 180°, or a path difference of $\lambda/2$, as the refractive index of the fluoride is less than that of glass. Thus if t is the required thickness of the coating and n' its retractive index, $2n't = \lambda/2$. Hence $t = \lambda/4n' = 6 \times 10^{-7}/(4 \times 1.38)$, assuming λ is 6×10^{-7} m and n' is 1.38; thus $t = 1.1 \times 10^{-7}$ m.

For best results n' should have a value equal to about $\sqrt{n}$, where n is the refractive index of the glass lens. The intensities of the two reflected beams are then equal, and hence complete interference occurs between them. No light is then reflected back from the lens. In practice, complete interference is not possible simultaneously for every wavelength of white light, and an average wavelength for λ, such as green-yellow, is chosen. "Bloomed" lenses effect a marked improvement in the clarity of the final image in optical instruments.

Lloyd's Mirror

In 1834 LLOYD obtained interference bands on a screen by using a plane mirror M, and illuminating it with light nearly at grazing incidence,

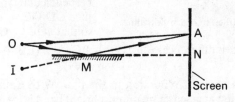

FIG. 29.12. Lloyd's mirror experiment.

coming from a slit O parallel to the mirror, Fig. 29.12. A point such as A on the screen is illuminated (i) by a ray OA and (ii) by a ray OM reflected along MA, which appears to come from the virtual image I of O in the mirror. Since O and I are close coherent sources interference bands are obtained on the screen.

Experiment showed that the band at N, which corresponds to the point of intersection of the mirror and the screen, was *dark;* since ON = IN, this band might have been expected, before the experiment was carried out, to be bright. Lloyd concluded that a phase change of 180°, equivalent to half a wavelength, occurred by reflection at the mirror surface, which is a denser surface than air (see p. 692).

Interference in Thin Films

The colours observed in a soap-bubble or a thin film of oil in the road are due to an interference phenomenon; they are also observed in thin transparent films of glass.

Consider a ray AO of monochromatic light incident on a thin parallel-sided film of thickness t and refractive index n. Fig. 29.13 is exaggerated for clarity. Some of the light is reflected at O along ON, while the remainder is refracted into the film, where reflection occurs at B. The ray BC then emerges into the air along CM, which is parallel to ON. The incident ray AO thus divides at O into two beams of different amplitude which are coherent, and if ON, CM are combined by a lens, or by the eye-lens, a bright or dark band is observed according to the path difference of the rays.

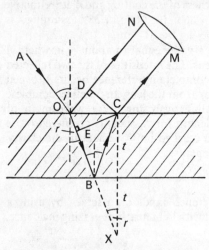

FIG. 29.13. Interference in thin films.

The time taken for light to travel a distance y in a medium of refractive index n is y/v, where v is the velocity of light in the medium. In this time, a distance $c \times y/v$ is travelled in air, where c is the velocity in air. But $n = c/v$. Hence the *optical* path of a length y in a medium of refractive index n is ny. The optical path difference between the two rays ON and OBCM is thus n (OB + BC) − OD, where CD is perpendicular to ON, Fig. 29.13. If CE is the perpendicular from C to OB, then OD/OE = sin i/sin $r = n$, so that nOE = OD.

∴   optical path difference $= n$ (EB + BC) $= n$ (EB + BX) $= n$. EX.
$$= 2\,nt \cos r,$$

where r is the angle of refraction in the film. With a phase change of 180° by reflection at a denser medium, a bright band is therefore obtained when $2\,nt \cos r + \lambda/2 = m\lambda$,

or $2\,nt \cos r = (m - \tfrac{1}{2})\,\lambda$.   .   .   .   (i)

For a dark band, $2\,nt \cos r = m\lambda$.   .   .   .   .   (ii)

Colours in Thin Films

The colours in thin films of oil or glass are due to interference from an extended source such as the sky or a cloud. Fig. 29.14 illustrates interference between rays from points O_1, O_2 respectively on the extended source. Each ray is reflected and refracted at A_1, A_2 on the film, and enter the eye at E_1. Although O_1, O_2 are non-coherent, the eye will see the same colour of a particular wavelength λ if $2\,nt \cos r = (m - \tfrac{1}{2})\,\lambda$. The separation of the two rays from A_1 or from A_2 must be less than the diameter of the eye-pupil for interference to occur, and this is the case only for thin films. The angle of refraction r is determined by the angle of incidence, or reflection, at the film. The particular colour seen thus depends on the position of the eye. At E_2, for example, a different colour will be seen from another point O_3 on the extended source. The variation of θ and hence r is small when the eye observes a particular area of the film, and hence a band of a particular colour, such as $A_1 A_2$, is the contour of paths of *equal inclination* to the film. The bands are localised at infinity, since the rays reaching E_1 or E_2 are parallel.

If a thin wedge-shaped film is illuminated by an extended source, as shown on p. 692 or in Newton's rings, the bands seen are contours of *equal thickness* of the film.

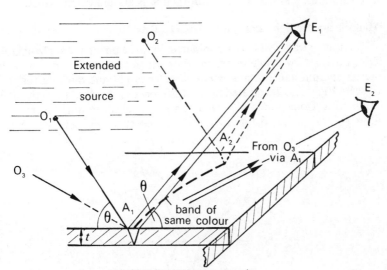

FIG. 29.14. Colours in thin films.

Vertical Soap Film Colours

An interesting experiment on thin films, due to C. V. Boys, can be performed by illuminating a vertical soap film with monochromatic light. At first the film appears uniformly coloured. As the soap drains to the bottom, however, a wedge-shaped film of liquid forms in the ring, the top of the film being thinner than the bottom. The thickness of the wedge is constant in a horizontal direction, and thus horizontal bright and dark bands are observed across the film. When the upper part of the film becomes extremely thin a black band is observed at the top (compare the dark central spot in Newton's rings experiment), and the film breaks shortly afterwards.

With white light, a succession of broad coloured bands is first observed in the soap film. Each band contains colours of the spectrum, red to violet. The bands widen as the film drains, and just before it breaks a black band is obtained at the top.

For normal incidence of white light, a particular wavelength λ is seen where the optical path difference due to the film $= (m - \frac{1}{2})\lambda$ and m is an integer. Thus a red colour of wavelength $7 \cdot 0 \times 10^{-7}$ m is seen where the optical path difference is $3 \cdot 5 \times 10^{-7}$ m, corresponding to $m = 1$. No other colour is seen at this part of the thin film. Suppose, however, that another part of the film is much thicker and the optical path difference here is $21 \times 3 \cdot 5 \times 10^{-7}$ m. Then a red colour of wavelength $7 \cdot 0 \times 10^{-7}$ m, $m = 11$, an orange colour of wavelength about

6.4×10^{-7} m, $m = 12$, a yellow wavelength about 5.9×10^{-7} m, $m = 13$, and other colours of shorter wavelengths corresponding to higher integral values of m, are seen at the same part of the film. These colours all overlap and produce a white colour. If the film is thicker still, it can be seen that numerous wavelengths throughout the visible spectrum are obtained and the film then appears uniformly white.

Monochromatic light and Thin Parallel Films

If a thin parallel film is illuminated by a beam of monochromatic light, obtained by using an extended or broad source such as a bunsen burner sodium flame, a number of *circular* bright and dark curves can be seen. Fig. 29.15 illustrated how interference is obtained from the light originating from points a, b which is refracted at an angle α into the film. This is similar to Fig. 29.14 if B represents an eye-lens.

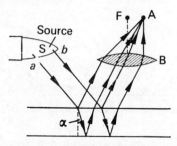

FIG. 29.15. Interference with extended source.

The emergent rays are combined by the eye-lens or a glass lens B, and a dark band is formed at A if $2nt \cos \alpha = m\lambda$, with the usual notation. If the light is incident on the film in every plane a circular band is obtained, whose centre is F, the focus of B. It is a band of 'equal inclination'.

When a *parallel* beam of monochromatic light is incident on the thin film, the angle of refraction r in the film and the thickness t are constant. The film thus appears uniformly bright at all points if the condition $2nt \cos r = (m + \frac{1}{2})\lambda$ is obeyed, and is uniformly dark if $2nt \cos r = m\lambda$. If the film is illuminated by a parallel beam of *white* light, the transmitted light appears to have dark bands across it when viewed through a spectroscope. The latter separates the colours, and a dark band is obtained where the condition $2nt \cos r = (m + \frac{1}{2})\lambda$ is satisfied for the particular wavelength, since we are now concerned with transmitted light.

EXAMPLE

What are Newton's rings and under what conditions can they be observed? Explain how they can be used to test the accuracy of grinding of the face of a lens. The face of a lens has a radius of curvature of 50 cm. It is placed in contact with a flat plate and Newton's rings are observed normally with reflected light of wavelength 5×10^{-7} m. Calculate the radii of the fifth and tenth bright rings. (C.)

First parts. See text.

Second part. With the usual notation, for a bright ring we have

$$2t = (m + \tfrac{1}{2})\lambda, \qquad \cdot \qquad \cdot \qquad \cdot \qquad \cdot \qquad \text{(i)}$$

where t is the corresponding thickness of the layer of air.

But, from geometry, $\qquad 2t = \dfrac{r^2}{a} \qquad \cdot \qquad \cdot \qquad \cdot \qquad \cdot \qquad \text{(ii)}$

where r is the radius of the ring and a is the radius of curvature of the lens face (p. 695).

$$\therefore \quad \frac{r^2}{a} = (m + \tfrac{1}{2})\lambda$$

$$\therefore \quad r^2 = (m + \tfrac{1}{2})\lambda a \qquad \cdot \qquad \cdot \qquad \cdot \qquad \text{(iii)}$$

The first ring corresponds to $m = 0$ from equation (iii). Hence the fifth ring corresponds to $m = 4$, and its radius r in metres is thus given by

$$r^2 = (4 + \tfrac{1}{2}) \times 5 \times 10^{-7} \times 50$$

$$\therefore \quad r = \sqrt{\frac{9 \times 5 \times 10^{-7} \times 50}{2}} = 1 \cdot 06 \times 10^{-3} \text{ m.}$$

The tenth ring corresponds to $m = 9$ in equation (iii), and its radius is thus given by

$$r^2 = 9\tfrac{1}{2} \times 5 \times 10^{-7} \times 50$$

$$\therefore \quad r = 1 \cdot 54 \times 10^{-3} \text{ m.}$$

DIFFRACTION OF LIGHT

In 1665 GRIMALDI observed that the shadow of a very thin wire in a beam of light was much broader than he expected. The experiment was repeated by Newton, but the true significance was only recognised more than a century later, after Huygens' wave theory of light had been resurrected. The experiment was one of a number which showed that light could bend round corners in certain circumstances.

We have seen how interference patterns, for example, bright and dark bands, can be obtained with the aid of two sources of light close to each other. These sources must be coherent sources, i.e., they must have the same amplitude and frequency, and always be in phase with each other. Consider two points on the *same wavefront*, for example the two points A, B, on a plane wavefront arriving at a narrow slit in a screen, Fig. 29.16. A and B can be considered as secondary sources of light, an aspect introduced by Huygens in his wave theory of light (p. 676); and as they are on the same wavefront, A and B have identical amplitudes

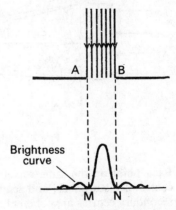

FIG. 29.16. Diffraction of light.

FIG. 29.17. Diffraction rings in the shadow of a small circular disc. The bright spot is at the centre of the geometrical shadow.

and frequencies and are in phase with each other. Consequently A, B, are coherent sources, and we can expect to find an interference pattern on a screen in front of the slit, provided the latter is small compared with the wavelength of light. For a short distance beyond the edges M, N, of the projection of AB, i.e., in the geometrical shadow, observation shows that there are some alternate bright and dark bands. See Fig. 29.20.

Thus light can travel round corners. The phenomenon is called *diffraction*, and it has enabled scientists to measure accurately the wavelength of light.

If a source of white light is observed through the eyelashes, a series of coloured images can be seen. These images are due to interference between sources on the same wavefront, and the phenomenon is thus an example of diffraction. Another example of diffraction was unwittingly deduced by POISSON at a time when the wave theory was new. Poisson considered mathematically the combined effect of the wavefronts round a circular disc illuminated by a distant small source of light, and he came to the conclusion that the light should be visible beyond the disc in the middle of the geometrical shadow. Poisson thought this was impossible; but experiment confirmed his deduction, and he became a supporter of the wave theory of light. See Fig. 29.17.

Diffraction at Single Slit

We now consider diffraction at a single slit in more detail. Suppose parallel light is incident on a narrow rectangular slit AB, Fig. 29.18.

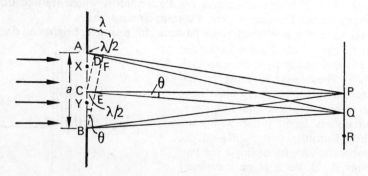

FIG. 29.18. Diffraction at single slit.

Each point on the same wavefront between A, B acts as a secondary centre of disturbance, and sends out wavelets beyond the slit. All the secondary centres are coherent, and their combined effect at any point such as P or Q can be found by summing the individual waves there, from the Principle of Superposition. The mathematical treatment is beyond the scope of this book. The general effect, however, can be derived by considering the two halves AC, CB of the wavefront AB. At a point P equidistant from A and B, corresponding secondary centres in AC, CB respectively, such as X and Y, are also equidistant from P. Consequently wavelets arrive in phase at P. When AB is of the order of a few wavelengths of light the resultant amplitude at P due to the whole wavefront AB is therefore large, and thus a bright band is obtained at P.

As we move from P parallel to AB, points are obtained where the secondary wavelets from the two halves of the wavefront become more and more out of phase on arrival and the brightness thus diminishes. Consider a point Q, where AQ is half a wavelength longer than CQ. A disturbance from A, and one from C, then arrive at Q 180° out of phase. This is also practically the case for all corresponding points such as X, Y on the two halves of the wavefront. In particular, CQ and BQ differ practically by $\lambda/2$, where C is the extreme point in the upper half of the wavefront and B is the extreme point on the lower half. Thus Q corresponds to the edge or minimum intensity of the central band round P, Fig. 29.19. As we move farther away from Q parallel to AB, the

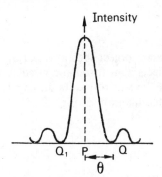

FIG. 29.19. Intensity variation – single slit.

intensity rises again to a much smaller maximum at R, where AR − BR = $3\lambda/2$, Fig. 29.18. To explain this, one can imagine the wavefront AB·in Fig. 29.18 divided into three equal parts. Two parts annul each other's displacements at R as just explained, leaving one-third of the wavefront, which produces a much less bright band at R than at P. Calculation shows that the maximum intensity of the band at R is less than 5 per cent of that of the central band at P. Other subsidiary maxima and minima diffraction bands are obtained if the slit is very narrow. See Fig. 29.20.

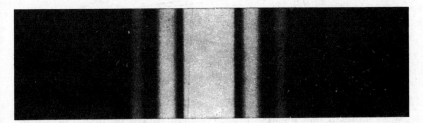

FIG. 29.20. Diffraction bands formed by a single small rectangular aperture.

Width of Central Band. Rectilinear Propagation

The angular width of the central bright band is 2θ, where θ is the angular width between the maximum intensity direction P and the mini-

mum at Q, Fig. 29.19. From Fig. 29.18, it can be seen that the line CQ
to the edge of the central band makes an angle θ with the direction CP
of the incident light given by

$$\sin \theta = \frac{AF}{AB} = \frac{AD + CE}{AB} = \frac{\lambda/2 + \lambda/2}{a} = \frac{\lambda}{a},$$

where $a =$ AB. When the slit is widened and a becomes large compared
with λ, then $\sin \theta$ is very small and hence θ is very small. In this case
the directions of the minimum and maximum intensities of the central
band are very close to each other, and practically the whole of the light
is confined to a direction immediately in front of the incident direction,
that is, no spreading occurs. This explains the *rectilinear propagation of
light*. When the slit width a is very small and equal to 2λ, for example,
then $\sin \theta = \lambda/a = 1/2$, or $\theta = 30°$. The light waves now spread round
through 30° on either side of the slit.

These results are true for any wave phenomenon. In the case of an
electromagnetic wave of 3 cm wavelength, a slit of these dimensions
produces sideways spreading. Sound waves of a particular frequency
256 Hz have a wavelength of about 1·3 m. Consequently, sound waves
spread round corners or apertures such as a doorway, which have
comparable dimensions to their wavelengths.

Diffraction in Telescope Objective

When a parallel beam of light from a distant object such as a star S_1
enters a telescope objective L, the lens collects light through a circular
opening and forms a diffraction pattern of the star round its principal
focus, F. This is illustrated in the exaggerated diagram of Fig. 29.21.

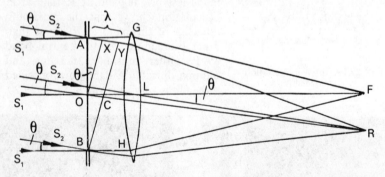

FIG. 29.21. Diffraction in telescope objective.

Consider an incident plane wavefront AB from the star S_1, and suppose
for a moment that the aperture is rectangular. The diffracted rays such
as AG, BH normal to the wavefront are incident on the lens in a direction
parallel to the principal axis LF. The optical paths AGF, BHF are equal.
This is true for all other diffracted rays from points between A, B which
are parallel to LF, since the optical paths to an image produced by a lens
are equal. The central part F of the star pattern is therefore bright.

Now consider those diffracted rays from all points between AB which enter the lens at an angle θ to the principal axis. This corresponds to a diffracted plane wavefront BY at an angle θ to AB. As described previously on p. 703, the wavefront AB can be divided into two halves, AO, OB. The rays from A and O in the two halves produce destructive interference if AX $= \lambda/2$, and likewise the extreme points O, B in the two halves produce destructive interference as OC $= \lambda/2$. Other corresponding points on the two halves also produce destructive interference. When the rays are collected and brought to a focus at R, darkness is thus obtained, that is, R is the edge of the central maximum of the star S_1. As explained on p. 704, other subsidiary maxima may be formed round F.

The angle θ corresponding to the edge R is given by

$$\sin \theta = \frac{\lambda/2 + \lambda/2}{D} = \frac{\lambda}{D},$$

where D is the diameter of the lens aperture. This is the case where the opening can be divided into a number of rectangular slits. For a circular opening such as a lens (or the concave mirror of the Palomar telescope), the formula becomes $\sin \theta = 1\cdot22\lambda/D$, and as θ is small, we may write $\theta = 1\cdot22\lambda/D$.

Resolving Power

Suppose now that another distant star S_2 is at an angular distance θ from S_1, Fig. 29.21. The maximum intensity of the central pattern of S_2 then falls on the minimum or edge of the central pattern of the star S_1, corresponding to R in Fig. 29.22 (i). Experience shows that the two stars

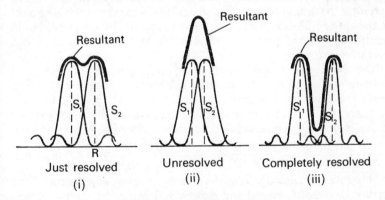

Just resolved (i)   Unresolved (ii)   Completely resolved (iii)

FIG. 29.22. Resolving power.

can then just be distinguished or *resolved*. Lord Rayleigh stated a criterion for the resolution of two objects, which is generally accepted: *Two objects are just resolved when the maximum intensity of the central pattern of one object falls on the first minimum or dark edge of the other.* Fig. 29.22 (i) shows the two stars just resolved. The resultant intensity in the middle dips to about 0·8 of the maximum, and the eye is apparently sensitive to the change here. Fig. 29.22 (ii) shows two stars S_1, S_2 unresolved, and Fig. 29.22 (iii) the same stars completely resolved.

The angular distance θ between two distant stars just resolved is thus given by $\sin \theta = \theta = 1 \cdot 22 \ \lambda/D$, where D is the diameter of the objective. This is an expression for the *limit of resolution*, or *resolving power*, of a telescope. The limit of resolution or resolving power increases when θ is smaller, as two stars closer together can then be resolved. Consequently telescope objectives of large diameter D give high resolving power. The Yerkes Observatory has a large telescope objective of about 1 metre. The angular distance θ between two stars which can just be resolved is thus given by

$$\theta = \frac{1 \cdot 22 \lambda}{D} = \frac{1 \cdot 22 \times 6 \times 10^{-7}}{100} = 7 \cdot 3 \times 10^{-7} \text{ radians,}$$

assuming 6×10^{-7} m for the wavelength of light. The Mount Palomar telescope has a parabolic mirror objective of aperture 5 metres, or 500 cm. The resolving power is thus five times as great as the Yerkes Observatory telescope. A large aperture D has also the advantage of collecting more light (p. 542). The Jodrell Bank radio telescope has a circular bowl of about 75 m, and for radio waves of 22 cm wavelength the resolving power, $\theta = 1 \cdot 22\lambda/D = 1 \cdot 22 \times 22/7500$ radians $= 4 \times 10^{-3}$ radians (approx.).

Magnifying Power of Telescope and Resolving Power

If the width of the emergent beam from a telescope is greater than the diameter of the eye-pupil, rays from the outer edge of the objective do not enter the eye and hence the full diameter D of the objective is not used. If the width of the emergent beam is less than the diameter of the eye-pupil, the eye itself, which has a constant aperture, may not be able to resolve the distant objects. Theoretically, the angular resolving power of the eye is $1 \cdot 22 \ \lambda/a$, where a is the diameter of the eye-pupil, but in practice an angle of 1 minute is resolved by the eye, which is more than the theoretical value.

Now the angular magnification, or magnifying power, of a telescope is the ratio a'/a, where a' is the angle subtended at the eye by the final image and a is the angle subtended at the objective (p. 533). To make the fullest use of the diameter D of the objective, the magnifying power should therefore be increased to the angular ratio given by (D in metres)

$$\frac{\text{resolving power of eye}}{\text{resolving power of objective}} = \frac{\pi/(180 \times 60)}{1 \cdot 22 \times 6 \times 10^{-7}/D} = 4 \ D \text{ (approx.)}$$

In this case the telescope is said to be in "normal adjustment". Any further increase in magnifying power will make the distant objects appear larger, but there is no increase in definition or resolving power.

Brightness of Images in Telescope

In a telescope, the eye is placed at the *exit-pupil* or *eye-ring*, the circle through which the emergent beam passes (p. 532). The *entrance pupil* of the telescope is the aperture or diameter of the objective. If the area of the latter is A, then the smaller area of the exit-pupil is A/M^2, where M is the angular magnification of the telescope in normal adjustment (see p. 535).

Consider a telescope used to observe (a) a small but finite area, or extended object, and (b) a point source, such as a star. Suppose that in each case the magnifying power is adjusted to make the exit pupil of the telescope equal to the eye pupil. In each case the luminous flux collected with the telescope is equal to the flux collected by an unaided eye multiplied by the ratio: area of objective/area of eye-pupil, which is M^2.

Geometrically, the telescope magnifies the finite object area by a factor M^2, but the image of the point object is still a point. Hence the area of the retinal image of the finite object is magnified by a factor M^2 when the telescope is used. On the other hand, since the eye-pupil is filled with light by the telescope, the retinal image of the point object with the telescope is the same as that without the telescope—it is the diffraction image for a point source. It will be seen that, for the finite object, the larger flux is spread over a larger image area, so that (apart from absorption losses in the telescope) the retinal illumination is unchanged. For the point object, however, the increased flux is spread over the same retinal area so that the brightness of the image is increased. On this account stars appear very much brighter when viewed by a telescope, whereas the brightness of the background, which acts as an extended object, remains about the same. Stars too faint to be seen with the naked eye become visible using a powerful telescope, and the number of stars seen thus increases considerably using a telescope.

Increasing Number of Slits. Diffraction Grating

On p. 703 we saw that the image of a single narrow rectangular slit is a bright central or principal maximum diffraction band, together with subsidiary maxima diffraction bands which are much less bright. Suppose that parallel light is incident on two more parallel close slits, and the light passing through the slits is received by a telescope focused at infinity. Since each slit produces a similar diffraction effect in the same direction, the observed diffraction pattern will have an intensity variation identical to that of a single slit. This time, however, the pattern is crossed by a number of interference bands, which are due to interference between slits (see *Young's experiment*, p. 688). The envelope of the intensity variation of the interference bands follow the diffraction pattern variation due to a single slit. In general, if I_s is the intensity at a point due to interference and I_d that due to diffraction, then the resultant intensity I is given by $I = I_d \times I_s$. Hence if $I_d = 0$ at any point, then $I = 0$ irrespective of the value of I_s.

As more parallel equidistant slits are introduced, the intensity and sharpness of the principal maxima increase and those of the subsidiary maxima decrease. The effect is illustrated roughly in Fig. 29.23. With several thousand lines per centimetre, only a few sharp principal maxima are seen in directions discussed shortly. Their angular separation depends only on the distance between successive slits. The slit width affects the intensity of the higher order principal maxima; the narrower the slit, the greater is the diffraction of light into the higher orders.

A *diffraction grating* is a large number of close parallel equidistant slits, ruled on glass or metal; it provides a very valuable means of studying spectra. If the width of a slit or clear space is a and the thickness of a ruled opaque line is b, the spacing d of the slits is $(a + b)$. Thus with a grating of 6000 lines per centimetre, the spacing $d = 1/6000$ centimetre $= 17 \times 10^{-5}$ cm, or a few wavelengths of visible light.

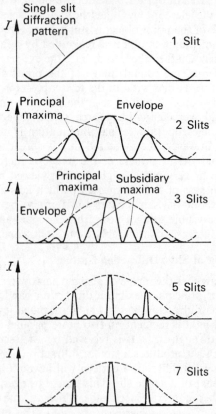

FIG. 29.23. Principal maxima with increasing slits.

Principal Maxima of Grating

The angular positions of the principal maxima produced by a diffraction grating can easily be found. Suppose X, Y are corresponding points in consecutive slits, where $XY = d$, and the grating is illuminated normally by monochromatic light of wavelength λ, Fig. 29.24. In a direction θ, the diffracted rays XL, YM have a path difference XA of $d \sin \theta$. The diffracted rays from all other corresponding points in the two slits have a path difference of $d \sin \theta$ in the same direction. Other pairs of slits throughout the grating can be treated in the same way. Hence bright or principal maxima are obtained when

$$d \sin \theta = m\lambda, \qquad . \qquad . \qquad . \qquad . \qquad \text{(i)}$$

where m is an integer, if all the diffracted parallel rays are collected by a telescope focused at infinity. The images corresponding to

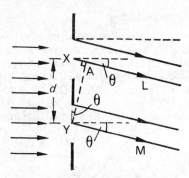

Fig. 29.24. Diffraction grating.

$m = 0, 1, 2, \ldots$ are said to be respectively of the zero, first, second ... orders respectively. The zero order image is the image where the path difference of diffracted rays is zero, and corresponds to that seen directly opposite the incident beam on the grating. It should again be noted that all points in the slits are secondary centres on the same wavefront and therefore coherent sources.

Diffraction Images

The *first order* diffraction image is obtained when $m = 1$. Thus
$$d \sin \theta = \lambda,$$
or
$$\sin \theta = \frac{\lambda}{d}.$$

If the grating has 600 lines per millimetre (600 mm^{-1}), the spacing of the slits, d, is $1/600$ mm $= 1/(600 \times 10^3)$ m. Suppose yellow light, of wavelength $\lambda = 5 \cdot 89 \times 10^{-7}$ m, is used to illuminate the grating. Then

$$\sin \theta = \frac{\lambda}{d} = 5 \cdot 89 \times 10^{-7} \times 600 \times 10^3 = 0 \cdot 3534$$

$$\therefore \quad \theta = 20 \cdot 7°$$

The *second order* diffraction image is obtained when $m = 2$. In this case $d \sin \theta = 2\lambda$.

$$\therefore \quad \sin \theta = \frac{2\lambda}{d} = 2 \times 5 \cdot 89 \times 10^{-7} \times 600 \times 10^3 = 0 \cdot 7068$$

$$\therefore \quad \theta = 45 \cdot 0°$$

If $m = 3$, $\sin \theta = 3\lambda/d = 1 \cdot 060$. Since the sine of an angle cannot be greater than 1, it is impossible to obtain a third order image with this diffraction grating.

With a grating of 1200 lines per mm the diffraction images of sodium light would be given by $\sin \theta = m\lambda/d = m \times 5 \cdot 89 \times 10^{-7} \times 12 \times 10^5 = 0 \cdot 7068\, m$. Thus only $m = 1$ is possible here. As all the diffracted light is now concentrated in one image, instead of being distributed over several images, the first order image is very bright, which is an advantage.

Diffraction with Oblique Incidence

When a diffraction grating is illuminated by a monochromatic parallel beam PX, QY at an angle of incidence i, each point in the clear spaces acts as a secondary disturbance and diffracted beams emerge from the grating, Fig. 29.25. For a diffracted beam such as AB, making an angle of diffraction θ on the same side of the normal as PX or QY, the path difference between two typical rays PXA, QYB is $d (\sin i + \sin \theta)$. For a diffracted beam such as CD on the other side of the normal, the path difference between typical

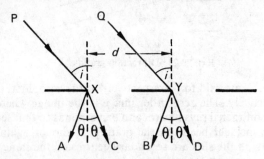

FIG. 29.25. Diffraction with oblique incidence.

rays PXC, QYD is $d (\sin i - \sin \theta)$. Thus, generally, a bright diffraction image is seen when $d (\sin i \pm \sin \theta) = m\lambda$, where m is an integer.

The zero order or central image is obtained in a direction opposite to the incident beam PX, QY. The first order diffraction image is obtained at angles θ on either side of this direction given respectively by $d (\sin i + \sin \theta) = \lambda$ and $d (\sin i - \sin \theta) = \lambda$. Diffraction images of higher order are obtained from similar formulae.

Reflection gratings can be used when light of particular wavelengths are absorbed by materials used in making transmission gratings. In this case the light is diffracted back into the incident medium at the clear spaces, and the diffraction images of various orders are given by $d (\sin i \pm \sin \theta) = m\lambda$.

Measurement of Wavelength

The wavelength of monochromatic light can be measured by a diffraction grating in conjunction with a spectrometer. The collimator C and telescope T of the instrument are first adjusted for parallel light (p. 445), and the grating P is then placed on the table so that its plane is perpendicular to two screws, Q, R, Fig. 29.26 (i). To level the table so that the plane of P is parallel to the axis of rotation of the telescope, the latter is first placed in the position T_1 directly opposite the illuminated slit of the collimator, and then rotated exactly through 90° to a position T_2. The table is now turned until the slit is seen in T_2 by reflection at P, and one of the screws Q, R turned until the slit image is in the middle of the field of view. The plane of P is now parallel to the axis of rotation of the telescope. The table is then turned through 45° so that the plane of the grating is exactly perpendicular to the light from C, and the telescope is turned to a position T_3 to receive the first diffraction image, Fig. 29.26 (ii). If the lines of the grating are not parallel to the axis of rotation of the telescope, the image will not be in the middle of the field of view. The third screw is then adjusted until the image is central.

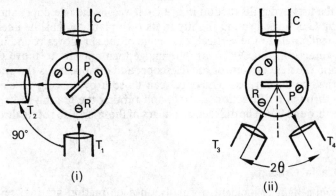

FIG. 29.26. Measurement of wavelength by diffraction grating.

The readings of the first diffraction image are observed on both sides of the normal. The angular difference is 2θ, and the wavelength is calculated from $\lambda = d \sin \theta$, where d is the spacing of the slits, obtained from the number of lines per centimetre of the grating. If a second order image is obtained for a diffraction angle θ, then $\lambda = d \sin \theta_1/2$.

Position of Image

If the grating lines are on the opposite side of the glass to the collimator C in Fig. 29.26 (ii), the light from C passes straight through the glass and the diffracted rays at the slits emerge into air. Suppose, however, that the grating is turned round so that the lines such as A, D face the collimator C, Fig. 29.27 (i). The rays are now diffracted into the glass and then refracted at B, F into the air at an angle θ to the normal. The optical path difference between the rays ABM, DFH from corresponding points A D, is then

$$n.AB + BL - n.DF = BL$$

since $AB = DF$. But $BL = BF \sin \theta = d \sin \theta$. Consequently the angular

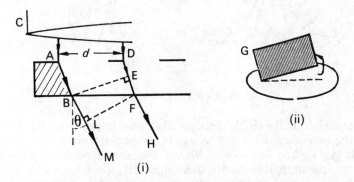

FIG. 29.27. Position of images with diffraction grating.

positions of the principal maxima diffraction images are given by $d \sin \theta = m\lambda$. Thus the images are observed at the same diffraction angles, no matter which side of the grating faces the collimator.

If the first order diffraction image is viewed in the telescope, and the grating G is turned round slightly in its own plane so that the lines are at a small angle to the vertical, the image of the slit moves round in the same direction, Fig. 29.27 (ii). The image then appears to move up or down in the field of view of the telescope, and disappears as the grating is turned round farther. The effect can be seen by viewing an electric lamp through a diffraction grating, and turning the grating in its own plane through 90°. The diffraction images of the lamp also rotate through 90°.

Spectra in Grating

If white light is incident normally on a diffraction grating, several coloured spectra are observed on either side of the normal, Fig. 29.28 (i). The first order diffraction images are given by $d \sin \theta = \lambda$, and as violet has a shorter wavelength than red, θ is less for violet than for red. Consequently the spectrum colours on either side of the incident white light are violet to red. In the case of a spectrum produced by dispersion in a glass prism, the colours range from red, the least deviated, to violet, Fig. 29.28 (ii). Second and higher order spectra are obtained with a diffraction grating on opposite sides of the normal, whereas only one spectrum is obtained with a glass prism. The angular spacing of the colours is also different in the grating and the prism.

If $d \sin \theta = m_1 \lambda_1 = m_2 \lambda_2$, where m_1, m_2 are integers, then a wavelength λ_1 in the m_1 order spectrum overlaps the wavelength λ_2 in the m_2 order. The

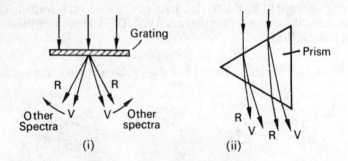

FIG. 29.28. Spectra in grating and prism.

extreme violet in the visible spectrum has a wavelength about $3 \cdot 8 \times 10^{-7}$ m. The violet direction in the second order spectrum would thus correspond to $d \sin \theta = 2\lambda = 7 \cdot 6 \times 10^{-7}$ m, and this would not overlap the extreme colour, red, in the first order spectrum, which has a wavelength about $7 \cdot 0 \times 10^{-7}$ m. In the second order spectrum, a wavelength λ_2 would be overlapped by a wavelength λ_3 in the third order if $2\lambda_2 = 3\lambda_3$. If $\lambda_2 = 6 \cdot 9 \times 10^{-7}$ m (red), then $\lambda_3 = 2\lambda_2/3 = 4 \cdot 6 \times 10^{-7}$ m (blue). Thus overlapping of colours occurs in spectra of higher orders than the first.

Dispersion by Grating

The *dispersion* of a grating, $d\theta/d\lambda$, is a measure of the change in angular position per unit wavelength change. Now $d \sin \theta = m\lambda$,

$$\therefore \quad d \cos \theta \frac{d\theta}{d\lambda} = m$$

$$\therefore \quad \frac{d\theta}{d\lambda} = \frac{m}{d \cos \theta} \qquad \cdot \qquad \cdot \qquad \cdot \qquad \text{(i)}$$

The dispersion thus increases with the order, m, of the image. It is also inversely proportional to the separation d of the slits, or, for a given grating width, directly proportional to the total number of slits on the grating. For a given order m, the dispersion increases when $\cos \theta$ is small, or when θ is large, which corresponds to the red wavelengths of the spectrum for normal incidence on the grating.

Resolving Power of Grating

For the mth order principal maximum of a grating, the path difference between diffracted rays from consecutive slits is $m\lambda$. The path difference AB between the extreme rays of the grating is thus $(N - 1) m\lambda$, where N is the total number of lines ruled on the grating, Fig. 29.29. The minimum intensity of the mth order principal maximum corresponds to a slightly different direction AC.

Now the discussion about the disturbances from various points across a wide slit (p. 702) can be applied to disturbances from various slits across a grating. It therefore follows that, for the minimum intensity, the path difference between disturbances from the first to the last slit is one wavelength, λ, more than that for the maximum intensity position. The path difference to the minimum is thus $(N - 1) m\lambda + \lambda$. The mth

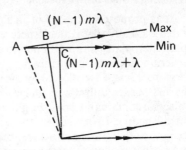

FIG. 29.29. Resolving power of diffraction grating.

order maximum of another wavelength λ', differing slightly from λ, is formed by extreme rays which have a path difference of $(N - 1) m\lambda'$ From Rayleigh's criterion, the two wavelengths λ' and λ are just resolved when the maximum of λ' falls on the first minimum of λ. In this case,

$$(N - 1) m\lambda' = (N - 1) m\lambda + \lambda.$$

$$\therefore \quad (N - 1) m (\lambda' - \lambda) = \lambda,$$

$$\therefore \quad \frac{\lambda}{\lambda' - \lambda} = (N - 1) m = Nm,$$

since 1 is negligible compared with N.

$$\therefore \quad resolving\ power = Nm$$

$$\therefore \quad resolving\ \ power = \frac{Nd \sin \theta}{\lambda} \text{ or } \frac{Nd (\sin i \pm \sin \theta)}{\lambda},$$

OPTICS, SOUND AND WAVES

the former being the expression for light incident normally on the grating and the latter if the angle of incidence is i. Either expression shows that for a given angle of incidence and diffraction, it is the *total width Nd of the grating which determines its resolving power*. The number of rulings in that width affects the dispersion in a given order but has no effect on the resolving power in that order. Thus a grating of 5 cm width and 6000 lines per cm has twice the resolving power of a grating 2·5 cm wide which also has 6000 lines per cm. If a grating is only 0·3 cm wide it has only about 2000 lines on it of the same spacing, whereas a grating 5 cm wide would have 12000 lines.

The two sodium lines or doublet have wavelengths $5·890 \times 10^{-7}$ and $5·896 \times 10^{-7}$ m respectively. The resolving power, R.P., required to distinguish them is given by:

$$\text{R.P.} = \frac{\lambda}{\lambda' - \lambda} = \frac{5·890 \times 10^{-7}}{0·006 \times 10^{-7}} = 1000 \text{ (approx.)}$$

Thus if a grating has 800 lines per cm, and the width covered by a telescope objective is 2·5 cm the sodium lines are clearly resolved in the first order images. If three-quarters of the grating is covered there are only 500 lines left, and the lines are now no longer resolved in the first order. They are just resolved in the second order images.

Resolving Power of Microscope

ABBE proposed a theory of image formation in a microscope which stated basically that if the structure of the illuminated object is regular (periodic), it acts like an illuminated diffraction grating. In this case the structure appears uniformly bright and unrecognisable if only the zero order image is collected by the microscope objective. If, in addition, the first order diffraction image is collected, the image plane in the objective contains alternate bright and dark strips or fringes in positions corresponding to images of the grating elements. The observer then recognises the grating structure, that is, the grating is "resolved". The more orders collected by the objective, the closer does the intensity distribution across the image plane resemble that transmitted by the object itself. The effect is analogous to the recognition of a note from a violin in Sound. This consists of a fundamental of the same frequency, together with overtones of higher frequency which gives the sound its timbre or quality. If only the fundamental is received, the note will not be recognisable as the note from the violin. The more overtones received in addition to the fundamental, the more faithful is the reproduction of the note.

An expression for the resolving power of a microscope can now be obtained. We require, for resolution, that a first order diffraction image is collected in addition to the zero order. Suppose that an object of regular structure is illuminated at an angle of incidence i by an oblique beam (Fig. 29.30). Then, if the first order image is just collected by the microscope objective,

$$d (\sin i \pm \sin a) = \lambda,$$

where a is the half-angle subtended by the objective at the object O and

d is the grating spacing of the object. The minimum value of d occurs when $i = \alpha$ and $d\,(\sin i + \sin \alpha) = \lambda$.

$$\therefore \quad 2d \sin \alpha = \lambda$$

$$\therefore \quad d = \frac{\lambda}{2 \sin \alpha}$$

This expression for d gives the grating spacing of the finest regular structure of the object which can just be resolved. If a medium such as oil of refractive index n is used in the object space beneath the objective,

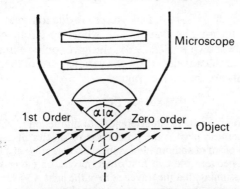

FIG. 29.30. Revolving power of microscope.

the least distance d or *limit of resolution* (or *"resolving power"*) is:

$$\text{limit of resolution} = \frac{\lambda}{2\,n \sin \alpha}$$

'$n \sin \alpha$' is called the *numerical aperture* of the objective.

The use of an oil-immersion objective was suggested by Abbe. The limit of resolution for the best optical microscopes is about 2×10^{-7} m. The eye can resolve about 0.01 cm. The largest useful magnifying power of a microscope is one which magnifies the limit of resolution of the objective to that of the eye, and is about 1000 with glass lenses and visible light. Higher resolving powers may be obtained with ultra-violet light, from $\lambda/2\,n \sin \alpha$. An *electron microscope*, which contains electron lenses and utilises electrons in place of light, has a limit of resolution less than 10^{-9} m owing to the much shorter wavelength of moving electrons compared with that of light waves. Much larger useful magnifying powers, such as 100 000, are thus obtained by using electron microscopes in place of optical microscopes.

Wavelengths of Electromagnetic Waves

In this book we have encountered rays which affect the sensation of vision (*visible rays*), rays which cause heat (*infra-red rays*, p. 456), and rays which cause chemical action (ultra-violet rays, p. 456). As these rays are all due to electric and magnetic vibrations they are examples of **electromagnetic** waves (see p. 719). Scientists have measured the wavelengths of these waves by a diffraction grating method, and results show a gradual transition in the magnitudes of the wavelength from one type of ray to another. Thus infra-red rays have a longer wavelength than visible rays, which in turn have a longer wavelength than ultra-violet rays. Radio waves are electromagnetic waves of longer wavelength

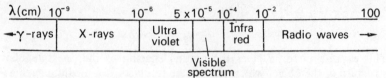

FIG. 29.31. Spectrum of electromagnetic waves (not to scale).

than infra-red rays, while X-rays and γ-rays are due to waves of shorter wavelength than ultra-violet waves. The whole spectrum of electromagnetic waves are shown in Fig. 29.31; this gives only an approximate value of the limits of the wavelength in the various parts of the spectrum, because these limits are themselves vague.

EXAMPLE

What is meant in optics by (a) interference, (b) diffraction? What part do each of these phenomena play in the production of spectra by a diffraction grating? A parallel beam of sodium light is incident normally on a diffraction grating. The angle between the two first order spectra on either side of the normal is 27° 42′. Assuming that the wavelength of the light is 5.893×10^{-7} m, find the number of rulings per mm on the grating. (N.)

First part. Briefly, interference is the name given to the phenomena obtained by the combined effect of light waves from two separate coherent sources; diffraction is the name given to the phenomena due to the combined effect of light waves from secondary sources on the same wavefront. In the diffraction grating, production of spectra is due to the interference between secondary sources on the same wavefront which are separated by a multiple of d, where d is the spacing of the grating rulings (p. 707).

Second part. The first order spectrum occurs at an angle $\theta = \frac{1}{2} \times 27° 42′$ $= 13° 51′$.

But
$$d \sin \theta = \lambda$$

$$\therefore \quad d = \frac{\lambda}{\sin \theta} = \frac{5.893 \times 10^{-7}}{\sin 13° 51′} \text{ m}$$

$\therefore$ number of rulings per metre $= \dfrac{1}{d} = \dfrac{\sin 13° 51′}{5.893 \times 10^{-7}}$

$$= 406 \text{ per millimetre}$$

POLARISATION OF LIGHT

We have shown that light is a wave-motion of some kind, i.e., that it is a travelling vibration. For a long time after the wave-theory was revived it was thought that the vibrations of light occurred in the same direction as the light wave travelled, analogous to sound waves. Thus light waves were thought to be longitudinal waves (p. 584). Observations and experiments, however, to be described shortly, showed that the vibrations of light occur in planes *perpendicular* to the direction along which the light wave travels, and thus light waves are *transverse* waves.

Polarisation of Transverse Waves

Suppose that a rope ABCD passes through two parallel slits, B, C, and is attached to a fixed point at D, Fig. 29.32 (i). Transverse waves can be set up along AB by holding the end A in the hand and moving it up and down in all directions perpendicular to AB, as illustrated by the arrows in the plane X. A wave then emerges along BC, but unlike the

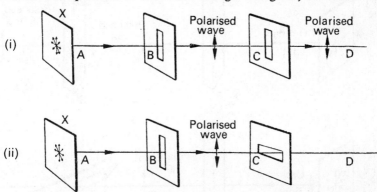

FIG. 29.32. Formation of plane-polarised waves.

waves along AB, which are due to transverse vibrations in every plane, it is due only to transverse vibrations parallel to the slit at B. This type of wave is called a *plane-polarised* wave. It shows a *lack of symmetry about the direction of propagation*, because a slit C allows the wave to pass through when it is parallel to B, but prevents it from passing when C is perpendicular to B, Fig. 29.32 (i), (ii). If B is turned so that it is perpendicular to the position shown in Fig. 29.32 (i), a polarised wave is again obtained along BC; but the vibrations which produce it are perpendicular to those shown between B and C in Fig. 29.32 (i).

Polarised Light

Years ago it was discovered accidentally that certain natural crystals affect light passing through them. *Tourmaline* is an example of such a crystal, *quartz* and *calcite* or *Iceland spar* are others (p. 720). Suppose two tourmaline crystals, P, Q, are placed with their axes, a, b, parallel, Fig. 29.33 (i). If a beam of light is incident on P, the light emerging from Q appears slightly darker. If Q is rotated slowly about the line of vision, with its plane parallel to P, the emergent light becomes darker and darker, and at one stage it disappears. In the latter case the axes a, b of the crystals are perpendicular, Fig. 29.33 (ii). When Q is rotated further the light reappears, and becomes brightest when the axes a, b are again parallel.

This simple experiment leads to the conclusion that light waves are *transverse* waves; otherwise the light emerging from Q could never be extinguished by simply rotating this crystal. The experiment, in fact, is analogous to that illustrated in Fig. 29.32, where transverse waves were set up along a rope and plane-polarised waves were obtained by means of a slit B. Tourmaline is a crystal which, because of its internal molecular structure, transmits only those vibrations of light parallel to its axis. Consequently plane-polarised light is obtained beyond the crystal P, and no light emerges beyond Q when its axis is perpendicular to P. Fig. 29.33 should be compared with Fig. 29.32.

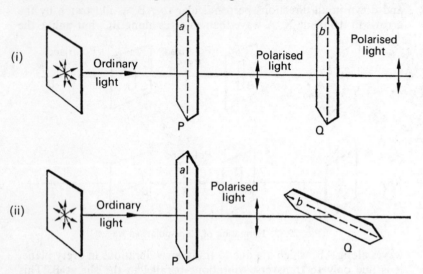

FIG. 29.33. Formation of plane-polarised light waves.

Vibrations in Unpolarised and Polarised Light

Fig. 29.34 (i) is an attempt to represent diagrammatically the vibrations of ordinary or unpolarised light at a point A when a ray travels in a direction AB. X is a plane perpendicular to AB, and ordinary (unpolarised) light may be imagined as due to vibrations which occur in

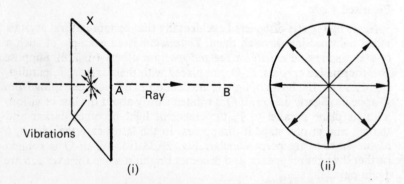

FIG. 29.34. (i). Vibrations occur in every plane perpendicular to AB.
(ii). Vibrations in ordinary light.

every one of the millions of planes which pass through AB and are perpendicular to X. As represented in Fig. 29.34 (ii), the amplitudes of the vibrations are all equal.

Consider the vibrations in ordinary light when it is incident on the tourmaline P in Fig. 29.33 (i). Each vibrations can be resolved into two components, one in a direction parallel to the axis a of the tourmaline P and the other in a direction m perpendicular to a, Fig. 29.35. Tourmaline absorbs the light due to the latter vibrations, known as the *ordinary* rays, allowing the light due to the former vibrations, known as the *extraordinary* rays, to pass through it. Thus plane-polarised light, due to the extraordinary rays,

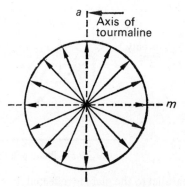

FIG. 29.35. Production of plane-polarised waves by tourmaline.

is produced by the tourmaline. Polaroid is a crystaline material, used in sun-glasses for example, which also has selective absorption.

Light waves are electromagnetic waves. Theory and experiment show that the vibrations of light are *electromagnetic* in origin; a varying electric vector E is present, with a varying magnetic vector B which has the same frequency and phase. E and B are perpendicular to each other, and are in a plane at right angles to the ray of light, Fig. 29.36. Experiments have shown that the electric force in a light wave affects a photographic plate and causes fluorescence, while the magnetic force, though present, plays no part in this effect of a light wave. On this account the vibrations of the electric force, E, are now chosen as the "vibrations of light", and the planes containing the vibrations shown in Fig. 29.35 (i), (ii) are those in which only the electric forces are present.

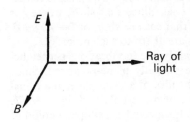

FIG. 29.36. Electromagnetic wave.

Polarised Light by Reflection

The production of polarised light by tourmaline is due to *selective absorption* of the "ordinary" rays. In 1808 MALUS discovered that polarised light is obtained when ordinary light is reflected by a plane sheet of glass (p. 719). The most suitable angle of incidence is about 56°, Fig. 29.37. If the reflected light is viewed through a tourmaline crystal which is slowly rotated about the line of vision, the light is practically extinguished at one position of the crystal. This proves that the light reflected by the glass is plane-polarised. Malus also showed that the light reflected by water is plane-polarised.

The production of the polarised light by the glass is explained as follows. Each of the vibrations of the incident (ordinary) light can be resolved into a component parallel to the glass surface and a component perpendicular to the surface. The light due to the components

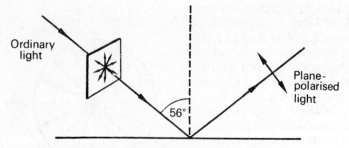

FIG. 29.37. Plane-polarised waves by reflection.

parallel to the glass is reflected, but the remainder of the light, due to the components perpendicular to the glass, is *refracted* into the glass. Thus the light reflected by the glass is plane-polarised.

Brewster's Law. Polarisation by Pile of Plates

The particular angle of incidence i on a transparent medium when the reflected light is almost completely plane-polarized is called the *polarising angle*. BREWSTER found that, in this case, $tan\ i = n$, where n is the refractive index of the medium (*Brewster's law*). Since $\sin i/\sin r$, where r is the angle of refraction, it then follows that $\cos i = \sin r$, or $i + r = 90°$. Thus the reflected and refracted beams are at 90° to each other.

The refracted beam contains light mainly due to vibrations perpendicular to that reflected and is therefore partially plane-polarised. Since refraction and reflection occur at both sides of a glass plate, the transmitted beam contains a fair percentage of plane-polarised light. A *pile of plates* increases the percentage, and thus provides a simple method of producing plane-polarised light. They are mounted inclined in a tube so that the ordinary (unpolarised) light is incident at the polarising angle, and the transmitted light it then fairly plane-polarised.

Polarisation by Double Refraction

We have already considered two methods of producing polarised light. The first observation of polarised light, however, was made by BARTHOLINUS in 1669, who placed a crystal of iceland spar on some words on a sheet of paper. To his surprise, two images were seen through the crystal. Bartholinus therefore gave the name of *double refraction* to the phenomenon, and experiments more than a century later showed that the crystal produced plane-polarised light when ordinary light was incident on it. See Fig. 29.38.

Iceland spar is a crystalline form of calcite (calcium carbonate) which cleaves in the form of a "rhomboid" when it is lightly tapped; this is a solid whose opposite faces are parallelograms. When a beam of unpolarised light is incident on one face of the crystal, its internal molecular structure produces two beams of polarised light, E, O, whose vibrations are perpendicular to each other, Fig. 29.39. If the incident direction AB is parallel to a plane known as the "principal section" of the crystal, one

FIG. 29.38. DOUBLE REFRACTION. A ring with a spot in the
centre, photographed through a crystal of Iceland spar. The
light forms two rings and two spots.

beam O emerges parallel to AB, while the other beam E emerges displaced
in a different direction. As the crystal is rotated about the line of vision the
beam E revolves round O. On account of this abnormal behaviour the
rays in E are called "extraordinary" rays; the rays in O are known as
"ordinary" rays (p. 719). Thus two images of a word on a paper, for
example, are seen when an Iceland spar crystal is placed on top of it; one
image is due to the ordinary rays, while the other is due to the extraordinary rays.

With the aid of an Iceland spar crystal Malus discovered the polarisation of light by reflection (p. 719). While on a visit to Paris he gazed

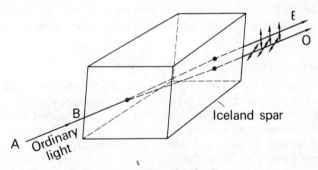

FIG. 29.39. Action of Iceland spar.

through the crystal at the light of the sun reflected from the windows of
the Palace of Luxemburg, and observed that only one image was obtained
for a particular position of the crystal when it was rotated slowly. The
light reflected from the windows could not therefore be ordinary (unpolarised) light, and Malus found it was plane-polarised.

Nicol Prism

We have seen that a tourmaline crystal produces polarised light, and that the crystal can be used to detect such light (p. 717). NICOL designed a crystal of Iceland spar which is widely used for producing and detecting

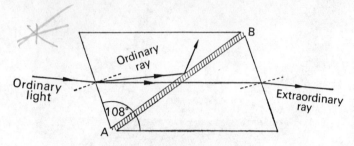

FIG. 29.40. Action of Nicol prism.

polarised light, and it is known as a *Nicol prism*. A crystal whose faces contain angles of 72° and 108° is broken into two halves along the diagonal AB, and the halves are cemented together by a layer of Canada balsam, Fig 29. 40. The refractive index of the crystal for the ordinary rays is 1·66, and is 1·49 for the extraordinary rays; the refractive index of the Canada balsam is about 1·55 for both rays, since Canada balsam does not produce polarised light. A critical angle thus exists between the crystal and Canada balsam for the ordinary rays, but not for the extraordinary rays. Hence total reflection of the former rays takes place at the canada balsam if the angle of incidence is large enough, as it is with the Nicol prism. The emergent light is then due to the extraordinary rays, and is polarised.

The prism is used like a tourmaline crystal to detect plane-polarised light, namely, the prism is held in front of the beam of light and is rotated. If the beam is plane-polarised the light seen through the Nicol prism varies in intensity, and is extinguished at one position of the prism.

Differences Between Light and Sound Waves

We are now in a position to distinguish fully between light and sound waves. The physical difference, of course, is that light waves are due to varying electric and magnetic forces, while sound waves are due to vibrating layers or particles of the medium concerned. Light can travel through a vacuum, but sound cannot travel through a vacuum. Another very important difference is that the vibrations of the particles in sound waves are in the same direction as that along which the sound travels, whereas the vibrations in light waves are perpendicular to the direction along which the light travels. Sound waves are therefore *longitudinal* waves, whereas light waves are *transverse* waves. As we have seen, sound waves can be reflected and refracted, and can give rise to interference phenomena; but no polarisation phenomena can be obtained with sound waves since they are longitudinal waves, unlike the case of light waves.

EXERCISES 29

Interference

1. Describe how to set up apparatus to observe and make measurements on the interference fringes produced by Young's slits. Explain how (i) the wavelengths of two monochromatic light sources could be compared, (ii) the separation of the slits could be deduced using a source of known wavelength. Establish any formula required.

State, giving reasons, what you would expect to observe (*a*) if a white light source were substituted for a monochromatic source, (*b*) if the source slit were then displaced slightly at right angles to its length in the plane parallel to the plane of the Young's slits. (*L.*)

2. Explain the formation of interference fringes by an air wedge and describe how the necessary apparatus may be arranged to demonstrate them.

Fringes are formed when light is reflected between the flat top of a crystal resting on a fixed base and a sloping glass plate. The lower end of the plate rests on the crystal and the upper end on a fixed knife-edge. When the temperature of the crystal is raised the fringe separation changes from 0·96 mm to 1·00 mm. If the length of the glass plate from knife-edge to crystal is 5·00 cm, and the light of wavelength $6·00 \times 10^{-7}$ m is incident normally on the wedge, calculate the expansion of the crystal. (*L.*)

3. Describe in detail how the radius of curvature of the spherical face of a planoconvex lens may be found by observations made on Newton's rings.

Two plane glass plates which are in contact at one edge are separated by a piece of metal foil 12·50 cm from that edge. Interference fringes parallel to the line of contact are observed in reflected light of wavelength $5·46 \times 10^{-7}$ m and are found to be 1·50 mm apart. Find the thickness of the foil. (*L.*)

4. Describe, with the aid of a labelled diagram, how the wavelength of monochromatic light may be found using Young's slits. Give the theory of the experiment.

State, and give physical reasons for the features which are common to this method and to *either* the method based on Lloyd's mirror *or* that based on Fresnel's biprism.

In an experiment using Young's slits the distance between the centre of the interference pattern and the tenth bright fringe on either side is 3·44 cm and the distance between the slits and the screen is 2·00 m. If the wavelength of the light used is $5·89 \times 10^{-7}$ m determine the slit separation. (*N.*)

5. Explain what is meant by the term *path-difference* with reference to the interference of two wave-motions.

Why is it not possible to see interference where the light beams from the headlamps of a car overlap?

Interference fringes were produced by the Young's slits method, the wavelength of the light being 6×10^{-7} m. When a film of material $3·6 \times 10^{-3}$ cm thick was placed over *one* of the slits, the fringe pattern was displaced by a distance equal to 30 times that between two adjacent fringes. Calculate the refractive index of the material. To which side are the fringes displaced?

(When a layer of transparent material whose refractive index is *n* and whose thickness is *d* is placed in the path of a beam of light, it introduces a path difference equal to $(n-1)d$.) (*O. & C.*)

6. Show how, with the aid of Huygens' idea of secondary wavelets, the wave theory of light will account for the laws of refraction and of reflexion at a plane surface.

Describe briefly Young's two-slit experiment and explain how it confirms the wave nature of light. (*L.*)

7. Describe, giving both theory and experimental detail, how you would find the radius of curvature of one surface of a convex lens by means of Newton's rings. You may assume that monochromatic light of a known wavelength is available.

Newton's rings are formed by reflexion between an equiconvex lens of focal length 100 cm made of glass of refractive index 1·50 and in contact with a plane glass plate of refractive index 1·60. Find the radius of the 5th bright ring using monochromatic light of wavelength 6000 Å.

Explain the changes which occur when oil of refractive index 1·55 fills the space between the lens and plate. (1 Å $= 10^{-10}$ m.) (*N.*)

8. Define *velocity*, *frequency* and *wavelength* for any wave motion, and deduce a relation between them. What do you understand by 'interference between waves' and 'coherent wave trains'? Explain why interference is not observed between the beams of two electric torches.

Deduce the relation connecting the refractive index of a material with the velocities of light in vacuo and in the material. State clearly the assumptions you make about wave fronts in order to do this.

Light passes through a single crystal of ruby 10·0 cm long and emerges with a wavelength of 6·94 $\times$ 10^{-7} m. If the critical angle of ruby for light of this wavelength if 34° 50′, calculate the number of wavelengths inside the crystal. (*C.*)

9. State the conditions necessary for the production of interference effects by two overlapping beams of light.

Describe fully *one* method for the production of interference fringes using light from a given monochromatic source. Show how with the aid of suitable measurements the wavelength of light emitted by the source may be determined with your apparatus.

Describe how the fringes produced by your apparatus would appear if a source of white light were employed instead of a monochromatic one. (*O. & C.*)

10. Explain how Newton's rings are formed, and describe how you would demonstrate them experimentally. How is it possible to predict the appearance of the centre of the ring pattern when (*a*) the surfaces are touching, and (*b*) the surfaces are not touching?

In a Newton's rings experiment one surface was fixed and the other movable along the axis of the system. As the latter surface was moved the rings appeared to contract and the centre of the pattern, initially at its darkest, became alternately bright and dark, passing through 26 bright phases and finishing at its darkest again. If the wavelength of the light was 5·461 $\times$ 10^{-7} m, how far was the surface moved and did it approach, or recede from, the fixed surface? Suggest one possible application of this experiment. (*O. & C.*)

11. Explain the formation of Newton's rings and describe how you would use them to measure the radius of curvature of the convex surface of a long-focus planoconvex lens.

The diameters of the *m*th and (*m* + 10)th bright rings formed by such a lens resting on a plane glass surface are respectively 0·14 cm and 0·86 cm. When the space between lens and glass is filled with water the diameters of the *q*th and (*q* + 10)th bright rings are respectively 0·23 cm and 0·77 cm. What is the refractive index of water? (*L.*)

12. What are the conditions essential for the production of optical interference fringes?

Explain how these conditions are satisfied in the case of (*a*) Young's fringes, and (*b*) thin film interference fringes. (*N.*)

13. Describe, in detail how you would arrange apparatus to observe, in monochromatic light, interference fringes formed by light reflected from two glass plates enclosing an air wedge. Show how the angle of the wedge could be obtained from measurements on the fringes.

Newton's rings are formed with light of wavelength $5·89 \times 10^{-7}$ m between the curved surface of a planoconvex lens and a flat glass plate, in perfect contact. Find the radius of the 20th dark ring from the centre if the radius of curvature of the lens surface is 100 cm. How will this ring move and what will its radius become if the lens and the plate are slowly separated to a distance apart of $5·00 \times 10^{-4}$ cm? (*L.*)

14. What are the necessary conditions for interference of light to be observable? Describe with the aid of a labelled diagram how optical interference may be demonstrated using Young's slits. Indicate suitable values for all the distances shown.

How are the colours observed in thin films explained in terms of the wave nature of light? Why does a small oil patch on the road often show approximately circular coloured rings? (*L.*)

Diffraction

15. Describe and give the theory of an experiment to compare the wavelengths of yellow light from a sodium and red light from a cadmium discharge lamp, using a diffraction grating. Derive the required formula from first principles.

White light is reflected normally from a soap film of refractive index $1·33$ and then directed upon the slit of a spectrometer employing a diffraction grating at normal incidence. In the first-order spectrum a dark band is observed with minimum intensity at an angle of $18° 0'$ to the normal. If the grating has 500 lines per mm, determine the thickness of the soap film assuming this to be the minimum value consistent with the observations. (*L.*)

16. Describe the phenomena which occur when plane waves pass (*a*) through a wide aperture, (*b*) through an aperture whose width is comparable with the wavelength of the waves.

How does the wave theory of light account for the apparent rectilinear propagation of light?

A diffraction grating has 6000 lines per cm. Calculate the angular separation between wavelengths $5·896 \times 10^{-7}$ m and $5·461 \times 10^{-7}$ m respectively after transmission through it at normal incidence, in the first-order spectrum. (*O. & C.*)

17. Describe two experiments to show the diffraction of light.

Describe how a diffraction grating may be used to measure the wavelength of sodium light, deriving any formulae employed. (*L.*)

18. What are the advantages and disadvantages of a diffraction grating as compared with a prism for the study of spectra?

A rectangular piece of glass 2 cm × 3 cm has 18000 evenly spaced lines ruled across its whole surface, parallel to the shorter side, to form a diffraction grating. Parallel rays of light of wavelength 5×10^{-7} m fall normally on the grating. What is the highest order of spectrum in the transmitted light?

What is the minimum diameter of a camera lens which can accept all the light of this wavelength in this order which leaves the grating on one side of the normal? *(O. & C.)*

19. In an experiment using a spectrometer in normal adjustment fitted with a plane transmission grating and using monochromatic light of wavelength 5.89×10^{-7} m, diffraction maxima are obtained with telescope settings of 153° 44′, 124° 5′, 76° 55′ and 47° 16′, the central maximum being at 100° 30′. Show that these observations are consistent with normal incidence and calculate the number of rulings per cm of the grating.

If this grating is replaced by an opaque plate having a single vertical slit 2.00×10^{-2} cm wide, describe and explain the diffraction pattern which may now be observed. Contrast the appearance of this pattern with that produced by the grating. *(N.)*

20. (*a*) What is meant by (i) *diffraction*, (ii) *superposition* of waves? Describe *one* phenomenon to illustrate each in the case of *sound waves*.

(*b*) The floats of two men fishing in a lake from boats are 22·5 metres apart. A disturbance at a point in line with the floats sends out a train of waves along the surface of the water, so that the floats bob up and down 20 times per minute. A man in a third boat observes that when the float of one of his colleagues is on the crest of a wave that of the other is in a trough, and that there is then one crest between them. What is the velocity of the waves? *(O. & C.)*

21. Give an account of the theory of the production of a spectrum by means of a plane diffraction grating. How does it differ from the spectrum produced by means of a prism?

Parallel light consisting of two monochromatic radiations of wavelengths 6×10^{-7} m and 4×10^{-7} m falls normally on a plane transmission grating ruled with 5000 lines per cm. What is the angular separation of the second-order spectra of the two wavelengths? *(C.)*

22. A pure spectrum is one in which there is no overlapping of light of different wavelengths. Describe how you would set up a diffraction grating to display on a screen as close an approximation as possible to a pure spectrum. Explain the purpose of each optical component which you would use.

A grating spectrometer is used at normal incidence to observe the light from a sodium flame. A strong yellow line is seen in the first order when the telescope axis is at an angle of 16° 26′ to the normal to the grating. What is the highest order in which the line can be seen?

The grating has 4800 lines per cm; calculate the wavelength of the yellow radiation.

What would you expect to observe in the spectrometer set to observe the first-order spectrum if a small but very bright source of white light is placed close to the sodium flame so that the flame is between it and the spectrometer? *(O. & C.)*

23. Describe how you would determine the wavelength of monochromatic light using a diffraction grating and a spectrometer. Give the theory of the method.

A filter which transmits only light between 6300 Å and 6000 Å is placed between a source of white light and the slit of a spectrometer; the grating has 5000 lines to the centimetre; and the telescope has an objective of focal length 15 cm with an eyepiece of focal length 3 cm. Find the width in millimetres of the first-order spectrum formed in the focal plane of the objective. Find also the angular width of this spectrum seen through the eyepiece. *(O.)*

Polarisation

24. What is meant by *plane of polarisation*? Explain why the phenomenon of polarisation is met with in dealing with light waves, but not with sound waves.

Describe and explain the action of (*a*) a nicol prism, (*b*) a sheet of Polaroid.

How can a pair of Polaroid sheets and a source of natural light be used to produce a beam of light the intensity of which may be varied in a calculable manner? (*L.*)

25. Explain what is meant by the statement that a beam of light is *plane polarised*. Describe *one* experiment in each instance to demonstrate (*a*) polarisation by reflexion, (*b*) polarisation by double refraction, (*c*) polarisation by scattering.

The refractive index of diamond for sodium light is 2·417. Find the angle of incidence for which the light reflected from diamond is completely plane polarised. (*L.*)

26. Give an account of the action of (*a*) a single glass plate. (*b*) a Nicol prism, in producing plane-polarised light. State *one* disadvantage of *each* method.

Mention *two* practical uses of polarising devices. (*N.*)

27. Describe how, using a long, heavy rope, you would demonstrate (*a*) a plane-polarised wave, and (*b*) a stationary wave.

Give a short account, with diagrams, of *three* ways in which plane-polarised light is obtained (other than by using 'polaroid'). State some uses of polarised light.

Two polaroid sheets are placed close together in front of a lamp so that no light passes through them. Describe and explain what happens when one sheet is slowly rotated, the other remaining in its original position. (*C.*)

28. Answer *two* of the following:

(i) How may it be shown that the radiation from (*a*) a sodium lamp, and (*b*) a radio transmitter (such as a broadcasting station or a microwave source) consists of waves?

(ii) Explain what is meant by the polarisation of light, and describe how you would demonstrate it. Why is light from most light sources unpolarised?

(iii) When a diffraction grating is illuminated normally by monochromatic light an appreciable amount of light leaves the grating in certain directions. Explain this phenomenon, and show how these directions may be predicted. (*O. & C.*)

29. What is meant by (*a*) polarised light, (*b*) polarising angle? Describe and explain two methods for producing plane-polarised light.

Calculate the polarising angle for light travelling from water, of refractive index 1·33, to glass, of refractive index 1·53. (*L.*)

30. A beam of plane-polarised light falls normally on a sheet of Polaroid, which is at first set so that the intensity of the transmitted light, as estimated by a photographer's light-meter, is a maximum. (The meter is suitably shielded from all other illumination.) Describe and explain the way in which you would expect the light-meter readings to vary as the Polaroid is rotated in stages through 180° about an axis at right angles to its plane.

How would you show experimentally (*a*) that calcite is doubly refracting, (*b*) that the two refracted beams are plane polarised, in planes at right angles to one another, and (*c*) that in general the two beams travel through the crystal with different velocities? (*O.*)

31. What is plane-polarised light?

Explain why two images of an object are seen through a crystal of Iceland Spar. What would be seen if the object were viewed through two crystals, one of which was slowly rotated about the line of vision?

How would you produce a plane-polarised beam of light by reflection from a glass surface? (*C.*)

32. What is meant by the polarisation of light? How is polarisation explained on the hypothesis that light has wave properties?

Describe how polarisation can be produced and detected by reflexion. Mention another way of obtaining polarised light and describe how you would determine which of the two methods is the more effective.

Describe briefly *two* uses of polarised light. (*N.*)

33. Give an account of the evidence for believing that light is a wave motion. What reason is there for believing light waves to be transverse waves?

Two dishes *A* and *B* each contain liquid to a depth of 3·000 cm. *A* contains alcohol, *B* a layer of water on which is a layer of transparent oil. The depths of the oil and water are adjusted so that for monochromatic light passing vertically through them, the number of wavelengths is the same in *A* and *B*. Find the depth of the water layer, if the refractive indices of alcohol, water and oil are respectively 1·363, 1·333 and 1·475. (*L.*)

chapter thirty

Further topics in Optics
Diffraction of light

Diffraction Due to Single Slit

As shown in the previous chapter, the light waves from a distant object arriving at the objective lens of a telescope are diffracted at the circular opening. Thus the quality of the image is limited by diffraction. The pupil of the eye is also a circular opening which diffracts light waves and hence the image on the retina is affected by diffraction. These two examples illustrate the importance of investigating more thoroughly diffraction.

In this section, *Fraunhofer diffraction* will be mainly discussed. This is the name given to diffraction of plane wavefronts (parallel rays), obtained from a distant object or by using a lens with a near object. Suppose a plane wavefront is incident on a narrow *rectangular slit* AB as shown in Fig. 30.1 (i). All points between A and B may be considered as coherent

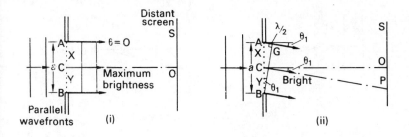

FIG. 30.1. Diffraction at rectangular slit.

sources, which are in phase, and send out waves beyond the slit. If a screen S is a long way from AB, we can consider that the waves reaching S have travelled towards it in parallel directions. Alternatively, a lens can be used to collect the parallel waves.

Diffraction Pattern. Central Band Intensity

The amplitude of the resultant wave at a point on the screen determines the intensity of the light there, since intensity $\propto$ amplitude^2. The mathematical treatment is discussed later on p. 731. Here a simpler treatment is adopted.

Consider first the direction corresponding to $\theta = 0$, where θ is the angle to the normal at A or B to the opening. In this direction, the waves from all the sources between A and B, such as X, C and Y, have no path difference to the screen. Hence they all arrive *in phase* at O, where CO is normal to AB. The resultant amplitude is therefore large. Hence a bright band is obtained at O. This is the midpoint of the central bright band on the screen.

Along the screen from O, the brightness of the band diminishes. Consider, for example, a point P near O which corresponds to a direction θ_1 to the normal, where $\sin \theta_1 = \lambda/2a$ and a is the slit width AB, Fig. 30.1 (ii). In this case, the waves from the extreme edges A,B have a path difference AG, where $AG = \lambda/2$, since $\sin \theta_1 = AG/AB = \lambda/2a$. Now a path difference of $\lambda/2$ is equivalent to a phase difference of waves of 180°. Hence the waves from A and B arrive in *antiphase* at P. The resultant amplitude of these waves is thus zero. The waves from all other points on the wavefront AB, such as X and Y, have a path difference *less* than $\lambda/2$. Thus although a resultant amplitude and intensity is obtained at P, the band is not as bright here as at the midpoint O.

Angular Width of Central Band

Now consider a point Q further from O than P, corresponding to a particular direction θ_2 from the normal such that $\sin \theta_2 = \lambda/a$, Fig. 30.2. In this case the path difference AF corresponding to waves from the edges A and B is λ, since $\sin \theta_2 = AF/AB = \lambda/a$.

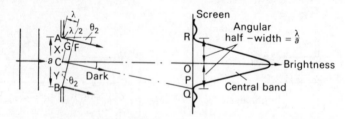

Fig. 30.2. Angular width of central band.

To find the resultant amplitude of the waves arriving at Q, divide the wavefront AB into two halves. The top point A of the upper half AC, and the top point C of the lower half CB, send out waves to Q which have a path difference $\lambda/2$, from above. Thus the resultant amplitude at Q is zero. All other pairs of corresponding points in the two halves of AB, for example, X and Y where $AX = CY$, also have a path difference $\lambda/2$. Hence it follows that the brightness at Q is zero. This point, then, is one edge of the central diffraction band on the screen. The other edge is R on the opposite side of O, where $OR = OQ$.

The *angular width* of the central diffraction band is thus $2\theta_2$, where $\sin \theta_2 = \lambda/a$. Since λ is usually very much smaller than a, the slit width, we may then write $\theta_2 = \lambda/a$. Fig. 30.2 shows roughly the variation of intensity or brightness of the central diffraction band due to a narrow rectangular slit. A photograph is shown on p. 703.

Secondary Bands

Secondary bright and dark bands are also obtained, as seen in this photograph. Consider, for example, a direction θ_3 from the normal beyond the edge of the central band, such that $\sin \theta_3 = 3\lambda/2a$, Fig. 30.3 (i). The path difference of the waves from the edges A,B is AF, and $AF = 3\lambda/2$ since $\sin \theta_3 = AF/AB = 3\lambda/2a$.

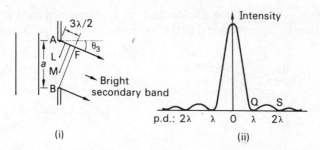

(i)

(ii)

FIG. 30.3. Secondary bands with single slit.

In this case, imagine the wavefront AB divided into *three* equal parts AL, LM, MB. Then, as seen in Fig. 30.3 (i), the extreme ends A and M of the upper two parts have a path difference λ. Hence, as explained for the band at Q, Fig. 30.2, the parts AL and LM of the wavefront together produce zero intensity at the screen. The third part MB produces an intensity on the screen much less than the intensity at O, the central band maximum intensity. This explains the secondary bright band.

At S on the screen, further away, another dark band is obtained. This corresponds to a direction θ_4 such that $\sin \theta_4 = 2\lambda/a$, that is, to a path difference of 2λ between the extreme points A and B on the wavefront AB. The dark band can be explained by dividing AB into four equal parts. The extreme points of the upper two parts have a path difference of λ in the direction θ_4, hence these parts produce zero intensity. Similarly for the lower two parts of AB. Fig. 30.3 (ii) shows diagrammatically the variation of intensity for path differences of waves coming from A,B respectively.

Phasor Addition

The intensity of a diffraction image can be investigated more thoroughly by addition of *phasors*. A phasor is a rotating vector which can be used to represent the displacement at a point due to a wave or the voltage in an a.c. circuit.

Suppose a plane wavefront reaches the slit AB, Fig. 30.4 (i). Waves from typical coherent sources such as S_1 and S_2 then arrive at a point P with an angular phase difference ϕ, which depends on the path difference S_2P-S_1P or x_2-x_1. Since a path difference λ is equivalent to an angle 2π in radians, then

$$\phi = \frac{2\pi}{\lambda} (x_2 - x_1).$$

The two waves of velocity v arriving at P may be represented respectively, in the usual notation, by

$$y_1 = a \sin \frac{2\pi}{\lambda} (vt - x_1)$$

and

$$y_2 = a \sin \frac{2\pi}{\lambda} (vt - x_2).$$

In a phasor diagram, the displacement y_1 is represented by a line OH whose length is proportional to the wave amplitude a, Fig. 30.4 (ii). The

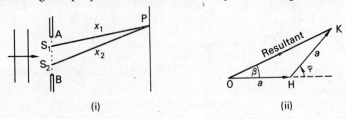

(i)   (ii)

FIG. 30.4. Phasor addition for resultant amplitude.

displacement y_2 is then represented by a line HK of equal length, since its amplitude is a, and this is drawn at the phase angle ϕ to OH. The *resultant amplitude* of the two waves is then represented by the line OK, and the phase angle between the resultant wave and that represented by OH is β in Fig. 30.4 (ii).

Intensity Values in Diffraction Image

(1) *Middle of central band.* The middle of the central diffraction band of a single slit corresponds to a direction $\theta = 0$ measured from a normal to the slit. See Fig. 30.1 (i). Since all the diffracted waves from the sources in the slit now have no path difference, they arrive in phase at the middle of the band. If we consider that there are n sources in the slit, and each of the n waves has an amplitude a, the phasor diagram is that shown in Fig. 30.5 ii (a) ($\phi = 0$). The resultant amplitude $A_0 = na$. Thus the intensity $\propto A_0^2 \propto n^2 a^2$.

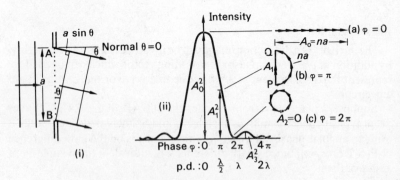

FIG. 30.5. Intensity values of bands.

(2) In a direction θ from the normal corresponding to a path difference $\lambda/2$ between the waves from the extreme points in the slit, $\sin \theta = \lambda/2a$ (p. 730). Thus, from one end A to the other end B, the phase difference of the waves from the n sources slowly increases from 0 to π (180°). The phasors are shown in Fig. 30.5 ii (b) $(\phi = \pi)$. The resultant amplitude A_1 is hence the length of the vector PQ, which is the diameter of the semi-circle formed.

The length of the semicircle is equal to na or A_0. If r is the radius of the semicircle, then $A_0 = \pi r$ and $A_1 = 2r$. Thus $A_1/A_0 = 2/\pi$. Hence the respective intensities I_1, I_0 are related by

$$\frac{I_1}{I_0} = \frac{A_1{}^2}{A_0{}^2} = \frac{4}{\pi^2} = 0.4 \text{ (approx.)}$$

(3) *Edge of band.* In a direction θ_2 where $\sin \theta_2 = \lambda/a$, the path difference from the sources at A and B of the slit is λ. Thus the phase difference of the waves is 2π (360°) and so the phase change for waves from successive sources is greater than before. Consequently the n phasors curl round more rapidly, Fig. 30.5 (ii) (c). The last phasor co-incides with the first as shown, since $\phi = 2\pi$, and so the resultant amplitude A_2 is zero. The intensity is hence zero. This corresponds to the edge of the central band.

Beyond the edge of the band, the phasors curl round more than before, since the phase angle ϕ exceeds 2π. A small resultant amplitude A_3 is thus obtained, as illustrated in Fig. 30.5 (ii). A weak secondary bright band is therefore formed.

General Intensity Formula

A general formula for the intensity I_θ in a direction θ from the normal can now be derived. In this case, the phase difference ϕ of the waves from the extreme ends of the slit is given by $\phi = 2\pi \, a \sin \theta/\lambda$, since the path

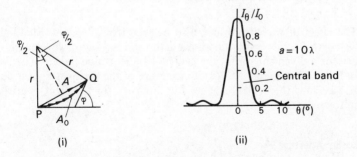

FIG. 30.6. Relative intensities of bright bands.

difference is $a \sin \theta$ where a is the slit width. The phasor diagram is shown in Fig. 30.6 (i). The arc PQ has a length $A_0 = r\phi$. The resultant amplitude

$A = PQ = 2r \sin \phi/2$. Hence

$$\frac{A}{A_0} = \frac{2 \sin \phi/2}{\phi} = \frac{\sin \phi/2}{\phi/2}$$

$$\therefore \quad \frac{I_\theta}{I_0} = \frac{\sin^2(\phi/2)}{(\phi/2)^2},$$

where $\phi/2 = \pi a \sin \theta/\lambda$. The ratio I_θ/I_0 is the relative intensity in a direction θ compared to that in the middle of the central diffraction band or maximum intensity. Fig. 30.6 (ii) shows roughly some of the relative intensity values for the case of $a = 10\lambda$. The intensities of the secondary maxima decrease rapidly, The first secondary has only about 5% of the intensity I_0. Thus most of the incident light is concentrated in the central diffraction band.

Angular Width of Central Band—Effect of λ and a

The angular width of the central bright band is 2θ, where $\sin \theta = \lambda/a$ (p. 730). When a is much greater than λ, the angular width is thus $2\lambda/a$.

Fig. 30.7 (i) shows the effect on the band when the width of the rectangular slit is decreased from a_1 to a_2. The effect of diffraction then increases and the width of the central band increases.

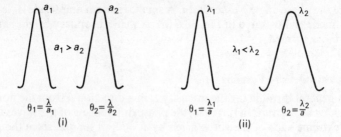

FIG. 30.7. Effect on central band of (i) slit width, (ii) wavelength.

The effect of wavelength can also be seen from $\theta = \lambda/a$. Blue light, wavelength λ_1, has a shorter wavelength than red light, wavelength λ_2. Thus, for a given slit width, the central band for blue light is narrower than for red light, Fig. 30.7 (ii). Since more of the incident light is concentrated into the central image, the intensity for the light is greater than for red light.

Circular Aperture

So far a rectangular opening or aperture has been considered. Almost all lenses used in optical instruments provide *circular* apertures.

Consider a circular and a rectangular aperture divided into strips of equal width, Fig. 30.8. In the case of the rectangular aperture, the contribution of each strip to the intensity of the diffraction image is equal in magnitude. For a circular aperture, however, the area of the slits decreases towards the left and right sides from the centre, as shown, and hence their contribution towards the intensity decreases. The directions of the maxima and minima in the diffraction pattern of the circular aperture are hence different from the case of the rectangular aperture.

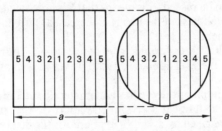

FIG. 30.8. Circular aperture.

The diffraction pattern produced by a circular aperture consists of a central bright circular disc surrounded by a series of concentric bright rings (secondary maxima) of diminishing intensity. It can be shown in advanced works that the first minimum of the diffraction pattern due to a circular aperture of diameter D occurs at an angle θ given by

$$\sin \theta = \frac{1 \cdot 22\lambda}{D}.$$

If λ is small compared to D, which is usually the case, then $\theta = 1 \cdot 22\lambda/D$. In the case of the rectangular aperture, the value of θ is given by $\theta = \lambda/a$.

Resolving Power

Although the resolving power of optical instruments was discussed in the previous chapter, it may not be out of place to briefly recapitulate here some of the main points in view of their importance.

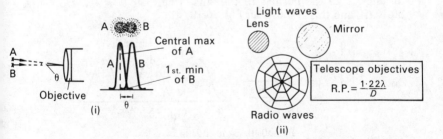

FIG. 30.9. (i) Resolving power. (ii) Telescope objectives.

1. *Rayleigh criterion*. This states that two distant points, viewed by an optical instrument, are *just* resolved when the central maximum of the diffraction image of one point A falls on the first minimum of the diffraction image of the other point B, Fig. 30.9 (i).

2. *Resolving power.* The smallest angle θ which the objects A and B subtend at the instrument when they are just resolved is taken as a measure of the resolving power R.P. of the instrument. Thus the resolving power of a telescope is given by $\theta = 1 \cdot 22\lambda/D$, where D is the diameter of the objective. Increasing D is beneficial on two counts. Firstly, the resolving power is then increased (this means that θ is now *smaller* than before); secondly, the brightness of the image is increased, since more light passes into the instrument.

Fig. 30.9 (ii) illustrates various types of telescopes. The resolving power expression $\theta = 1 \cdot 22\lambda/D$ applies to radio waves as well as to light waves (see also p. 706).

At night, the pupil of the eye has a diameter about 3 mm. Suppose the headlamps of a distant car are $1 \cdot 5$ metre apart, so that they subtend an angle α at the eye given by $\alpha = 1 \cdot 5/x$, where x is the distance from the car to the observer. The smallest angle θ when the two lamps are just resolved is given by $\theta = 1 \cdot 22\lambda/D$. Assuming $\lambda = 6 \times 10^{-7}$ m, the observer will see the two lamps just resolved when

$$\frac{1 \cdot 5}{x} = \frac{1 \cdot 22 \times 6 \times 10^{-7}}{3 \times 10^{-3}}$$

Solving for x, we find $x = 6000$ metres (approx.). Thus the distance is about 6 km.

Diffraction Due to Two Slits

We now consider diffraction due to two parallel narrow rectangular slits, separated by a distance d and illuminated by plane wavefronts incident normally to the slits, Fig. 30.10 (i). The two waves from corresponding points A,B in the two slits add constructively when their path difference is λ. The waves from all the corresponding points in the slits, separated by the same distance d as A,B, also have a path difference of λ in the same direction. Hence a bright band is formed. The direction is at an angle θ to the normal to the slits, where $\sin \theta = \lambda/d$, from Fig. 30.10 (i), and this is shown in Fig. 30.10 (ii).

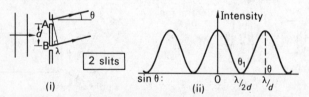

FIG. 30.10. Bands due to two slits.

In a direction of smaller angle θ_1 where $\sin \theta_1 = \lambda/2d$, waves from corresponding points in the two slits such as A,B have a path difference $\lambda/2$. Thus the intensity is zero in this direction. Fig. 30.10 (ii) shows roughly the variation of intensity of the band pattern. The middle of the central band at 0 is brightest, since this corresponds to zero path difference between the waves.

These results can be derived from the phasor diagram for the waves from the two slits (p. 731). Suppose the amplitude of each wave is a and their phase difference is ϕ in a particular direction. From the cosine formula for the resultant of the two phasors shown in Fig. 30.4 (ii), the resultant amplitude A is given by

$$A^2 = a^2 + a^2 + 2a^2 \cos \phi = 2a^2(1 + \cos \phi).$$

The position of maximum amplitude or intensity of the bright band corresponds to $\phi = 0$. Here $A^2 = 4a^2$ from above. The intensity diminishes from $\phi = 0$ to $\phi = \pi$, where $A^2 = 0$. This corresponds to the position of the centre of the dark band. As ϕ increases from π to 2π, the intensity increases to a maximum again. The mean value of the intensity over a bright band and neighbouring dark band is thus the mean value of A^2 when ϕ varies from 0 to 2π. In this range, the mean value of $\cos \phi = 0$. Hence, from above, the mean value of the intensity A^2 is $2a^2$. The two individual waves together have an intensity equal to $a^2 + a^2$, or $2a^2$. Thus we see that the energy lost in the dark bands has reappeared in the bright bands. The law of conservation of energy is therefore obeyed.

It should be noted that we have just discussed the theory of "Young's interference bands" (p. 688). Diffraction and interference are examples of the same phenomenon, namely, the resultant effect produced by overlapping waves. In the case of two slits, the resultant intensity of the bands is the product of two terms. One term is the intensity due to the interference between the two slits just considered, and the other is the intensity due to diffraction at a single slit, which modulates the resultant intensity. This is discussed more fully later (p. 738). The modulation effect can be seen by using a laser beam to illuminate two close slits. Numerous bright and dark bands are then obtained on a white screen beyond the slits, and the intensity of the bright bands decreases from the central band outwards.

Diffraction Due to Three Slits

With three slits having the same separation d, maximum intensity is again obtained in directions θ corresponding to path differences λ, 2λ, ... For example, a bright band occurs in a direction θ where $\sin \theta = \lambda/d$, Fig. 30.11 (i).

In a direction θ_1 of smaller angle than θ, and such that the path difference between corresponding rays in successive slits is $\lambda/2$, the waves from

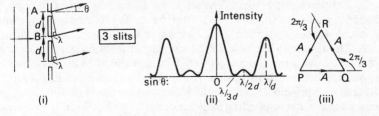

FIG. 30.11. Bands due to three slits.

corresponding points in the first two slits are 180° out of phase. But the waves from the third slit are in phase with those from corresponding points in the first slit. Hence the resultant wave amplitude is roughly one-third that in the direction θ, and the resultant image intensity about one-ninth, since intensity $\propto$ amplitude^2. The weak image is shown roughly in Fig. 30.11 (ii); it corresponds to a direction θ_1 given by sin $\theta_1 = \lambda/2d$.

At a smaller angle θ_2 than θ_1 given by sin $\theta_2 = \lambda/3d$, the waves from corresponding points in successive slits, distance d apart, have a path difference of $\lambda/3$. This is equivalent to a phase difference of $2\pi/3$ (120°). Suppose each of the waves from the slits has equal amplitude A and is represented by phasors PQ, QR and RQ, Fig. 30.11 (iii). Then, by vector addition,

$$\text{resultant} = \overrightarrow{PQ} + \overrightarrow{QR} + \overrightarrow{RP} = 0.$$

Hence, as shown in Fig. 30.11 (ii), the intensities are zero for directions θ corresponding to sin $\theta = \lambda/3d, 2\lambda/3d$, and so on. Thus the band pattern consists of bright and dark bands. The intensity of the background varies in a similar way to that due to a single slit (p. 737).

Diffraction Pattern

Fine curtains, chiffon and other closely woven materials form a network of numerous small parallel close openings in two perpendicular directions. When placed in front of a lens and illuminated by parallel light, these materials produce a pattern of rectangular rows of spots on a screen. This is a diffraction pattern.

The spots rotate about the centre when the material is rotated, keeping their rectangular pattern. If finer material is used, the spots become broader and farther apart. Using coloured filters, the spots become slightly narrower and closer together for blue light than for red light.

Diffraction Grating

A diffraction grating may have several hundred equidistant ruled lines per millimetre. The theory of the grating was discussed in the last chapter. Here we can restate the main points:

1. The grating produces principal maxima, or strong images, in directions θ given by d sin $\theta = m\lambda$, where m, the order of the spectra, is 0, 1, 2, . . . This is illustrated in Fig. 30.12 (ii).

2. The intensity of these images is modulated by the variation of intensity I_d due to a single slit. As shown in Fig. 30.12 (i), the zero values of I_d occur where sin $\theta = \lambda/a, 2\lambda/a, \ldots$, where a is the width of the *slit*.

3. The resultant intensity I of the principal maxima is given by the product of two terms, I_d and I_s, where I_s is the intensity of the interference bands due to waves from the slits. Thus I varies as shown in Fig. 30.12 (iii). Note that no principal maxima are obtained where the single slit curve has zero intensity, corresponding to directions sin $\theta = \lambda/a, 2\lambda/a, \ldots$. Thus if $d = 5a$, for example, the 5th order principal maximum is missing, since sin $\theta = m\lambda/d = 5\lambda/5a$ in this case.

Finally, it may again be noted that the *sharpness* of the grating principal images in different orders depends on the number of lines N in the grating for a given separation d. The first order image, for example, corresponds to a direction θ such that $d \sin \theta = \lambda$, so that the path difference between waves from corresponding points in the first and last slit is $(N - 1)\lambda$. When the angle of diffraction is increased to $\theta + \Delta\theta$, so that the path

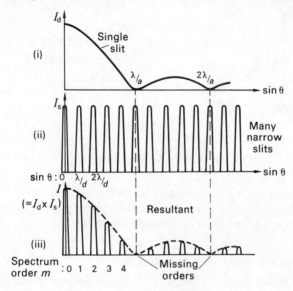

Fig. 30.12. Intensities of grating images.

difference increases from $(N - 1)\lambda$ to $N\lambda$, the intensity falls to zero since this is an extra path difference of λ (compare the theory for a single slit, p. 730). The angle $\Delta\theta$ corresponds to a change of λ in $(N - 1)\lambda$, or $N\lambda$ when N is large, which is a fractional change of $1/N$. When N is large, therefore, $\Delta\theta$ is small and hence the image is sharp. The *resolving power* of the grating is shown on p. 713 to be equal to Nm, where m is the order of the spectrum.

Concave Reflection Grating

The plane diffraction grating requires two lenses in order to obtain spectra in sharp focus; one lens produces a parallel beam from the source and the other brings the diffracted parallel beam from the grating to a focus. With this grating the spectra of different wavelengths, in say the ultra-violet and visible spectra, would not be in focus at the same time.

In 1883 Rowland showed that gratings ruled on *concave spherical surfaces* produced sharp spectra of different wavelengths without the need for lenses. This has also the advantage of eliminating chromatic aberration due to a lens and of avoiding absorption of ultra-violet light by the glass. The theory of the grating is beyond the scope of this work and the interested reader is referred to Dr. R. S. Longhurst's advanced book

Geometrical and Physical Optics (Longmans). This shows that if the diameter GC of the circle in Fig. 30.13 (i) is equal to the radius of curvature r of the concave grating G, the spectra of different wavelengths from a source at S on the circle will all be sharply focused at the circumference

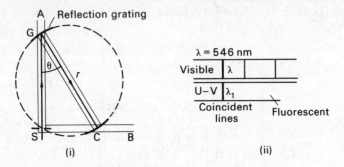

FIG. 30.13. Concave reflection grating.

of the circle. A photographic camera at C will thus record the wavelengths diffracted in a direction normal to the concave grating G. A large radius of curvature of G produces appreciable dispersion.

Normal Spectra. Ultra-violet Wavelengths

Rowland used perpendicular rails SA,SB along which G and C moved, the grating and camera being joined by a rigid rod. This method enabled the angle of incidence θ, angle SGC, to be varied, so that successive wavelengths and successive spectra appeared at C. For the mth order spectrum of wavelength λ, the path difference $= m\lambda$. Now the path difference for rays diffracted normal to the grating at G is $d \sin \theta$, where d is the spacing of the grating.

$$\therefore \ d \sin \theta = m\lambda.$$

Thus $\sin \theta = m\lambda/d$. But, from Fig. 30.13 (i), SC $= r \sin \theta$, since GC $= r$ and angle GSC $= 90°$. Hence, for a given order m,

$$\text{SC} = \frac{m\lambda r}{d}$$

$$\therefore \ \text{SC} \propto \lambda$$

This result shows that the linear separation of two lines along SB is proportional to their difference in wavelength. Such a spectrum is called a *normal spectrum*. Hence if the photographic plate C is curved to a radius $r/2$, the spectra will not only appear in focus at C but their wavelengths can be easily read directly from a linear scale.

Overlapping orders are also in sharp focus round C. This enabled Rowland to determine accurately wavelengths in the ultra-violet, which are shorter than those in the visible spectrum. Suppose, for example, that the second order image of an ultra-violet line wavelength λ_1 coincides with the first order green wavelength $\lambda = 546$ nm, Fig. 30.13 (ii). Then $d \sin \theta = 2 \lambda_1 = \lambda$. Hence $\lambda_1 = \lambda/2 = 273$ nm. Rowland employed this "method of coincidences" to determine wavelengths in different parts of the spectrum, using known wavelengths in the visible spectrum as standards.

FRESNEL DIFFRACTION

Half-period Zones

The diffraction phenomena resulting from plane wavefronts (parallel beams) are classed as *Fraunhofer diffraction*, see p. 729. The diffraction phenomena resulting from curved wavefronts (close sources) are classed as *Fresnel diffraction*. When the sources are at great distances, this class of diffraction phenomena merges into *Fraunhofer diffraction*.

Fresnel first showed that a wavefront from a point source could be divided into a number of *zones*, the effective phase difference between adjacent zones being π, which is equivalent to a path difference of $\lambda/2$.

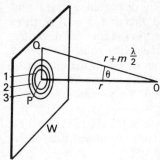

FIG. 30.14. Fresnel zones.

To illustrate the idea, suppose that the effect of a plane wavefront W at a point O is required, Fig. 30.14. The nearest point P on W from O is called the *pole*, and if P is a distance r from O, spheres of centre O and radii $r + \lambda/2, r + \lambda, r + 3\lambda/2, \ldots$ are drawn. These intersect W in circles, producing zones $1, 2, 3, \ldots$ known as *half-period zones* because the extreme points on each consecutive zone are 180° out of phase with one another. As we proceed outwards along a radius from P, for example, the phase of vibration of each point in zone 1 increases, and at the beginning of zone 2 it just reaches 180°. Moving further outwards along the radius in zone 2 the phase increases until, at the beginning of zone 3, it just reaches 360°. On account of the 180° phase change it can now be seen that, if $a_1, a_2, a_3, a_4 \ldots a_m$ are the respective amplitudes of vibration due to zones $1, 2, 3, 4 \ldots m$ at O, the resultant amplitude of vibration A_m at O is given by

$$A_m = a_1 - a_2 + a_3 - a_4 + \ldots + (-1)^{m+1} a_m.$$

Resultant Amplitude

The magnitude of the amplitude of vibration at O due to a particular zone depends on three factors: (i) the area of that zone, (ii) its distance from O, (iii) the angle θ between the normal to the wavefront and the direction of the diffracted ray. Suppose Q is a point on the extreme edge of the mth zone, that is, $OQ = r + m\lambda/2$, Fig. 30.14. Then, by Pythagoras'

Theorem, the radius PQ or R_m of that zone is given by

$$R_m{}^2 = OQ^2 - OP^2 = \left(r + \frac{m\lambda}{2}\right)^2 - r^2 = m\lambda r,$$

to a very good approximation, since the term in λ^2 is very small compared to $m\lambda r$. If R_{m-1} is the radius of the $(m-1)$th zone, it follows that

$$R^2{}_{m-1} = (m-1)\lambda r.$$

$$\therefore \text{ area of } m\text{th zone} = \pi R_m{}^2 - \pi R_{m-1}^2 = \pi(R_m{}^2 - R_{m-1}^2)$$
$$= \pi[m\lambda r - (m-1)\lambda r] = \pi\lambda r.$$

Since π, λ, are constants, the area of each zone is the same.

The amplitude of vibration due to a source is inversely proportional to its distance from the point concerned, see p. 607. The distance of the mth zone from O is $r + m\lambda/2$; and since λ is small compared with r, the distance is practically equal to r. Thus the net effect at O due to the area of the zone and its distance away $\propto$ area $\div$ distance $\propto \pi\lambda r \div r \propto \pi\lambda$. Since π and λ are constants, it follows that each zone produces the same effect at O due to the two factors concerned. Stokes investigated how the amplitude at O varied with the third factor θ, the angle between the normal to the wavefront and the direction of the diffracted ray. He found it varied as $(1 + \cos\theta)$. Thus in a backward direction, $\theta = 180°$, there was no effect, but there was a maximum effect in the forward direction, $\theta = 0°$. As we move outwards from P through each zone, the angle θ increases, and hence the amplitude diminishes slowly as the number of each zone increases. Thus the net effect of the three factors, area, r, θ, is a slow diminution of amplitude as the number of the zone increases.

We have now to find the sum A_m of $a_1 - a_2 + a_3 - \ldots + (-1)^{m+1}a_m$. Here m is infinite for an unbounded plane wave but limited in value for a wave passing through a small circular aperture. As stated previously, there is a gradual increase in phase angle of the vibrations in the wavefront if we move outwards from P through the various zones. The succession of small amplitudes in zone 1 can thus be represented vectorially by the arc ABC in Fig. 30.15; the phase angle is zero at A, 90° at B, and 180° at C, the edge of zone 1. AC thus represents the amplitude a_1 of zone 1. As we proceed from the edge of zone 1 through zone 2, a vector diagram CDE is obtained. The amplitude a_2 is represented by CE, which is opposite in sign to AC, p. 741. If there were only two zones, the resultant amplitude would be $a_1 - a_2$ or AE. Proceeding to zone 3, the vector curve EFG is obtained, where EG $= + a_3$. Similarly, GK $= - a_4$. Dealing with the amplitudes of each zone in this way, it can be seen that *the vector curve spirals towards the centre*, T, *when m is large*. The resultant amplitude A_m is equal to AY if the mth zone finishes at Y, and equal to AZ if the mth zone finishes at Z. This corresponds to resultants (AT + TY) or (AT − TZ) respectively, which can be denoted by $a_1/2 + a_m/2$ when m is odd or by

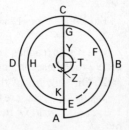

FIG. 30.15. Vector diagram.

$a_1/2 - a_m/2$ when m is even. In either case the small term $a_m/2$ can be neglected when m is very large, which occurs for an infinite plane wave, and the total amplitude due to all the zones is then given by

$$A_m = \frac{a_1}{2}.$$

Thus the effect is equal to half that due to the first zone only.

The result we have just derived vectorially can also be obtained by writing the series $A_m = a_1 - a_2 + a_3 - \ldots$ in the form

$$A_m = \frac{a_1}{2} + \left(\frac{a_1}{2} - a_2 + \frac{a_3}{2}\right) + \left(\frac{a_3}{2} - a_4 + \frac{a_5}{2}\right) + \ldots + (-1)^{m+1}\frac{a_m}{2}.$$

The terms in the brackets are practically zero as a_2 is about the arithmetic mean of a_1 and a_3, for example, and so on for the other brackets. When m is very large, $A_m = a_1/2$.

Rectilinear Propagation of Light. Shadows

To illustrate the size of the half-period zones, consider a plane wave at a distance of 10 cm from the point O. If it is sodium light of mean wavelength $5\cdot893 \times 10^{-7}$ m, the radius R_1 of the first half-period zone is given, from p. 742, by

$$R_1 = \sqrt{\left(r + \frac{\lambda}{2}\right)^2 - r^2} = \sqrt{\lambda r} = \sqrt{5\cdot893 \times 10^{-7} \times 0\cdot1} = 2\cdot4 \times 10^{-4}\,\text{m}.$$

Thus the radius of the first zone is a fraction of a millimetre, and hence the zone is confined to a tiny area round the pole. For a wavefront of width about 1 metre, the number m of the extreme zone is of the order given by

$$R_m = 0\cdot5 = \sqrt{m\lambda r} = \sqrt{m \times 5\cdot893 \times 10^{-7} \times 0\cdot1};$$

from which $\qquad m = \dfrac{0\cdot5^2 \times 10^7}{5\cdot893 \times 0\cdot1} = 4 \times 10^6 \text{ (approx.)}.$

The 4 millionth zone will contribute a very small amplitude at O, and hence, from the expansion for A_m given above, the total amplitude at O can be regarded as due to half of the first zone. The light energy thus travels to O from a tiny area around the pole, and this explains the rectilinear propagation of light. A more direct method is by the consideration of the extent of the diffraction of light into the geometrical shadow of an opaque object, discussed on p. 731.

With sound of frequency 500 Hz, corresponding to a wavelength in air of about 0·7 metre, the first half-period zone for a point 3 metres away has a radius $= \sqrt{\lambda r} = \sqrt{0\cdot7 \times 3} = 1\cdot5$ m. Thus (i) an obstacle would have to be very large to produce a sound "shadow", (ii) an aperture of normal size would not even cover the first zone. It can now be seen that sound energy, unlike light energy, is not confined to a small region, and

its effect spreads round obstacles of normal size. The rectilinear propagation of light is thus basically due to the shortness of the wavelength
of light.

Small Circular Aperture and Obstacle

If a circular aperture is so small that it is equal in area to the first half-
period zone, a bright light is observed at the point O, due to the amplitude a_1, see Fig. 30.14, p. 741. If the aperture covers two zones, the
amplitude at O is now $(a_1 - a_2)$, which is very small, and little light is now
seen at O. For three zones, the resultant amplitude is $(a_1 - a_2 + a_3)$,
which gives bright light again. Similar maxima and minima of light
intensity are obtained by keeping the size of the small aperture fixed, and
moving towards it from a point whose distance away is large compared
with the radius of the aperture.

Similarly, with a small circular obstacle, only a few of the first zones
may be blocked out. The amplitude at a distant point on the axis may
then be considered as $a/2$, where a is the amplitude due to the first zone
round the obstacle. Off the axis, consideration of the zone effect shows
that the amplitude reduces quickly to zero. Hence a fairly sharp spot of
light should be observed at a point on the axis, in the geometrical shadow.
This was actually observed by Arago, p. 702, in early discussions of the
wave theory, and is shown in photograph in Fig. 29.17, p. 702.

Zone Plate

If circles are drawn on a transparent plate whose radii are proportional to $\sqrt{m}$, where m is a whole number, half-period zones are obtained

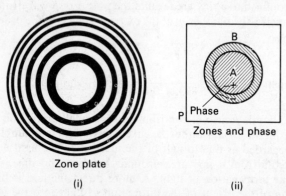

Zone plate

(i)   (ii)

FIG. 30.16 (i) Zone plate—increased illumination due to blacked out
zones. (ii) Demonstration of phase effect on resultant.

(p. 742). When alternate zones such as 2, 4, 6, . . . are blacked out and
the plate is illuminated, the amplitude A beyond the plate at the point P
on the axis is given by $A = a_1 + a_3 + a_5 + \ldots$, which are all positive
terms, Fig. 30.16 (i). The plate, suitably reduced in size, then produces a

bright image of an illuminated object at some point on its axis, and thus acts like a converging lens. R. W. Wood developed a zone plate coated with suitable materials in which the phase was *reversed* in alternate zones. The resultant amplitude A was then $a_1 + a_2 + a_3 + a_4 + \ldots$ The zone plate thus produced a very bright image at the appropriate point on its axis, its effective focal length being $R_1{}^2/\pi\lambda$, where R_1 is the radius of the first zone.

The opposite phases of waves coming from the zones can be demonstrated by using 3 cm microwaves. Following the same calculation given before for visible light, the radius of the first and second zones can be found when $\lambda = 3$ cm, assuming a particular distance r for the microwave transmitter. These two zones, a disc A and a ring B, are then cut from a metal plate P, Fig. 30.16 (ii). With A removed, or B removed, a strong signal can be received at a particular place in front of P when the transmitter is behind at r. With A and B *both* removed, the signal received is practically zero, showing that the waves from A and B have opposite phases and their amplitudes are nearly equal.

If a is the amplitude when only the first zone or disc A is removed, then $a/2$ is the amplitude when the whole of the plate P is removed (p. 743). Hence the intensity reduces to one-quarter.

Diffraction at Straight Edge

When monochromatic light passes over an opaque obstacle AB with a straight edge A, a few interference bands parallel to A are observed on a screen M at X, just above the geometrical shadow, A_1, of A, Fig. 30.16 (i). Below A_1 the shadow is not sharp, but falls away gradually in intensity.

The appearance of the bands can be explained by considering the cylindrical wavefront RQ which reaches the obstacle from a source S in front of a slit parallel to D, Fig. 30.17 (i). We divide RQ into half-period

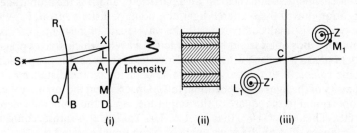

FIG. 30.17. Diffraction at straight edge (*not to scale*).

zones in a different manner from that for plane or spherical wavefronts. Cylinders of radius $r + m\lambda/2$, with their axes passing through A_1, cut the wave surface into two sets of half-period zones in the form of *strips*, one set above and one below the centre, Fig. 30.17 (ii). This leads to an amplitude phase diagram of the type shown in Fig. 30.17 (iii), which in its generalised form is known as *Cornu's spiral*. The spirals in opposite quadrants refer respectively to the upper and lower halves of the wavefront. The difference in shape from the spiral in Fig. 30.15 results from the greater rapidity with which the areas of consecutive zones and therefore the amplitude contributions diminish for a cylindrical wavefront.

In the Cornu spiral, the total amplitude of the upper and lower halves of the wavefront are CZ and CZ′ respectively. The amplitude of the whole wave is ZZ′. The point A_1 on the geometrical shadow receives only half the wavefront, equivalent to an amplitude CZ, or half ZZ′; the intensity at A_1 is thus one-quarter of that due to the whole wave. As we move into the geometrical shadow, light is received from less than half the wavefront. At M in the shadow for example, the amplitude would correspond to ZM_1, and this diminishes rapidly as M_1 approaches Z. Outside the shadow, at L for example, the amplitude corresponds to ZL_1; and it can be seen that, as L_1 proceeds from the origin C along the spiral to Z′, ZL_1 goes through a series of maxima and minima values, with a corresponding fluctuation in intensity. The intensity variations for positions near the geometrical shadow are considerable but they rapidly die away in the outside region, Fig. 30.17 (i).

Scattering of Light by Small Particles. Rayleigh's Law

When the size of a particle is very large compared with the wavelength of light, and it is illuminated by a beam of light, a defined geometrical shadow is produced. If the size of the particle is of the order of the wavelength of light, however, diffraction occurs round the edge of the particle and the light is now "scattered". The intensity of the scattered light in this case is proportional to $1/\lambda^4$ where λ is the wavelength, among other factors, a relationship first derived by Lord Rayleigh 1871, using a method of dimensions.

Rayleigh's Formula. Suppose A_0 is the amplitude of vibration of a monochromatic beam incident on a particle P, and A is the amplitude of vibration due to the scattered light at a point X distant r from P. When the particle is small compared with the wavelength of the light, every part round it sends out a secondary (diffracted) wavelet which arrives practically in phase with each other at X. Thus the amplitude A is proportional to the *volume* V of the particles. A also depends on the distance r. The intensity of the light is proportional to $1/r^2$, p. 565, and the intensity is also proportional to the square of the amplitude, A^2, at the point considered, p. 607. Thus $A^2 \propto 1/r^2$, that is, $A \propto 1/r$. Taking also into account the dependence of A on the wavelength λ, it can now be stated that

$$A = A_0 \frac{V\lambda^x}{r}, \qquad \cdot \qquad \cdot \qquad \cdot \qquad \text{(i)}$$

where x is some unknown power. From (i),

$$\frac{A}{A_0} = \frac{V\lambda^x}{r}.$$

Now A/A_0 has no dimensions. Thus $V\lambda x/r$ has no dimensions. The volume V has dimensions $[L^3]$, the distance r has dimensions $[L]$; consequently, as wavelength has dimensions $[L]$,

$$3 + x - 1 = 0, \text{ or } x = -2.$$

Hence, from (i), $\qquad A = A_0 \dfrac{V\lambda^{-2}}{r} = A_0 \dfrac{V}{r\lambda^2}$

$\therefore$ intensity of scattered light, $I_s, \propto A^2 \propto A_0{}^2 \dfrac{V^2}{r^2\lambda^4}$

Thus, for given values of V, A_0, and r,

$$I_s \propto \frac{1}{\lambda^4} \qquad . \qquad . \qquad . \qquad . \qquad \text{(ii)}$$

When the size of the particle is large compared with the wavelength of light, the diffracted wavelets from the points on the particle arrive out of phase, and hence some interference occurs. The amplitude of vibration is then no longer proportional to $1/\lambda^2$, and the fourth-power law of Rayleigh does not hold.

Colour of Sky, Sunset, and Smoke. The wavelength of the blue end of the spectrum is of the order of 0·0004 mm, that of the red end of the spectrum is 0·0007 mm. From the fourth-power law, it follows that blue light is scattered by fine particles about ten times as much as red light. When we gaze at the sky, we receive light from the sun which is scattered by molecules of air and water-vapour in the atmosphere. Since blue light is scattered much more than red light, the light reaching the eye is mainly blue. This explanation of the blue of the sky was first given by Lord Rayleigh.

The sun viewed directly appears to be more yellow than white because some blue light is scattered. When the sun sets, the light from it passes through a thicker amount of atmosphere than when it is high in the heavens, as at midday, and hence more of the wavelengths are scattered and the sun appears red in the sky. For a similar reason, the sun appears orange-red in a fog or a mist. If the smoke from a cigarette or wood fire is first observed, it appears blue, as this light is scattered more than the others by fine particles. When the volume of smoke increases, and the particles coalesce and form large particles, the colour of the smoke changes to white as the fourth-power law is not obeyed. From the above, infra-red rays are scattered only very slightly by fine particles, and hence infra-red photographs can be taken through fog and mist, see p. 456.

An experiment due originally to Tyndall illustrates vividly the effect of particles on scattering. If a solution of hypo is added to a dilute solution of sulphuric acid, fine particles of sulphur are first formed in the solution. When the solution is illuminated by light from a projection lantern, the scattered light appears blue and the transmitted light red. As the particles combine and form large groups, the transmitted light diminishes in intensity and the scattered light becomes white.

The light from the sky is partially polarised. This can be shown by rotating a Nicol prism, when a variation in brightness is observed. The change in brightness is a maximum, that is, the polarisation is most complete, when the sky is viewed in a direction perpendicular to the sun's rays.

Bands of Equal Inclination

In wedge films, the bands of the interference pattern are produced by interference along paths of equal thickness, and they are hence classified as *bands of equal thickness*. In 1849 Haidinger discovered the existence of interference bands when a thick parallel-sided plate was used. They are *bands of equal inclination* in contrast to those with the wedge films, because each band is produced by interference along paths for which the angle of incidence on the plate is the same. Further, as we shall explain below shortly for a thick plate, the bands are formed at infinity, and to see them the eye must be focused on infinity, or a telescope focused on infinity is used.

Consider a ray AO_1 of monochromatic light incident at an angle θ on a plane-parallel thick plate, Fig. 30.18. On account of successive reflection, the emergent beam consists of parallel rays at $O_1, O_2, O_3, \ldots$, as shown. The path difference between the parallel rays at O_1, O_2 is $2nt \cos r$, where r is the angle of refraction in the plate. The path difference between any two successive parallel rays is also $2nt \cos r$. Thus if all the emergent rays

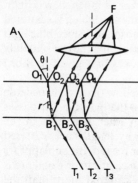

are brought to a focus F by a lens they interfere, and a bright band is obtained when $2nt \cos r = (m + \frac{1}{2})\lambda$, allowing for a phase change of 180° by reflection. It follows from $2nt \cos r$ that the path difference decreases as θ increases; the largest path difference occurs when $r = 0$, at normal incidence. This is the reverse to the case of Newton's rings.

For a given band, θ is constant. Thus the band is the arc of a circle whose centre is the foot of the perpendicular from F to the principal axis of the lens. Other circular bands are obtained corresponding to different values of θ. It should be noted that the

FIG. 30.18. Bands with thick plates.

bands are located at infinity since they are due to parallel rays, and can interfere only when they are brought together by a lens.

Similar arguments to the above show that the transmitted beam of light, B_1T_1, B_2T_2, $B_3T_3 \ldots$ (Fig. 30.18), also give rise to circular interference bands. Since there is no net phase change in this case, a bright band is obtained when $2nt \cos r = m\lambda$; the band system is thus complementary to that due to reflected light. The full theory, involving summation of the multiple internal reflections, shows that the bright bands in the transmitted system are narrower than the intervening dark bands; the reverse is true for the reflected system.

Interference in Air-film between Parallel Plates

When a thick parallel-sided air-film between two plates A, B is illuminated by a monochromatic beam of light, multiple reflections occur in the
air-film. Fig. 30.19 illustrates the multiple reflections due to a ray CD
incident at an angle θ on the first plate, A; the beam of parallel rays
emerges at the same angle θ to B, as shown. Replacing r by θ, the angle in
the air-film, and putting $n = 1$ in our formula $2nt \cos r = m\lambda$ for a bright
transmitted ring, it then follows that $2t \cos \theta = m\lambda$ is the condition here
for a bright band. Other points on the source also emit rays incident at the
same angle θ on the air-film, and thus more emergent rays, parallel to
those shown, are obtained. If they are all collected by a lens, a circular
bright ring is formed in the focal plane, any radius of which subtends an
angle θ at the centre of the lens. The circular bands are bands of equal
inclination (p. 700).

Fabry–Perot Interferometer

If the plates A, B are unsilvered, the bright rings are fairly broad and
only slightly narrower than the intervening dark rings. When the plates
are partially silvered, however, the relative strength of the internally

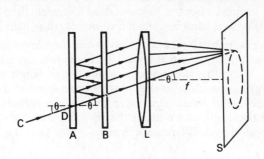

FIG. 30.19. Fabry–Perot principle.

reflected beams is increased, and there is a remarkable change in the
appearance of the ring system. The bright rings now become much
narrower and sharper in intensity, and the dark rings between them
become much broader. The bright rings are now seen very clearly, that is,
the silvering has considerably increased their visibility.

This is the principle of the Fabry–Perot interferometer. It has been used
for investigations into the "fine structure" of spectral lines. In the case of
the sodium line for example, two ring systems are clearly observed with
the interferometer, showing that the sodium line is actually a doublet (p.
696). The interferometer has also been used to measure wavelengths, and
it has enabled the length of the standard metre to be evaluated in terms of
the wavelength of the red cadmium line. Details of the instrument are
given in *Advanced Practical Physics* by Worsnop and Flint (Methuen).

Refractive Index of Gases. Jamin Refractometer

About 1860 Jamin designed an instrument producing interference bands which could be used to measure the refractive index of a gas. This instrument, called a *refractometer*, consists of two equal thick vertical plates X, Y silvered on their back surfaces, with two tubes A, B of equal length between them, Fig. 30.20. When X is illuminated by light from a

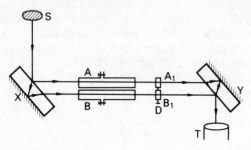

FIG. 30.20. Jamin refractometer.

broad source S, parallel reflected beams from the front and back surfaces travel through A, B. If X, Y are parallel the total optical light paths of the two beams after reflection at the back and front are equal. When the plates are slightly inclined about a horizontal axis straight line bands are observed; the central band is white if a source of white light is used.

To measure the refractive index of a gas, A and B are first evacuated and the gas is passed slowly into A say, until it is at the pressure and temperature desired. While this is happening, a number of bands m cross the field of view and are counted as they pass the cross-wires of the telescope T. The optical path in A has increased from t, when a vacuum existed, to nt, where t is the length of A and n is the refractive index of the gas. The increase in optical path difference is thus $nt - t$, or $(n - 1)t$, and hence

$$(n - 1)t = m\lambda,$$

where λ is the wavelength.

$$\therefore n = 1 + \frac{m\lambda}{t},$$

and n can be calculated when m, t, λ are known.

If the number of fringes cross the field of view too quickly to be counted in an experiment, an arrangement incorporating two plates A_1, B_1 of equal thickness is used. This is called a *compensator*. The plates are inclined at a small angle, and can rotate about a common axis by a screw D. When a source of white light is used, the central band of the system is white, and this provides a "reference" line. After the gas is introduced, the central band is displaced. By turning the compensator slightly, thus altering the optical path, the central band can be brought back to the cross-wires again. The screw D is previously calibrated with sodium light by counting the bands which pass for a known rotation. The increased path difference due to the gas is thus known from the rotation of the compensator in terms of the wavelength of sodium light, and can be calculated. The dial on the screw can be calibrated to read refractive indices directly if desired.

Rayleigh Refractometer

Rayleigh designed a refractometer which is more sensitive than Jamin's. Two parallel narrow slits, S_1, S_2, are illuminated by a monochromatic source S at the focus of a lens L, and tubes A, B, of equal length are traversed respectively by parallel beams, Fig. 30.21. As in Young's experiment, interference bands are produced when the beams are brought

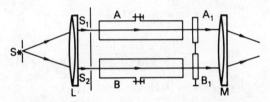

Fig. 30.21. Rayleigh refractometer.

together at the focus of the lens M of a powerful telescope. The central fringe of the system can be found by using a source of white light, and as in the Jamin refractometer, the change in optical path when gas is introduced in A can be obtained by using the compensator A_1, B_1 and a calibrated dial. The Rayleigh refractometer is extremely sensitive, and it has been applied to detect and measure the presence of very small quantities of gas.

FERMAT'S PRINCIPLE

Fermat's Principle

In the seventeenth century, Fermat stated that the path taken by a ray of light in travelling between two points was such as to make the time of transit a minimum. Later, it became known that there were cases when the time taken was a maximum, and thus in its general form Fermat's Principle can be stated as follows: *The path taken by a light ray in travelling between two points is one in which the time has a stationary value.* This implies that the path taken by a light ray between two points is equal, to a first approximation, to other paths closely adjacent to the actual path. As we shall now show, the laws of light are all embodied in Fermat's principle.

Rectilinear Propagation. Laws of Reflection and Refraction

In an isotropic medium the velocity of light is the same in all directions. Now the shortest distance between two points is the straight line joining them, and thus the minimum time is taken when a light ray travels along the straight line joining the two points. This explains the rectilinear propagation of light by Fermat's principle.

Reflection. Consider a ray AO reflected along OB by a plane mirror, Fig. 30.22 (i). Let AM = b, BN = d, where M, N are the foot of the perpendiculars from A, B respectively to the mirror, and let OM = x. Then, if MN = a, ON = a − x.

If c is the velocity of light in air, the time T for the light to travel from A to B is given by

$$T = \frac{1}{c}(AO + OB) \qquad . \qquad . \qquad . \qquad (i)$$

Now AO = $(b^2 + x^2)^{\frac{1}{2}}$, OB = $[d^2 + (a − x)^2]^{\frac{1}{2}}$.

$$\therefore \frac{d(AO)}{ax} = x(b^2 + x^2)^{-\frac{1}{2}} = \frac{x}{(b^2 + x^2)^{\frac{1}{2}}}, \qquad . \qquad (ii)$$

$$\text{and} \qquad \frac{d(OB)}{dx} = \frac{-(a − x)}{[d^2 + (a − x)^2]^{\frac{1}{2}}} \qquad . \qquad (iii)$$

For a stationary value, $dT/dx = 0$. Hence, from (i)

$$\frac{d(AO)}{dx} + \frac{d(OB)}{dx} = 0.$$

Substituting from (ii) and (iii),

$$\therefore \frac{x}{(b^2 + x^2)^{\frac{1}{2}}} − \frac{(a − x)}{[d^2 + (a − x)^2]^{\frac{1}{2}}} = 0.$$

From Fig. 30.22 (i), it can be seen that $\sin i = x/(b^2 + x^2)^{\frac{1}{2}}$, and $\sin r = (a − x)/[d^2 + (a − x)^2]^{\frac{1}{2}}$.

$$\therefore \sin i − \sin r = 0,$$
$$\text{or} \qquad i = r.$$

This is the law of reflection.

Refraction. Consider a light ray AO in air refracted along OB in a medium in which the velocity of light is v. Fig. 30.22 (ii). The time taken, T, is given by

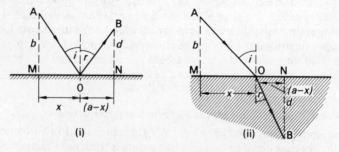

FIG. 30.22. Reflection and refraction by Fermat Principle.

$$T = \frac{AO}{c} + \frac{OB}{v}$$

$$\therefore \frac{dT}{dx} = \frac{1}{c}\frac{d(AO)}{dx} + \frac{1}{v}\frac{d(OB)}{dx} \qquad . \qquad . \qquad \text{(iv)}$$

Now, as before, $AO = (b^2 + x^2)^{\frac{1}{2}}$, $OB = [d^2 + (a - x)^2]^{\frac{1}{2}}$, and differentiation of each with respect to x results in $\sin i$ and $- \sin r$ respectively, as already shown. For a stationary time, $dT/dx = 0$. Hence, from (iv),

$$\frac{\sin i}{c} - \frac{\sin r}{v} = 0.$$

$$\therefore \frac{\sin i}{\sin r} = \frac{c}{v} = \text{constant}.$$

This is Snell's law of refraction.

EXERCISES 30

Single Slit. Resolving Power

1. A parallel beam of light, wavelength 600 nm (6×10^{-7} m) is incident normally on a rectangular slit 0·8 mm wide. Calculate the angular width of the central bright band. If a lens of +2D (focal length 50 cm) is placed behind the slit to focus the emerging beam on to a white screen, find the distance on the screen from the middle of the central band to the edge of the band.

2. Draw a sketch of the variation in intensity of the bright and dark bands obtained with a single slit illuminated by a parallel beam of monochromatic light and received on a distant screen.
Draw a sketch of the diffracted plane wavefronts which travel in the direction of the *edge* of the central band (first minimum), showing the path difference.
Describe and explain the colours observed at the edges of the central bright band when white light is used.

3. A rectangular slit of width a is illuminated by a parallel beam of monochromatic light of wavelength λ. What is the angular width of the central bright band when $a = 10\lambda$? If the slit width is slowly reduced until it is equal to λ, describe and explain the effect observed on the central band.

4. Calculate the smallest angular resolution of a double star by (i) the Isaac Newton reflector telescope, objective diameter 2·5 m, (ii) Mount Yerkes lens telescope, objective diameter 1·3 m, assuming $\lambda = 5·5 \times 10^{-7}$ m. What are the advantages of a reflector telescope over a lens telescope?

5. If the Jodrell Bank radio-telescope objective or bowl is 80 m diameter, calculate the resolving power for reception of radio waves of frequency 150 MHz ($c = 3 \times 10^8$ m s^{-1}).

6. Two headlamps on a car are just resolved at a distance of 5·0 km. What is the separation of the two lamps if the average wavelength $\lambda = 6 \times 10^{-7}$ m and the diameter of the pupil of the eye is 3 mm?

7. Find the diameter of a radio-telescope bowl for reception of wavelengths 22 cm long, if the smallest angle of resolution for distant galaxies is to be 1°.

8. A ruler with millimetre graduations is viewed by an observer. (i) At what distance will the graduations be just resolved when viewed through a red filter ($\lambda = 7 \times 10^{-7}$ m, red light)? (ii) If a blue filter now replaces the red filter, will the graduations be better resolved or not? (iii) If $\lambda = 4 \times 10^{-7}$ m for blue light, at what distance will the graduations be just resolved? Assume a pupil diameter of 4 mm.

9. A diffraction grating has a width of 2 cm and 50 lines per mm; another has a width of 3 cm and 30 lines per mm. Each grating is used in turn to observe the sodium doublet, lines with wavelength 589·6 nm ($5·896 \times 10^{-7}$ m) and 589·0 nm ($5·890 \times 10^{-7}$ m). (i) Which grating is able to just resolve the lines in the first order spectrum? (ii) Can both gratings resolve the lines in the second-order spectrum?

10. A grating has 100 lines per mm and a width of 4 cm. What is the resolving power of the grating for the first and second order bright images?
If alternate slits of the grating are blacked out, what effect has this on (i) the intensities of the bright images, (ii) the grating resolving power?

11. Calculate the closest distance between two point objects when they are just resolved 25 cm from the eye, assuming a pupil diameter of 3 mm and a wavelength of $5·5 \times 10^{-7}$ m.

Diffraction Grating

12. Describe *two* experiments to illustrate the diffraction of light. How in general do the effects produced depend on the size of the obstacle or aperture relative to the wavelength of light use?
A diffraction grating consists of a number of fine equidistant wires each of diameter $1·00 \times 10^{-3}$ cm with spaces of width $0·75 \times 10^{-3}$ cm between them. A parallel beam of infra-red radiation of wavelength $3·00 \times 10^{-4}$ cm falls normally on the grating, behind which is a converging lens made of material transparent to the infra-red. Find the angular positions of the second order maxima in the diffraction pattern. (*L.*)

13. Explain why the slits of a diffraction grating are (*a*) narrow (*b*) close together (*c*) identical (*d*) equally spaced. Draw a diagram to show plane waves falling normally on a grating and being diffracted. Using the diagram explain in which directions diffracted maxima will be seen.
A parallel beam of monochromatic light of 6000 Å is incident normally on one face of a 10° glass prism the other face of which has a diffraction grating scratched on it, the lines being parallel to the refracting edge of the prism. It is found that one of the first order diffracted beams is parallel to the incident light. Explain this and deduce the number of lines per cm on the grating. Also find the position of the other first order diffracted beam, illustrating your answer with a diagram. Assume the refractive index of glass to be 1·52. (*N.*)

14. What is meant, in optics, by (*a*) interference, (*b*) diffraction?
Parallel monochromatic light of wavelength $5·9 \times 10^{-7}$ m from a collimator with a vertical slit is incident normally on a plane transmission diffraction grating with 200 vertical rulings per cm and a pattern is focused on a screen by means of a lens of 1 metre focal length. Describe the parts played by diffraction

and interference in the formation of this pattern and calculate from first principles the linear separation of the bright lines at the centre of the pattern. (*N.*)

15. Explain what is meant by *diffraction* and describe a diffraction grating. Describe also how the wavelength of sodium light may be measured using a diffraction grating and a spectrometer.

In such an experiment the grating has 750 lines per millimetre and forms a first order diffracted beam at an angle of $26° 15'$ with the direction of the incident beam which is normal to the grating. Calculate the wavelength of the sodium light. Prove any formula you use.

How does the spectrum of white light produced by a diffraction grating differ from that produced by a glass prism? (*L.*)

16. Describe the arrangement you would use to demonstrate experimentally, by projection on to a screen, the spectra formed by a diffraction grating on which white light is incident normally. Mention the chief features that would be observed, and explain briefly how these depend on the spacing of the rulings on the grating.

For a certain grating illuminated normally with light of wavelength 600 nm $(6 \times 10^{-7} \text{ m})$, first-order bright lines are observed at $17·5°$ to the normal on each side of the normal. Calculate (*a*) the separation of the rulings on the grating, (*b*) the directions in which the first-order bright lines will be found if the light were *incident on the grating* at an angle of $17·5°$ to the normal. (*O.*)

17. (i) State two advantages of a *concave reflection grating* over a plane diffraction grating. (ii) Draw a diagram of a concave reflection grating in use to measure wavelengths and explain any special features of the arrangement you draw. (iii) How has the grating been used to measure ultra-violet wavelengths?

Interference

18. The diagram shows a vertical section of an arrangement used to demonstrate Young's interference experiment. S_1 and S_2 are opaque screens containing one and two long narrow horizontal slits respectively. A source of monochromatic light is placed to the left of S_1. Describe and explain the changes in the fringes observed at the screen S_3 brought in turn by:

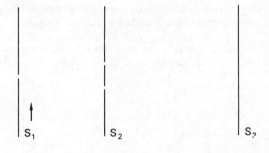

FIG. 30A.

(*a*) displacing S_1 vertically upwards in the direction shown by the arrow; (*b*) increasing the width of the slit in S_1; (*c*) making the width of one slit in S_2 twice that of the other; (*d*) covering one of the slits in S_2; (*e*) placing a very thin piece of glass over, and to the right of one of the slits in S_2. (*O. and C.*)

19. Discuss critically the conditions which must be fulfilled in order to demonstrate interference in (*a*) light, and (*b*) sound.

A radio astronomy aerial at the Equator is mounted on a high cliff, overlooking the sea, facing due East where an intense radio star rises. If the aerial is connected to a receiver which measures total power at one particular radio wavelength (of about one metre), describe as fully as you can how you expect the power to vary as the radio star rises. You may assume that the sea acts as a perfect plane reflector. (*C.*)

20. In order that interference fringes may be obtained from two sets of light waves, the two sources must be *coherent*. Explain the meaning of *coherent*. Explain why the two interfering beams are coherent in interference experiments in which they arise from (*a*) two slits (e.g. Young's experiment), (*b*) an extended source (e.g. Newton's rings).

A parallel beam of white light passes normally through an air film contained between parallel glass plates, and is then examined in a spectroscope. The visible spectrum is found to contain three dark bands. If the thickness of the air film is $1{\cdot}2 \times 10^{-4}$ cm, calculate the wavelengths at the centres of the dark bands. The wavelength limits of the visible spectrum may be taken as $4{\cdot}0 \times 10^{-7}$ m and $7{\cdot}0 \times 10^{-7}$ m. (*O. and C.*)

21. Describe the "Young's slits" experiment for demonstrating the interference of monochromatic light. Specify the measurements which must be made in this experiment if it is desired to deduce the wavelength of the light, and show how the wavelength is calculated from them.

After performing the experiment and noting that the width of the primary slit is an important factor, two pupils A and B comment as follows:

A: This slit should be as narrow as possible so that the secondary slits are illuminated by the central diffraction maximum from it. If it is too wide there will be considerable brightness variations in the interference pattern.

B: My reason for demanding a narrow primary slit is based upon the fact that a wide slit is equivalent to a number of narrow ones. Each constituent narrow slit produces its own sharp set of interference fringes; but, as these sets are displaced relative to one another on the screen, the overall pattern produced by a wide primary slit would obviously be blurred and show undesirable brightness variations.

Discuss briefly the comments made by A and B; and state, giving your reason, which of the two arguments you find the more convincing. (*O.*)

22. ABCD is part of a rectangular glass block of refractive index $1{\cdot}50$ with face BC horizontal. S is an opaque screen in contact with face AB, having thin

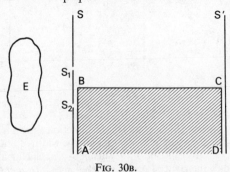

Fig. 30B.

parallel slits close together at S_1 and S_2. The slits are horizontal and equidistant from B. S' is a translucent screen in contact with face CD. E is an extended source of monochromatic light. (*Note.* This is not Young's slits apparatus.)

Explain the following observations: (*a*) When there is no glass block between S and S' no fringes are visible on S'. (*b*) If the glass block is in position as shown between A and S' and slit S_1 open and S_2 covered up, fringes are observed above C. (*c*) If slit S_2 is now uncovered fringes will be observed also below C. (*d*) The fringes below C persist if slit S_1 is closed. (*e*) The fringe width in the system below C is 0·67 times that in the system above C.

If the block is raised slightly so that B approaches S_1, what changes would you expect in the fringe systems above and below C? (*O.*)

23. Describe a method of measuring the wavelength of monochromatic light. Two partially silvered, parallel glass plates separated by a small thickness of air are used to produce interference effects. A parallel beam of white light is passed normally through the plates and then focused on the slit of a spectrometer. Describe and explain what is seen in the spectrometer.

Explain the effect of gradually increasing the distance between the plates from zero, assuming that they remain parallel. (*C.*)

chapter thirty-one

Electromagnetic Waves

Hertzian Transmitter

In 1887 Hertz demonstrated the existence of waves in space generated by a *spark gap*. They were called *electromagnetic waves* because they consist of travelling electric and magnetic fields.

As we have seen in previous chapters, waves are produced by oscillators. When electrons in metals are made to oscillate at high frequencies of the order of a million hertz or more, electromagnetic waves are radiated from them. Fig. 31.1 (i) shows a circuit in which this occurs. A 5 kV

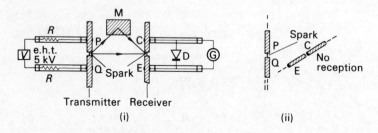

FIG. 31.1. Hertzian dipole transmitter of radio waves.

supply V is connected across two highly-polished brass rods P, Q with a very small gap between them. This arrangement is known as a *Hertzian dipole*. A similar dipole arrangement, C, E, is placed near and parallel to P, Q. A Scalamp (mirror) galvanometer G and a high-frequency diode D, which rectifies alternating current, is connected between the rods C, E.

When the high-tension supply to P and Q is switched on, sparks occur in the gap between P, Q. At the same time a small deflection is observed in the galvanometer G. We conclude that some energy is radiated into space from P, Q and received in the circuit C, E. The spark actually consists of bursts of current in the air gap which produce high-frequency oscillations of current in the rods P, Q.

Using this apparatus, it can be shown that (i) the intensity of the waves decreases with distance from the source P, Q, (ii) the waves can be reflected and can interfere with each other—a metal plate M, held near P and Q as shown in Fig. 31.1 (i), produces an increased or decreased deflection in G, showing that the waves are reflected to C, E and interfere with the waves reaching C, E directly, (iii) the radiation is polarised—when C, E are turned into the horizontal plane, the deflection in G reduces to a low value, Fig. 31.1 (ii).

Radio Waves

Since polarisation occurs, it follows that *electromagnetic waves are transverse waves*. Light waves are shown to be transverse waves by using Polaroids (p. 772). The circuit in Fig. 31.1 (i) can produce electromagnetic waves of about 30 cm wavelength. Commercial transmitters may produce radio waves of wavelength say 300 metres. Maximum reception is obtained in a small transistor receiver when the line of the aerial inside is normal to a line joining the distant transmitter to the aerial. Further, when the aerial is made to rotate in a plane normal to this line, the strength of the reception diminishes, thus showing polarisation. Radio waves are hence similar in nature to Hertzian waves—they are both electromagnetic waves.

Microwaves

Experiments with 3 cm electromagnetic waves, or *microwaves*, can be carried out in the laboratory to show the properties of electromagnetic waves. A transmitter T contains a special valve and circuit for generating the microwaves, and the receiver or detector R contains a diode for rectifying the waves and a microammeter A for showing the resulting small current which flows, Fig. 31.2 (i). A probe P or linear detector can be connected directly to the meter in R for more accurate detection. Further, by using an audio-frequency modulator on the transmitter, and an amplifier and loudspeaker connected to R, the detected waves can produce audible signals.

By using a metal plate as a reflector, it can be shown that the angle of incidence = the angle of reflection for microwaves. A paraffin-wax 60° prism demonstrates refraction of microwaves, and total internal reflection is obtained with a 90°, 45°, 45° prism of the same material. A wax lens can produce a parallel beam of microwaves. Fig. 31.2 (ii) illustrates experiments which show also that microwaves have properties of interference, diffraction and polarisation. Thus electromagnetic waves have the same properties as light waves.

Speed of Electric Pulse in Coaxial Cable

Before considering the transmission and speed of electromagnetic waves in space, it is useful to investigate the speed of an *electrical pulse* along a coaxial cable due to a direct voltage applied for a short time.

Fig. 31.3 shows a circuit consisting of a coil of coaxial cable A, B about 200 metres long, with a "pulse generator" V connected across A and B. V generates short pulses at the rate of 200 kHz or 2×10^5 Hz. A resistor S, about 75 Ω, is joined in parallel with A, B as shown (this ensures that the cable is terminated correctly at the pulse generator in order to reduce interference due to multiple reflections). A resistance R, about 75 Ω, can be connected across C, D at the other end of the cable. An oscilloscope, with its timebase set to 1 μs/cm, is connected between A and B to monitor the voltage here.

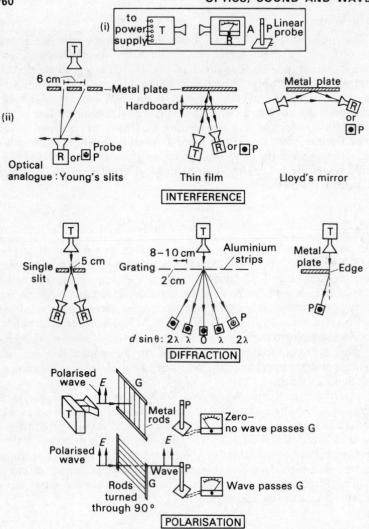

FIG. 31.2. Microwave demonstration of interference, diffraction, polarisation.

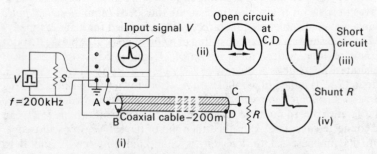

FIG. 31.3. Pulse along coaxial cable transmission line.

With the cable disconnected from the oscilloscope, the input pulse trace is seen on the screen, Fig. 31.3 (i). When the cable is now connected to V and the other end of the cable left on open circuit, a smaller second pulse is obtained after the first or input pulse, Fig. 31.3 (ii). If the ends C and D are short-circuited, the second pulse appears inverted. Fig. 31.3 (iii). If C and D are shunted by the resistance R, the second pulse is considerably diminished, Fig. 33.3 (iv).

Determination of Speed

These experiments show that the pulse applied at A, B has travelled along the cable to the other end C, D, where it is reflected. From the appearance of the screen traces, it follows that (a) when C and D are on open circuit the phase of the reflected pulse is unchanged at C, D, (b) when C and D are short-circuited the phase of the reflected pulse at C, D changes by 180°, (c) with a suitable resistance R at C, D, the pulse arriving at C, D is practically absorbed here.

The speed of the pulse along the cable can be estimated by measuring the separation on the screen of the input and returning pulses in Fig. 31.3 (ii). If this is 2·0 cm, then, with a time-base of 1 μs/cm, this represents a time of 2·0 μs or $2 \cdot 0 \times 10^{-6}$ s. Now the distance travelled along the cable and back = 2 × 200 m = 400 m.

$$\therefore \text{ speed of pulse } = \frac{400 \text{ m}}{2 \cdot 0 \times 10^{-6}} = 2 \times 10^8 \text{ m s}^{-1}$$

Thus the pulses along wires travel at speeds less than, but comparable to, the speed of light, which is about $3 \cdot 0 \times 10^8$ m s^{-1}.

Speed of Pulse Along Inductors and Capacitors

The speed of a pulse along a cable depends on the *inductance per unit length*, and on its *capacitance per unit length*. The influence of these factors on the speed can be shown with the apparatus in Fig. 31.4, where a large number of inductors and capacitors are used.

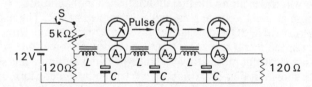

FIG. 31.4. Transmission line—inductors (L) and capacitors (C).

Many air-cored inductors L and electrolytic capacitors C of 1000 μF are arranged in a row as shown. Electrically the arrangement represents a "cable"; here the inductance and capacitance are "lumped" at places, whereas in a cable they are spread along uniformly. The "cable" is terminated with a 120 Ω resistor which eliminates reflection effects.

The 5 kΩ potentiometer is varied until, on closing the switch S to start the pulse from the 12 V battery, a suitable voltage such as 0·4 V appears across the first capacitor C as shown by a voltmeter. A number of 100–0–100 μA meters, A_1, A_2, ..., are placed in series with each inductor to follow the pulse, as shown in Fig. 31.4. When S is now closed, the pointers are deflected in turn as the pulse travels from one end to the other. When S is opened the returning pulse can be seen, as the deflections of the meters reverse.

When the experiment is repeated with a soft-iron core inside each inductor, the pulse is now observed to travel much more slowly. Thus increasing L decreases the speed. Experiment also shows that the speed is decreased when C, the capacitance, is increased. This explains why the speed of the pulse in a coaxial cable is high; both L and C are very small in this case.

Velocity of Pulse Along Transmission Line

We can now derive a general formula for the speed of a pulse in a coaxial cable or transmission line. Consider two long parallel wires PQ and RS, Fig. 31.5 (i). These two wires together form a transmission line.

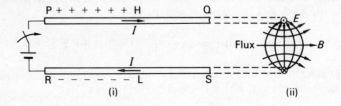

Fig. 31.5. Speed of pulse along transmission line.

We shall assume that the wires have an inductance L *per unit length*, a capacitance C *per unit length*, and a negligible resistance. If there are no losses of energy in the dielectric or the conductors, a voltage applied at one end will travel along the line undiminished in magnitude.

Suppose a direct voltage V is suddenly applied at PR. The displaced charge, positive in PQ and negative in RS, spreads along the wires with a constant speed v and reaches a position HL say at some time t. Simultaneously, a current pulse I travels along the wires with the same speed v. Thus, as shown in Fig. 31.5 (ii), both electric and magnetic flux, or fields E and B respectively, move along the transmission line in the surrounding medium.

The magnetic flux Φ *per unit length*, which is due to the current, is moving with velocity v. In a time dt, the increase or change of flux = $\Phi \times v.\mathrm{d}t$, since the length travelled = $v.\mathrm{d}t$. Hence the rate of change of flux = Φv = induced voltage. But in the absence of resistance, the induced voltage just balances V, the applied voltage.

$$\therefore \ V = \Phi v$$

Now $\Phi = LI$, since L may be defined as "the flux change per unit current change" for the unit length of line.

$$\therefore \quad V = LIv \qquad . \qquad . \qquad . \qquad . \qquad (1)$$

The capacitance of a length l of line is Cl. Hence

$$I = \mathrm{d}Q/\mathrm{d}t = \mathrm{d}(ClV)/\mathrm{d}t = CV . \mathrm{d}l/\mathrm{d}t.$$

But $\mathrm{d}l/\mathrm{d}t = v$.

$$\therefore \quad I = CVv \qquad . \qquad . \qquad . \qquad . \qquad (2)$$

From (1), $V/I = Lv$. From (2), $V/I = 1/Cv$.

$$\therefore Lv = \frac{1}{Cv}$$

$$\therefore \quad v = \frac{1}{\sqrt{LC}} \qquad . \qquad . \qquad . \qquad . \qquad (3)$$

The ratio V/I is called the *characteristic impedance*, Z_0, of the line. Hence, when there are no losses,

$$Z_0 = Lv = \sqrt{\frac{L}{C}} \qquad . \qquad . \qquad . \qquad . \qquad (4)$$

It can be shown that the characteristic impedance of a transmission line with no energy losses is purely resistive, that is, there is no reactive component. Thus if the line is terminated by a resistance of this value, all the energy sent down the line is absorbed in this resistance. The line is then said to be *matched*. No reflection of energy now occurs (see Fig. 31.3 (iv)).

From (1) and (2), note also that

$$v = \frac{I}{CV} = \frac{V}{LI}$$

$$\therefore LI^2 = CV^2$$

Thus $\frac{1}{2}LI^2 = \frac{1}{2}CV^2$. Now $\frac{1}{2}LI^2$ is the magnetic energy per unit length of the line and $\frac{1}{2}CV^2$ is the electric energy per unit length. Thus half the energy is stored in the magnetic field as the pulse travels, and the other half in the electric field. As seen later, this is also the case when electromagnetic waves travel in free space or in a medium.

Dimensions

The velocity formula $v = 1/\sqrt{LC}$ can be checked by dimensional analysis. We may use four primary dimensions: mass M, length L, time T, charge Q. From the energy formula $W = QV$, we have $V = W/Q$. Hence $C = Q/V = Q^2/W$. Now the dimensions of W (force × distance) $= \mathrm{ML^2T^{-2}}$.

$$\therefore \text{ dimensions of } C \text{ per unit length } (C_1) = \frac{Q^2 \mathrm{M^{-1}L^{-2}T^2}}{\mathrm{L}}$$

$$= \mathrm{Q^2 M^{-1} L^{-3} T^2} \qquad . \qquad . \qquad . \qquad . \qquad (1)$$

From the energy relation $W = \frac{1}{2}LI^2 = \frac{1}{2}L(Q/t)^2$, we have

$$\text{dimensions of } L \text{ per unit length } (L_1) = \frac{ML^2T^{-2} \times T^2}{Q^2 \times L}$$

$$= MLQ^{-2} \quad . \quad . \quad . \quad . \quad (2)$$

From (1) and (2),

$$\text{dimensions of } \frac{1}{\sqrt{L_1 C_1}} = \frac{1}{\sqrt{L^{-2}T^2}} = LT^{-1}$$

$$= \text{dimensions of velocity}$$

Velocity of Pulse Along Parallel Plates

The formula $v = 1/\sqrt{LC}$ can be used to obtain the velocity of a pulse along two parallel rectangular plates in free space. As for the case of the transmission line, when a pulse voltage is applied charge spreads along the plates, and electric flux and magnetic flux travel in the medium.

The capacitance of a parallel-plate capacitor is given by $C' = \varepsilon_0 A/d$, where ε_0 is the permittivity of a free space, A is the common area and d is the separation of the plates. If l is the length of the plates, the capacitance per unit length, C, is given by

$$C = \frac{C'}{l} = \frac{\varepsilon_0 A}{ld} = \frac{\varepsilon_0 bl}{ld} = \frac{\varepsilon_0 b}{d}, \quad . \quad . \quad . \quad (1)$$

where b is the breadth of the plates, Fig. 31.6 (i).

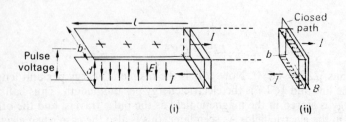

FIG. 31.6. Speed of pulse along rectangular plates.

Consider a section of unit length of the plates, Fig. 31.6 (ii). We can imagine the top and bottom plates each to consist of a very large number of *close parallel wires* as shown, which carry currents totalling I, the current travelling along the plates. Thus the two plates are now effectively the top and bottom of short rectangular current-carrying coils. From Ampere's circuital law for flux density B in the magnetic field round the currents, $\oint B.dl = \mu_0 I$, where the integral or sum is taken over a closed path. Choosing a path which goes through the middle of the plates or coil, where the field is B, and round the outside, where the field is practically zero, then

$$B.b = \mu_0 I$$

$$\therefore \quad B = \frac{\mu_0 I}{b}$$

$\therefore$ flux per unit length, $\Phi = B \times \text{area} = B \times (d \times 1) = \dfrac{\mu_0 I d}{b}$

$\therefore$ L per unit length $= \dfrac{\Phi}{I} = \mu_0 \dfrac{d}{b}$ . . . . . (2)

From (1) and (2), it follows that

$$v = \frac{1}{\sqrt{LC}} = \frac{1}{\sqrt{\mu_0 \varepsilon_0}}$$

This is the speed of the pulse along the plates. The same result is obtained for the case of a pulse travelling along a coaxial cable (p. 763). With an air core cable, it can be shown that $L = \mu_0/2\pi \ln(b/a)$ and $C = 2\pi\varepsilon_0/\ln(b/a)$, where b and a are the respective outer and inner radii of the cable. Thus again we find that

$$v = \frac{1}{\sqrt{LC}} = \frac{1}{\sqrt{\mu_0 \varepsilon_0}}.$$

The formula for the velocity shows that *the speed depends only on the properties μ_0 and ε_0 of the medium between the conductors.* This suggests that the plates or the coaxial cable act as "guides" for the travelling pulse, and that the moving electric and magnetic fields carry the energy in the medium between them.

Moving Fields. Electromagnetic Wave

We can now discuss some important points in connection with moving fields.

(1) *Moving B-field.* Consider a straight metal conductor XY, moving perpendicular to a horizontal magnetic field of flux density B in a downward direction with a constant velocity v, Fig. 31.7 (i). The free

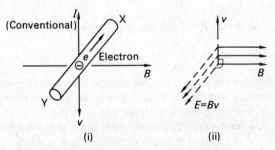

FIG. 31.7. Moving B-field produces E-field.

electrons moving down with velocity v are equivalent to an upward current I in the conventional direction, as shown. Thus applying Fleming's left hand rule for conventional current, the force on the electrons is towards X.

The force on an electron of charge e is given by $F = Bev$. This force would also be produced by an electric field E such that $F = Ee = Bev$. In this case, therefore,

$$E = Bv \quad . \qquad . \qquad . \qquad . \quad (1)$$

E is in the direction XY as, conventionally, E is the direction which a positive charge would tend to move.

The downward movement of XY through the stationary field B is equivalent to a field B moving upward with a velocity v past the stationary conductor. From (1), it follows that *a magnetic field moving with a velocity v is equivalent to a moving electric field E given by $E = Bv$.* It should be noted that B, E and v are mutually perpendicular, as shown in Fig. 31.7 (ii).

(2) *Moving E-field.* Consider a positive charge q moving with a velocity v in a constant direction PR, Fig. 31.8. This is equivalent to a

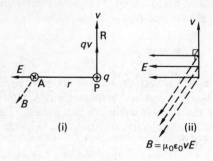

Fig. 31.8. Moving *E*-field produces *B*-field.

current I in a conductor of length l, where $I.l = qv$. Now the magnetic field B at a point A distance r from a current I in a direction normal to I is given by the Biot–Savart law:

$$B = \frac{\mu_0}{4\pi} \cdot \frac{I.l \sin 90°}{r^2} = \frac{\mu_0 qv}{4\pi r^2}$$

But the electric intensity at the same point is given by $E = q/4\pi\varepsilon_0 r^2$. Substituting,

$$\therefore \quad B = \mu_0 \varepsilon_0 vE \quad . \qquad . \qquad . \qquad . \quad (2)$$

Thus *a moving electric field E is equivalent to a moving magnetic field B whose magnitude is given by* (2). B, E and v are mutually perpendicular, as for the case of a moving magnetic field, Fig. 31.8 (ii).

(3) *Velocity of moving fields.* Consider the motion of a magnetic field B in free space. From (1) above, this is equivalent to a moving electric field E given by $E = Bv$. But, from (2) above, a moving electric field E is

equivalent to a moving magnetic field B' given by $B' = \mu_0\varepsilon_0 vE$. Hence, substituting for E from $E = Bv$,

$$B' = \mu_0\varepsilon_0 v^2 B \qquad . \qquad . \qquad . \qquad . \qquad (3)$$

If $B' = B$, then the system can sustain itself. From (3), this will only occur if $\mu_0\varepsilon_0 v^2 = 1$. Hence the speed v of the moving field is given by

$$v = \frac{1}{\sqrt{\mu_0\varepsilon_0}} \qquad . \qquad . \qquad . \qquad . \qquad (4)$$

This is the only speed for which our equations are consistent. The same result for v is obtained for a moving electric field.

We now see that electric and magnetic fields can only travel in free space with the speed $1/\sqrt{\mu_0\varepsilon_0}$. The moving B- and E-fields together form *an electromagnetic wave.* Generally, the speed is given by $v = 1/\sqrt{\mu\varepsilon}$ for a medium of permeability μ and permittivity ε. For a vacuum, $\mu_0 = 4\pi \times 10^{-7}$ H m^{-1} and $\varepsilon_0 = 8\cdot854 \times 10^{-12}$ F m^{-1}. On substituting these values, we find that, to a good approximation,

$$v = 3 \times 10^8 \text{ m s}^{-1}.$$

This is the same value as found experimentally for the velocity of light in free space, denoted by the symbol c. We therefore infer that *light is an electromagnetic wave,* and this fact is utilised later.

A.C. Voltage and Transmission Line

So far we have considered a pulse or direct voltage applied to a transmission line, and the resulting travelling pulse of electric and magnetic fields. If an alternating voltage, however, is applied to an infinitely-long

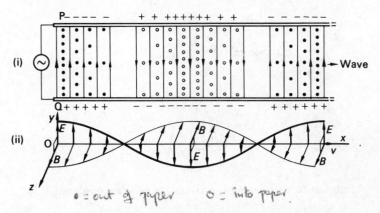

FIG. 31.9. B- and E-fields in electromagnetic wave.

transmission line, the two travelling fields both alternate with time at any given point along the line. This is the characteristic of a *wave*, as explained on p. 587. For a sinusoidal plane-progressive wave travelling in the x-direction, the instantaneous values of E and B, the electric and magnetic field values, are given by

$$E = E_0 \sin 2\pi \, (ct - x)/\lambda$$

and

$$B = B_0 \sin 2\pi \, (ct - x)/\lambda.$$

E_0 and B_0 are the respective peak values or amplitudes, c is the speed of the wave in free space, and λ is the wavelength.

Fig. 31.9 (i) illustrates the electric and magnetic field lines at some instant between two infinitely-long parallel rectangular plates P, Q. The electric field lines are directed from one conductor to the other. The magnetic field lines, perpendicular to the electric field lines, are represented by small dots if they point towards the reader, and by small circles if they point away from the reader into the paper, as shown in Fig. 31.9 (ii). The densities of the electric and magnetic lines indicate the magnitudes of E and B respectively at the various points along the line. Note that E and B reach their maximum and minimum values simultaneously, that is, E and B are in phase in the plane progressive waves.

Standing (Stationary) Waves and Aerials

The moving voltage, charge and current also travel as plane-progressive waves along the infinitely-long transmission line. When the line has a finite length, however, reflection occurs at the ends. A *stationary* or *standing* wave of voltage and current can then be set up along the line, as for the case of reflection of sound waves discussed on p. 641.

Aerials are transmission lines of finite length. Consider the case, for example, of a vertical metal rod such as the half-wave dipole (p. 758). In Fig. 31.10, two equal rods A and B have a high-frequency generator G

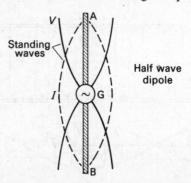

FIG. 31.10. Standing waves of V and I in half-wave dipole aerial.

between them, which drives electrons along the metal. When the frequency is adjusted (see later), standing waves of current I and voltage V are set up, as shown. The ends A and B, where the electrons cannot move are *current nodes*. They are *voltage antinodes* because the high density electrons here exert maximum "pressure" or voltage. A current antinode appears in the middle at G, since the electrons are most free to move here. This is the position of a voltage node.

The half-wave dipole aerial corresponds to the case in Sound of a stretched string fixed at both ends, or to a resonance tube closed at both ends and with a source in the middle.

Propagation of Electromagnetic Waves from Aerials

The aerial has the same function in a radio transmitter as the sounding board in a violin or the resonance tube with a tuning fork. It helps the radio-frequency oscillator G to radiate its energy into space. When the length of the Hertz aerial is adjusted to be $\lambda/2$, where $\lambda = c/f = 3 \times 10^8/f$ and f is the frequency of G, standing waves of voltage and current are set up. The radio-frequency energy of the oscillator is then radiated most strongly into space.

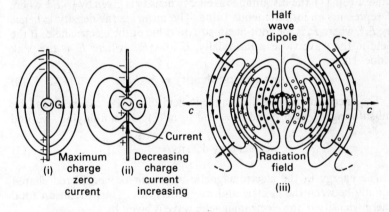

FIG. 31.11. Radiation from dipole aerial.

Fig. 31.11 (i) and (ii) show the electric field lines close to the moving charges of the dipole aerial. They pass through points whose distance r from the aerial is much less than the wavelength λ. The electric field E and the magnetic field B which are near to the aerial are out of phase. The magnitudes of B and E each decrease as $1/r^2$. These fields form the *induction field*, which is relatively unimportant in propagation.

Further away from the aerial, where the distance r is very much greater than the wavelength, the electric and magnetic flux have detached themselves from the dipole and formed complete loops in space, Fig. 31.11 (iii). In this case B and E are in phase with each other, and their magnitude varies as $1/r$. This is called the *radiation field* of the aerial; it predominates over the induction field at large distances. As explained previously, B and E now constitute an electromagnetic wave travelling in space with a speed c. In Fig. 31.11 (iii), the density of the electric lines and that of the magnetic lines, represented by dots and small circles, follow a variation similar to the field between a transmission line shown in Fig. 31.9 (i).

Energy in Field and Energy Density

The energy stored in a capacitor C charged to a p.d. V is $\frac{1}{2}CV^2$, with which we assume the reader is familiar. This is the energy stored entirely in the electric field between the plates of the capacitor (see p. 578). For convenience, consider a parallel plate capacitor of area A and plate separation d, with a vacuum between the plates.

Then, with the usual notation,

$$C = \varepsilon_0 A / d \quad \text{and} \quad V = E.d.$$

Substituting, then

$$\text{energy per unit volume in field} = \frac{\frac{1}{2}CV^2}{A.d} = \frac{1}{2}\varepsilon_0 E^2 \quad . \quad (1)$$

The energy per unit volume is called the *energy density* in the field.

For a varying field such as in an electromagnetic wave, E varies with time. From (1), the instantaneous energy density is given by $\frac{1}{2}\varepsilon_0 E^2$, where E represents an instantaneous value. The mean energy density is hence $\frac{1}{2}\varepsilon_0 E_r^2$, where E_r is the root-mean-square value of the electric field. If the field in the wave varies sinusoidally, $E_r^2 = \frac{1}{2}E_0^2$, where E_0 is the peak value. Hence

$$\text{mean energy density} = \frac{1}{4}\varepsilon_0 E_0^2 \quad . \qquad . \qquad (2)$$

The energy density in the magnetic field of a wave can be found by substituting $E_0 = B_0 c$ and $\varepsilon_0 = 1/c^2\mu_0$ in (2). On simplifying, we find

$$\text{mean energy density} = \frac{1}{4}\frac{B_0^2}{\mu_0} \quad . \qquad . \qquad (3)$$

The energy in the electromagnetic wave can be considered shared equally between the electric and magnetic fields. Thus the mean total energy density in the electromagnetic wave is given by

$$\text{mean total energy density} = \frac{1}{2}\varepsilon_0 E_0^2 = \frac{1}{2}\frac{B_0^2}{\mu_0} \quad . \qquad . \qquad (4)$$

Intensity of Electromagnetic Wave

The *intensity*, I, of a wave is defined as the mean energy per second flowing across unit area normal to the direction of the wave. See *Sound*, p. 607. The speed of the electromagnetic wave in a vacuum is c. Hence in one second, the wave occupies a volume $= c \times 1 = c$. Thus, from our previous result in (4),

$$\text{intensity } I = \frac{1}{2}\varepsilon_0 E_0^2 c \quad . \qquad . \qquad . \qquad (5)$$

If a small dipole radiates a power P equally in all directions, the mean intensity at a distance $r = P/(\text{area of sphere of radius } r) = P/4\pi r^2$. When r is large, we can imagine that a small area of the sphere is substantially plane. In this case we can equate I in (5) to $P/4\pi r^2$.

$$\therefore \frac{1}{2}\varepsilon_0 E_0^2 c = \frac{P}{4\pi r^2}$$

Using the relation $\varepsilon_0 = 1/c^2\mu_0$ from the expression for the velocity c (p. 767), where $\mu_0 = 4\pi \times 10^{-7}$ H m^{-1} and $c = 3 \times 10^8$ m s^{-1}, we obtain

$$E_0 = \frac{(60P)^{\frac{1}{2}}}{r} \qquad \qquad (6)$$

In this expression E_0 is in volt metre^{-1}, P is in watt and r is in metre.

Example. Assuming that all the energy from a small 100 W lamp is radiated as electromagnetic waves, find (i) the intensity 1 metre away, (ii) the electromagnetic wave energy density, (iii) the magnitudes of E_0 and B_0 in the wave. (Assume $\varepsilon_0 = 8\cdot85 \times 10^{-12}$ F m^{-1}, $c = 3 \times 10^8$ m s^{-1}.)

(i) 1 metre away, intensity $I = \dfrac{100\text{W}}{4\pi \times 1^2 \, \text{m}^2} = 8\cdot0$ W m^{-2} (approx.)

(ii) I = energy per unit volume $\times c$ = energy density $\times c$

$$\therefore \text{ energy density} = \frac{I}{c} = \frac{8\cdot0}{3 \times 10^8} = 2\cdot7 \times 10^{-8} \text{ J m}^{-3}$$

(iii) Energy density $= \frac{1}{2}\varepsilon_0 E_0^2$

$$\therefore E_0 = \sqrt{\frac{2 \times 2\cdot7 \times 10^{-8}}{8\cdot85 \times 10^{-12}}} = 78 \text{ V m}^{-1} \text{ (approx.)}$$

$$\therefore B_0 = \frac{E_0}{c} = \frac{78}{3 \times 10^8} = 2\cdot6 \times 10^{-7}\text{T}$$

Polarisation of Light

The polarisation of light, described in Chapter 29, is due to the transverse nature of electromagnetic waves. In the discussion here, we assume the direction of light vibrations to be the direction of the electric vector E in the wave. The "plane of polarisation" is usually defined as the plane containing the light ray and E.

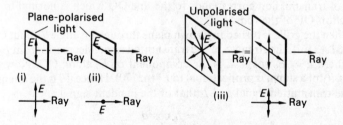

FIG. 31.12. Electric vectors in polarised and unpolarised light.

Fig. 31.12 (i) shows *plane-polarised* light due to vertical vibrations or a vertical vector E. Fig. 31.12 (ii) shows plane-polarised light due to horizontal vibrations; in this case a dot is used to indicate the vibrations or vector E perpendicular to the paper. Fig. 31.12 (iii) shows *ordinary* or *unpolarised* light.

The associated field vectors E in ordinary light act in all directions in a plane perpendicular to the ray and vary in phase. Since each vector can be resolved into components in perpendicular directions, the total effect is equivalent to two perpendicular vectors equal in magnitude. Thus ordinary or unpolarised light can be represented in a diagram by an arrow and a dot, as shown in Fig. 31.12 (iii).

Intensity of Transmitted Light

A microwave (3 cm) transmitter produces polarised electromagnetic waves. This can be seen by placing a receiver with meter directly in front of the transmitter, so that an appreciable deflection is produced on the meter. When the transmitter is rotated in its own plane, at one stage the meter deflection diminishes to zero.

As shown in Fig. 31.2, p. 760, a grille absorbs microwaves when the metal rods are parallel to the electric vector. The incident field E_0 produces forced oscillations of the electrons along the rods, so that the energy is absorbed, Fig. 31.13 (i). When the grille is turned through 90°,

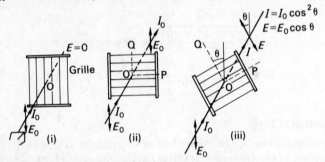

FIG. 31.13. Transmission of electric field through grill—$I = I_0 \cos^2\theta$.

hardly any forced oscillations occur and the strength of the transmitted signal is then practically undiminished, Fig. 31.13 (ii). The "easy" direction of transmission corresponds to the line OQ which is normal to the direction OP of the rods.

When the grille is turned in its own plane through an angle θ about O in Fig. 31.13 (iii), the strength of the transmitted signal decreases from E_0 to E, where $E = E_0 \cos \theta$. E is the component of E_0 along OQ. Now the *intensity* of a signal is proportional to E^2 (p. 770). Hence if I is the intensity of the transmitted signal and I_0 that of the incident signal,

$$I = I_0 \cos^2\theta$$

When $\theta = 45°$, $\cos \theta = 1/\sqrt{2}$. Thus the intensity transmitted is reduced to one-half of its original value by the grille.

Polaroid Transmission of Light

The $\cos^2\theta$ law for the transmitted intensity was discovered by Malus in 1810 using polarising devices with light. A Polaroid A produces polarised light whose electric vector vibrates in a particular direction, say AY in Fig. 31.14 (i). If another Polaroid B is placed in front of A so that its "easy" direction of transmission BX is perpendicular to AY, the overlapping areas appear black because no light is transmitted.

If B is turned so that BX makes an angle θ with AY, Fig. 31.14 (ii), the intensity I of the transmitted light is related to that of the incident light by $I = I_0 \cos^2\theta$, as for the case of 3 cm electromagnetic waves passing through a grille. Here I_0 is the intensity of the polarised light passing through A. The intensity of the unpolarised light *incident* on A is $2I_0$, since, from Fig. 31.12, one of the perpendicular vectors is absorbed by A.

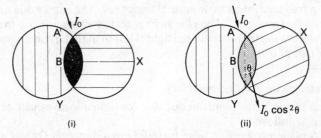

FIG. 31.14. Transmission of light through Polaroid—$I = I_0 \cos^2\theta$.

Scattering of Light

The scattering of light, discussed on p. 746, can be explained from the fact that light is a transverse electromagnetic wave. Suppose ordinary or unpolarised light is incident on a molecule of air at O, Fig. 31.15. The

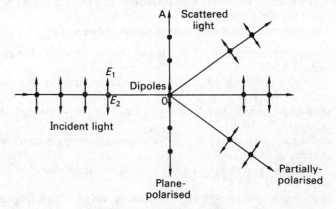

FIG. 31.15. Scattering of light.

electric vector in the wave displaces the electrons inside the molecule and induces a dipole. This dipole oscillates in the direction of the electric vector.

Ordinary or unpolarised light can be represented by two perpendicular component vectors, as explained on p. 771. In Fig. 31.15, the two vectors concerned, E_1 and E_2, are represented by an arrow and a dot (see Fig.

31.12). The axes of the induced dipoles at O are parallel to E_1 and E_2, and hence lie in and perpendicular to the paper. There is no radiation along OA due to E_1, since a dipole does not radiate along its axis. The only radiation in this direction is due to E_2. Hence the light scattered along OA is *plane-polarised*. The two oscillating dipoles both produce radiation in other directions so that the scattered light here is only partially polarised.

The blue of the sky can also be explained by considering the oscillation of electrons in molecules. The natural frequency of oscillation of the electron is of the order 10^{15} Hz, which is the ultra-violet range of wavelengths. When sunlight is incident on molecules in the atmosphere, the blue light in it produces appreciable forced oscillations of the electrons lower in frequency than their natural frequency. The energy due to these oscillations is scattered. The blue light is scattered much more than the red light since the frequency of red light is less than that of blue light from $c = f\lambda$ (see p. 690).

Electromagnetic Waves—Summary

Some of the main points about electromagnetic waves can be summarised as follows:

(1) A plane-progressive wave travelling in a direction Ox has a magnetic field B and a perpendicular field E both perpendicular to Ox.

(2) If c is the speed in a vacuum, then $E = Bc$ and $B = \mu_0\varepsilon_0cE$.

(3) $c = 1/\sqrt{\mu_0\varepsilon_0}$ for a vacuum. In other media, speed $v = 1/\sqrt{\mu\varepsilon}$.

(4) The fields B and E in the wave are in phase. In the Ox direction, $B = B_0 \sin 2\pi (ct - x)/\lambda$ and $E = E_0 \sin 2\pi (ct-x)/\lambda$ for a sinusoidal wave.

(5) The wave is transverse. The vibrations of light are drawn in the direction of the electric vector E.

(6) The electromagnetic spectrum of waves consists of:

γ-rays, wavelength 10^{-11} m; origin—nuclear energy changes of order 1 MeV.

X-rays, wavelength 10^{-10} m; origin—inner electron changes of order 0·1 MeV.

ultra-violet rays, wavelength 10^{-8} m; origin—outer electron changes of order 10–1000 eV.

visible rays, wavelength 10^{-7} m; origin—outermost electron energy changes of order 1 eV.

infra-red rays, wavelength 10^{-6} m; origin—molecular changes of order 0·1 eV.

radio waves, wavelength 10^{-2} to 1000 m; origin—high-frequency oscillations of electrons, energy change of order 10^{-6} eV.

There are no definite limits to the range of wavelengths in the spectrum of electromagnetic waves. The values above are approximate only.

Wave-particle theory shows that the energy in electromagnetic waves of frequency v can be carried by particles of energy hv, where h is Planck's constant, and this aspect of waves is treated in the atomic physics section of *Advanced Level Physics* (Heinemann).

EXERCISES 31

(*Where necessary, assume* $c = 3 \times 10^8$ m s^{-1}, $\varepsilon_0 = 8 \cdot 85 \times 10^{-12}$ F m^{-1})

1. A coaxial cable has an inductance per metre $L = 9 \times 10^{-7}$ H m^{-1} and a capacitance per metre $C = 4 \times 10^{-11}$ F m^{-1}. Find the speed of a pulse along the cable.

2. In question **1** the cable has a length of 300 m and the pulse is applied at one end. If an oscilloscope with a suitable time-base is connected at this end, draw sketches of the trace seen when the other end of the cable is (*a*) open, (*b*) short-circuited.

What is a suitable time-base for the oscilloscope: 1μs/cm, $100\ \mu$s/cm or 10 ms/cm?

3. A cable 200 m long has a capacitance per unit length of 2×10^{-11} F m^{-1}. A pulse applied at one end is reflected back in $2 \cdot 5$ microseconds. Calculate the inductance per unit length of the cable.

4. Draw a circuit arrangement, with coils and capacitors, for demonstrating a pulse travelling along a "transmission line". Why are large inductances and large capacitances necessary? What effect is observed when (i) the inductance is reduced to 1/1600th of its value, (ii) the capacitance is increased to 25 times its value?

5. Assuming that the energy in a capacitor is stored entirely in its electric field, estimate the energy density in the field when a 50 μF capacitor is charged to a p.d. of 100 V, assuming the volume of the dielectric used is 10^{-5} m^3.

6. In the following question on electromagnetic waves, answer A if (i) only is correct, B if (ii) only is correct, C if (i) and (ii) only are correct, D if (i) and (iii) only are correct, E if (i), (ii) and (iii) are correct:
   (i) The B- and E-fields are always in phase.
   (ii) The speed in a medium is $v = \sqrt{\mu\varepsilon}$.
   (iii) In a vacuum, $B = Ec$ and $E = \mu_0\varepsilon_0 cB$.

7. Calculate the speed of an electromagnetic wave in a vacuum, given that $\mu_0 = 4\pi \times 10^{-7}$ H m^{-1} and $\varepsilon_0 = 8 \cdot 854 \times 10^{-12}$ F m^{-1}. What is the speed along a coaxial cable filled with material of permittivity 25 times that of a vacuum and permeability equal to that in a vacuum?

8. The energy density (energy per unit volume) of an electromagnetic wave in a vacuum is $\frac{1}{2}\varepsilon_0 E_0{}^2$, where ε_0 is the permittivity of a vacuum and E_0 is the peak value of the electric field. By using units or dimensions for ε_0 and E_0, show that this expression is dimensionally correct. (*Note.* ε_0 has units F m^{-1} and E_0 has units V m^{-1}.)

9. Write down an expression for the *intensity* I of a plane-progressive electromagnetic wave. Show that the intensity in watt metre^{-2} is about $2 \cdot 7 \times 10^{-3}\ E^2$, where E is the root-mean-square (r.m.s.) value of the electric field. ($E = E_0/\sqrt{2}$, where E_0 is the peak value.)

10. A plane-progressive electromagnetic wave in free space has a peak electric field of 60 V m^{-1}. Calculate the magnitude of the peak magnetic field. Draw a sketch showing the relative directions of the two fields and of the wave.

11. Find the respective peak values of E and B in the electromagnetic energy from the sun on a hot day when the intensity reaching the earth is estimated at 100 W m^{-2}.

12. An indoor aerial, 0·50 metre long, is positioned along the electric field direction of an electromagnetic wave of intensity 5×10^{-16} W m^{-2}. Calculate the magnitude of the peak voltage in microvolts induced between the ends of the aerial.

13. Calculate the mean intensity at a distance of 2 m from a small lamp of 100 W. Assuming that all the energy radiated from the lamp is electromagnetic, estimate the magnitudes E_0 and B_0 of the peak values of the fields in the wave at this distance from the lamp.

14. A beam of plane-polarised light is reduced to one-quarter of its intensity after passing through a Polaroid. Calculate the angle between the planes of polarisation of the incident light and the Polaroid.

15. A small light source is placed above a horizontal surface. Beneath the source are placed a pair of horizontal Polaroid sheets and beneath these again is placed a horizontal sheet of plane glass. The lower Polaroid sheet is rotated in a horizontal plane with respect to the upper until maximum illumination is obtained on the horizontal surface. When the glass is removed, one of the Polaroid sheets has to be rotated through 38° 0′ with respect to the other to reduce the illumination to the original value. Determine the percentage of light transmitted by the glass, explaining the basis of your calculation. (*L.*)

16. Draw sketches showing the direction of vibration or electric vector in (i) plane-polarised light, (ii) ordinary or unpolarised light, (iii) partially plane-polarised light. Explain why unpolarised light can be represented by two perpendicular electric vectors, and why a Polaroid transmits 50% of the incident light intensity.

17. A Polaroid reduces the intensity of unpolarised light by 50%. If unpolarised light of intensity 8 W m^{-2} is incident on two Polaroids held in front of each other, find the intensity of the transmitted light when the "easy" directions of the Polaroids are (i) parallel, (ii) at an angle of 60° to each other.

18. With the aid of a diagram, explain how a metal grille can be used to find the direction of the electric and magnetic vectors in a beam of microwaves.

How can this arrangement be used to investigate the law relating I and θ, where I is the intensity of the transmitted beam and θ is the angle between the incident electric vector and the normal direction to the rods in the grille?

Answers

OPTICS

EXERCISES 16 (p. 400)

4. $4a$; $2na$.

EXERCISES 17 (p. 417)

1. (i) 15 cm, 1·5, (ii) 12 cm, 3. **3.** 6 cm, 0·4. **6.** 4/21 m. **7.** Object distance = 10 cm, $r = 40$ cm; concave. **9.** $2R$. **10.** 2 radians, or 114°. **11.** Inverted. **12.** 4·5 cm behind mirror; 0·25 mm; 5/38. **13.** (a) 240 cm, (b) 1·3 cm.

EXERCISES 18 (p. 437)

1. 35·3°. **2.** 41·8°. **3.** (i) 26·3°. (ii) 56·4°. **4.** (b) 3 cm from bottom. **6.** (i) 41·8, (ii) 48·8°, (iii) 62·5°. **9.** 1·47. **11.** (b) 12 cm above mirror. **15.** 1·60. **17.** 1·41. **20.** Nearer by $(n-1)d/n$.

EXERCISES 19 (p. 451)

1. 42°. **2.** (i) 1·52, (ii) 52·2°. **3.** 60°, 55° 30′, 1·648. **4.** 4° 48′. **5.** angle i on second face = 60·7°, $c = 41·8°$. **7.** 43° 35′. **8.** 37° 45′; 10° 8′; 180°. **10.** 55°, 1·53. **12.** 27·9°.

EXERCISES 20 (p. 468)

1. Crown: (i) 3·07°, (ii) 3·14°, (iii) 3·11°; flint: (i) 2·58°, (ii) 2·66°, (iii) 2·62°. **2.** 0·023, 0·031. **3.** 3·92°, 0·021°. **4.** 0·54 mm; 0·54 mm. **5.** 1·75. **8.** 0·144°. **11.** (a) 49° 12′, (b) 50° 38′, (c) 1° 26′. **12.** 6·67°, 0·83°. **13.** 1170 km s^{-1}.

EXERCISES 21 (p. 504)

1. (i) 40 cm virtual, (ii) 80 cm real. **2.** 7·2, 18 cm from nearest point on sphere. **3.** (a) 1·51, (b) 7·5 cm. **4.** $v = 6r$, where r is radius. **5.** (i) 12 cm, $m = 1$, (ii) 12 cm, $m = 3$. **6.** $6\frac{2}{3}$ cm, $\frac{3}{5}$. **7.** $13\frac{1}{3}$, 40 cm. **8.** (i) $5\frac{1}{7}$ cm, (ii) $22\frac{1}{2}$ cm. **9.** (i) $9\frac{3}{5}$ cm, (ii) $16\frac{2}{3}$ cm. **10.** 10 cm. **11.** 1·4. **12.** 12·0, 18·7 cm. **13.** 80 cm. **14.** (a) 40 cm, (b) 160 cm. **15.** (a) $11\frac{1}{9}$, (b) $5\frac{5}{9}$ cm; $3\frac{19}{27}$ cm. **16.** $27\frac{1}{4}$ cm from lens. **18.** 1·44. **19.** radii = 9·9, 24·8 cm; $n = 1·51$. **20.** 4 cm. **21.** $12\frac{6}{7}$, $37\frac{6}{7}$ cm above water surface. **22.** 1·4. **23.** (a) 20·5 cm, (b) 12·85 cm, (c) 1·63. **24.** (a) 120 cm from converging lens, (b) 92·2 cm, (c) 2·2. **25.** (a) Beside object O, (b) 72·5 cm from O.

EXERCISES 22 (p. 522)

1. Diverging, $f = 200$ cm; $22\frac{2}{3}$ cm. **2.** Converging, $f = 28\frac{4}{7}$ cm; $35\frac{5}{17}$ to 25 cm. **3.** near pt.: 50 cm, far pt.: 200 cm, $r = 16\frac{2}{3}$, $+100$ cm. **5.** 220 cm. **6.** Diverging, $f = 20$ cm. Infinity to 20 cm from eye. **8.** Diverging, $f = 30$ cm; 30 cm. **9.** 15·3 cm; 39·7 cm, diverging. **10.** 40 cm, converging. **12.** 64 cm; $106\frac{2}{3}$ cm. **14.** 0·98, 1·02 cm; 3·8. **15.** 10 cm.

EXERCISES 23 (p. 548)

1. 8:5:2. **2.** (a) infinity, (b) least distance of distinct vision. **3.** (a) 4, (b) 4·8.
4. long sight, $f = +38\frac{8}{9}$ cm. **5.** 30 cm from scale; $f = 6$ cm. **8.** $f_e = 4$ cm,
dia. $= 0.5$ cm; $r = 4.8$ cm; 0·013 cm. **9.** 3·2, 8·8 cm. **10.** 4·0, 21·0 cm. $M = 6.0$.
11. (a) 250, (b) 0·12 cm. **12.** 89° or 91°. **13.** 8·7 cm, 46·7. **14.** 0·55 cm; 2:1. **15.** 2;
22·5 cm. **16.** (a) 22·4, (b) 4·9 cm diameter. **17.** 6.25×10^{-3} rad.

EXERCISES 24 (p. 573)

7. 22·5 m; 0·02 m. **8.** 25; 6×10^4. **9.** 25 rev s^{-1}, 3×10^4 cycles. **12.** 187·5 lux
(m-candle), 67·5 cd. **13.** (i) $3\frac{1}{8}$, (ii) 1·6 lux. **14.** 125 cd. **15.** 2, 10 lux. **16.** 0·5.
17. 1, 0·83 lux; 1·41, 1·2 lux. **18.** 2·63 metres. **19.** 64 lumen m^{-2}. **20.** 63·6°
22. 90·3%.

WAVES AND SOUND

EXERCISES 25 (p. 604)

2. $L=$ sound, $T=$ remainder. **3.** (i) 133 cm, (ii) 400 Hz. **4.** (i) 100 Hz,
(ii) 1·7 m, (iii) 170 m s^{-1}, (iv) π, (v) 0·2 sin $(400\pi t + 20\pi x/17)$. **6.** 380, 391 Hz.
7. 625·9 Hz. **10.** (i) $5\pi/3$, (ii) $y = 0.03$ sin $2\pi(250t - 25x/3)$, (iii) 6 cm, (iv) 0·01
sin $50\pi x/3$. cos $500\pi t$ (x in m).

EXERCISES 26 (p. 635)

1. (a) 333, (b) 360·4 m s^{-1}; 342 m s^{-1}. **2.** 349 m s^{-1}. **3.** 332 m s^{-1}. **4.** 53·1°.
6. 680 Hz. **9.** 1366 m. **11.** 1115 Hz. **13.** 1·83 s. **14.** 6·7 m s^{-1}. **15.** 6.44×10^6
m s^{-1}. **16.** 1174, 852 Hz. **17.** (b) (i) 66 cm, (ii) (1) 550, (2) 545, (3) 600 Hz. **21.** 10·8
db. **22.** 25 W. **23.** 5×10^{-17} W m^{-2}.

EXERCISES 27 (p. 671)

1. (i) $\lambda/2$, (ii) $\lambda/4$, (iii) $\lambda/2$; 567 Hz. **2.** 20 cm. **3.** (a) 0·322, (b) 0·645 m. **4.** (b)
Amp.: (i) max, (ii) 0, (iii) max, (iv) half-max. **5.** 267 Hz. **8.** (i) 15·95 cm, (ii) 4·5 Hz.
9. +5·2°C. **12.** −3·2, +6·7%. **13.** 239 Hz. **15.** (a) 2 m, (b) 100 Hz. **16.** touch
1/6th from end. **18.** 2·08 kgf. **19.** 100, 300, 500 Hz.

EXERCISES 28 (p. 685)

2. 39°. **3.** 18·6°. **4.** 47° 10′; 41° 48′ with vertical at oil surface. **8.** 4.0×10^{-7} m.

EXERCISES 29 (p. 723)

2. 6.25×10^{-7} m. **3.** 2.27×10^{-5} m. **4.** 0·34 mm. **5.** 1·5. **7.** 1·64 mm; oil—
centre bright, fringes closer. **8.** 2.52×10^5. **10.** 1.42×10^{-6} m, receded. **11.** $1\frac{1}{4}$.
13. inwards, 0·133 cm. **15.** 2.32×10^{-7} m. **16.** 2° 35′. **18.** 3; 1·2 cm. **19.** 6,800.
20. (b) 5 m s^{-1}. **21.** 13·3°. **22.** 3; 5.895×10^{-7} m. **23.** 2·35 mm, 0·0785 rad.
25. 67·5°. **29.** 49·0°. **32.** 2·37 cm.

EXERCISES 30 (p. 753)

1. 7.5×10^{-4} rad, 0·375 mm. **3.** 0·1 rad. **4.** (i) 2.7×10^{-7} rad (ii) 5.2×10^{-7} rad. **5.** 0·03 rad. **6.** 1·22 m. **7.** 15·38 m. **8.** (i) 4·68 m (ii) 8·20 m.

9. (i) 2 cm wide grating (ii) Yes. **10.** 4000, 8000. (i) 1/4 (ii) 1/2. **11.** 0·056 mm. **12.** ± 20·1°. **13.** 1505 cm^{-1}, 10° to normal on other side. **14.** 1·18 cm. **15.** 5·897 × 10^{-7} m. **16.** (a) 2 × 10^{-4} cm (b) along normal and at 37° to normal. **18.** (a) Bands shift downward—optical paths from S$_1$ to the central fringe *via* each slit in S$_2$ must be equal. (b) Decreases contrast while increasing intensity— light from each part of S$_1$ slit gives rise to its own band system in different positions. (c) Decreases band contrast, i.e. dark bands become less dark. Effect due to Young's two equal slits plus background due to extra width. (d) Bands disappear—no two-slit interference. (e) Displaces bands due to change in optical path. **20.** 6 × 10^{-7}, 4·8 × 10^{-7}, 4·0 × 10^{-7} m. **22.** (a) Different points on E are non-coherent and produce their own fringes in different positions— hence uniform illumination on S′. (b) Direct light from S$_2$ interferes with light reflected from surface BC (as Lloyd's mirror). (c) As in (b), direct light from S$_2$ interferes with light reflected internally at surface BC. (d) Fringes in (c) remain as no effect here due to S$_1$. (e) Fringe width ∝ λ, and λ$_{\text{glass}}$ = (1/n) × λ$_{\text{air}}$ = 0·67 × λ$_{\text{air}}$. Fringes above C become wider (sources closer together); below C they become narrower (sources wider apart). C is centre of band system in each case.

ELECTROMAGNETIC WAVES

EXERCISES 31 (p. 775)

1. 1·7 × 10^8 m s^{-1}. **2.** 1 μs/cm. **3.** 1·95 × 10^{-6} H m^{-1} **4.** (i) speed in- creases 40 times; (ii) reduced to 1/5th. **5.** 2·5 × 10^4 J m^{-3}. **6.** A. **7.** 2·998 × 10^8 m s^{-1}, 6 × 10^7 m s^{-1}. **10.** 2 × 10^{-7} T. **11.** 275 V m^{-1}, 9 × 10^{-7} T. **12.** 0·3 μV. **13.** 2 W m^{-2}; 39 V m^{-1}, 1·3 × 10^{-7} T. **14.** 60°. **15.** 62%. **17.** (i) 4 W m^{-2}; (ii) 1 W m^{-2}.

Index

880 66 22 Appleton.

Ext : 363